Climate Change and Agriculture in the United States:

Effects and Adaptation

USDA Technical Bulletin 1935

Climate Change and Agriculture in the United States: Effects and Adaptation

This document may be cited as:

Walthall, C.L., J. Hatfield, P. Backlund, L. Lengnick, E. Marshall, M. Walsh, S. Adkins, M. Aillery, E.A. Ainsworth, C. Ammann, C.J. Anderson, I. Bartomeus, L.H. Baumgard, F. Booker, B. Bradley, D.M. Blumenthal, J. Bunce, K. Burkey, S.M. Dabney, J.A. Delgado, J. Dukes, A. Funk, K. Garrett, M. Glenn, D.A. Grantz, D. Goodrich, S. Hu, R.C. Izaurralde, R.A.C. Jones, S-H. Kim, A.D.B. Leaky, K. Lewers, T.L. Mader, A. McClung, J. Morgan, D.J. Muth, M. Nearing, D.M. Oosterhuis, D. Ort, C. Parmesan, W.T. Pettigrew, W. Polley, R. Rader, C. Rice, M. Rivington, E. Rosskopf, W.A. Salas, L.E. Sollenberger, R. Srygley, C. Stöckle, E.S. Takle, D. Timlin, J.W. White, R. Winfree, L. Wright-Morton, L.H. Ziska. 2012. *Climate Change and Agriculture in the United States: Effects and Adaptation.* USDA Technical Bulletin 1935. Washington, DC. 186 pages.

This document was produced as part of of a collaboration between the U.S. Department of Agriculture, the University Corporation for Atmospheric Research, and the National Center for Atmospheric Research under USDA cooperative agreement 58-0111-6-005. NCAR's primary sponsor is the National Science Foundation.

Images courtesy of USDA and UCAR.

This report is available on the Web at: http://www.usda.gov/oce/climate_change/effects.htm

Printed copies may be purchased from the National Technical Information Service. Call 1-800- 553-NTIS (6847) or 703-605-6000, or visit http://www.ntis.gov.

February 2013

Climate Change and Agriculture in the United States: Effects and Adaptation

Table of Contents

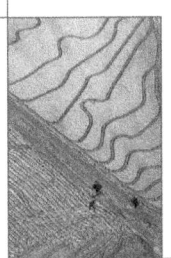

Climate Change and Agriculture in the United States:
Effects and Adaptation

Executive Summary

Key Messages

Increases of atmospheric carbon dioxide (CO_2), rising temperatures, and altered precipitation patterns will affect agricultural productivity. Increases in temperature coupled with more variable precipitation will reduce productivity of crops, and these effects will outweigh the benefits of increasing carbon dioxide. Effects will vary among annual and perennial crops, and regions of the United States; however, all production systems will be affected to some degree by climate change. Agricultural systems depend upon reliable water sources, and the pattern and potential magnitude of precipitation changes is not well understood, thus adding considerable uncertainty to assessment efforts.

Fig. 1. Storm gathers over farmland. Image courtesy UCAR.

Livestock production systems are vulnerable to temperature stresses. An animal's ability to adjust its metabolic rate to cope with temperature extremes can lead to reduced productivity and in extreme cases death. Prolonged exposure to extreme temperatures will also further increase production costs and productivity losses associated with all animal products, e.g., meat, eggs, and milk.

Projections for crops and livestock production systems reveal that climate change effects over the next 25 years will be mixed. The continued degree of change in the climate by midcentury and beyond is expected to have overall detrimental effects on most crops and livestock.

Climate change will exacerbate current biotic stresses on agricultural plants and animals. Changing pressures associated with weeds, diseases, and insect pests, together with potential changes in timing and coincidence of pollinator lifecycles, will affect growth and yields. The potential magnitude of these effects is not yet well understood. For example, while some pest insects will thrive under increasing air temperatures, warming temperatures may force others out of their current geographical ranges. Several weeds have shown a greater response to carbon dioxide relative to crops; understanding these

physiological and genetic responses may help guide future enhancements to weed management.

Agriculture is dependent on a wide range of ecosystem processes that support productivity including maintenance of soil quality and regulation of water quality and quantity. Multiple stressors, including climate change, increasingly compromise the ability of ecosystems to provide these services. Key near-term climate change effects on agricultural soil and water resources include the potential for increased soil erosion through extreme precipitation events, as well as regional and seasonal changes in the availability of water resources for both rain-fed and irrigated agriculture.

The predicted higher incidence of extreme weather events will have an increasing influence on agricultural productivity. Extremes matter because agricultural productivity is driven largely by environmental conditions during critical threshold periods of crop and livestock development. Improved assessment of climate change effects on agricultural productivity requires greater integration of extreme events into crop and economic models.

The vulnerability of agriculture to climatic change is strongly dependent on the responses taken by

1

humans to moderate the effects of climate change. Adaptive actions within agricultural sectors are driven by perceptions of risk, direct productivity effects of climate change, and by complex changes in domestic and international markets, policies, and other institutions as they respond to those effects within the United States and worldwide. Opportunities for adaptation are shaped by the operating context within which decision-making occurs, access to effective adaptation options, and the capacity of individuals and institutions to take adaptive action as climate conditions change. Effective adaptive action across the multiple dimensions of the U.S. agricultural system offers potential to capitalize on emerging opportunities and minimize the costs associated with climate change. A climate-ready U.S. agriculture will depend on the development of geographically specific, agriculturally relevant, climate projections for the near and medium term; effective adaptation planning and assessment strategies; and soil, crop and livestock management practices that enhance agricultural production system resilience to climatic variability and extremes. Anticipated adaptation to climate change in production agriculture includes adjustments to production system inputs, tillage, crop species, crop rotations, and harvest strategies. New research and development in new crop varieties that are more resistant to drought, disease, and heat stress will increase the resilience of agronomic systems to climate change and will enable exploitation of opportunities that may arise.

Over the last 150 years, U.S. agriculture has exhibited a remarkable capacity to adapt to a wide diversity of growing conditions amid dynamic social and economic changes. These adaptations were made during a period of relative climatic stability and abundant technical, financial and natural resources. Future agricultural adaptation will be undertaken in a decision environment characterized by high complexity and uncertainty driven by the sensitivity of agricultural system response to climatic variability, the complexity of interactions between the agricultural systems, non-climate stressors and the global climate system, and the increasing pace and intensity of climatic change. New approaches to managing the uncertainty associated with climate change, such as integrated assessment of climate change effects and adaptation options, the use of adaptive management and robust decision-support strategies, the integration of climate knowledge into decisionmaking by producers, technical advisors, and agricultural research and development planning efforts, and the development of resilient agricultural production systems will help to sustain agricultural production during the 21st century.

Climate change poses unprecedented challenges to U.S. agriculture because of the sensitivity of agricultural productivity and costs to changing climate conditions.

Introduction

Agriculture in the United States produces approximately $300 billion a year in commodities with livestock accounting for roughly half the value. Production of these commodities is vulnerable to climate change through the direct (i.e., abiotic) effects of changing climate conditions on crop and livestock development and yield (e.g., changes in temperature or precipitation), as well as through the indirect (i.e., biotic) effects arising from changes in the severity of pest pressures, availability of pollination services, and performance of other ecosystem services that affect agricultural productivity. Thus, U.S. agriculture exists as a complex web of interactions between agricultural productivity, ecosystem services, and climate change.

Climate change poses unprecedented challenges to U.S. agriculture because of the sensitivity of agricultural productivity and costs to changing climate conditions. Adaptive action offers the potential to manage the effects of climate change by altering patterns of agricultural activity to capitalize on emerging opportunities while minimizing the costs associated with negative effects. The aggregate effects of climate change will ultimately depend on a complex web of adaptive responses to local climate stressors. These adaptive responses may range from farmers adjusting planting patterns and soil management practices in response to more variable weather patterns, to seed producers investing in the development of drought-tolerant varieties, to increased demand for Federal risk management programs, to adjustments in international trade as nations respond to food security concerns. Potential adaptive behavior can occur at multiple levels in a highly diverse international agricultural system including production, consumption, education, research, services, and governance. Understanding the complexity of such interactions is critical for developing effective adaptive strategies.

The U.S. agricultural system is expected to be fairly resilient to climate change in the short term due to the system's flexibility to engage in adaptive behaviors such as expansion of irrigated acreage, regional shifts in acreage for specific crops, crop rotations, changes to management decisions such as choice and timing of inputs and cultivation practices, and altered trade patterns compensating for yield changes caused by changing climate patterns. By midcentury, when temperature increases are expected to exceed 1°C to 3°C and precipitation extremes intensify, yields of major U.S. crops and farm returns are projected to decline. However, the simulation studies underlying such projections often fail to incorporate production constraints caused by changes of pest pressures, ecosystem services and conditions that limit adaptation that can significantly increase production costs and yield losses.

Crop Response to Changing Climate

Plant response to climate change is dictated by a complex set of interactions to CO_2, temperature, solar radiation, and precipitation. Each crop species has a given set of temperature thresholds that define the upper and lower boundaries for growth and reproduction, along with optimum temperatures for each developmental phase. Plants are currently grown in areas in which they are exposed to temperatures that match their threshold values. As temperatures increase over the next century, shifts may occur in crop production areas because temperatures will no longer occur within the range, or during the critical time period for optimal growth and yield of grain or fruit.

For example, one critical period of exposure to temperatures is the pollination stage, when pollen is released to fertilize the plant and trigger development of reproductive organs, for fruit, grain, or fiber. Such thresholds are typically cooler for each crop than the thresholds and optima for growth. Pollination is one of the most sensitive stages to temperatures, and exposure to high temperatures during this period can greatly reduce crop yields and increase the risk of total crop failure. Plants exposed to warm nighttime temperatures during grain, fiber, or fruit production also experience lower productivity and reduced quality. Increasing temperatures cause plants to mature and complete their stages of development faster, which may alter the feasibility and profitability of regional crop rotations and field management options, including double-cropping and use of cover crops. Faster growth may create smaller plants, because soil may not be able to supply water or nutrients at required rates, thereby reducing grain, forage, fruit, or fiber production. Increasing temperatures also increase the rate of water use by plants, causing more water stress in areas with variable precipitation. Estimated reductions in solar radiation in agricultural areas over the last 60 years are projected to continue due to increased cloud cover and radiative scattering caused by atmospheric aerosols. Such reductions may partially offset the temperature-induced acceleration of plant growth. For vegetables, exposure to temperatures in the range of 1°C to 4°C above optimal for biomass growth moderately reduces yield, and exposure to temperatures more than 5°C to 7°C above optimal often leads to severe, if not total, production losses.

While many agricultural enterprises have the option to respond to climate changes by shifting crop selection, development of new cultivars in perennial specialty crops commonly requires 15 to 30 or more years, greatly limiting that sector's opportunity to adapt by shifting cultivars unless cultivars can be introduced from other areas.

An increase in winter temperatures also affects perennial cropping systems through interactions with plant chilling requirements. All perennial specialty crops have a winter chilling requirement (typically expressed as hours below 10°C and above 0°C) ranging from 200 to 2,000 cumulative hours. Yields will decline if the chilling requirement is not completely satisfied because flower emergence and viability will be low. Projected air temperature increases for California, for example, may prevent the chilling requirements for fruit and nut trees by the middle to the end of the 21st century. In the Northeast United States, perennial crops with a lower 400-hour chilling requirement will continue to be met for most of the Northeast during this century, but crops with prolonged cold requirements (1,000 or more hours) could demonstrate reduced yields, particularly in southern sections of the Northeast. Climate change affects winter temperature variability, as well; mid-winter warming can lead to early bud-burst or bloom of some perennial plants, resulting in frost damage when cold winter temperatures return.

Increasing carbon dioxide (CO_2) in the atmosphere is a positive for plant growth, and controlled experiments have documented that elevated CO_2 concentrations can increase plant growth while decreasing soil water-use rates. The effects of elevated CO_2 on grain and fruit yield and quality, however, are mixed; reduced nitrogen and protein content observed in some nitrogen-fixing plants causes a reduction in grain and forage quality. This effect reduces the ability of pasture and rangeland to support grazing livestock. The magnitude of the growth stimulation effect of elevated CO_2 concentrations under field conditions, in conjunction with changing water and nutrient constraints, is uncertain. Because elevated CO_2 concentrations disproportionately stimulate growth of weed species, they are likely to contribute to increased risk of crop loss from weed pressure.

The effects of elevated CO_2 on water-use efficiency may be an advantage for areas with limited precipitation. Other changing climate conditions may either offset or complement such effects. Warming temperatures, for instance, will act to increase crop water demand, increasing the rate of water use by crops. Crops grown on soils with a limiting soil water-holding capacity are likely to experience an increased risk of drought and potential crop failure as a result of temperature-induced increases in crop water demand, even with improved water-use efficiencies. Conversely, declining trends of near-surface winds over the last several decades and projections for future declines of winds may decrease evapotranspiration of cropping regions.

Crops and forage plants will continue to be subjected to increasing temperatures, increasing CO_2, and more variable water availability caused by changing precipitation patterns. These factors interact in their effect on plant growth and yield. A balanced understanding of the consequences of management actions and genetic responses to these factors will form the basis for more resilient production systems to climate change. Due to the complexities of these relationships, integrated research and development of management practices, plant genetics, hydrometeorology, socio-economics, and agronomy is necessary to enable successful agricultural adaptation to climate change.

Livestock Response to Changing Climate

Animal agriculture is a major component of the U.S. agricultural system. Changing climatic conditions affect animal agriculture in four primary ways: (1) feed-grain production, availability and price; (2) pastures and forage crop production and quality; (3) animal health, growth and reproduction; and (4) disease and pest distributions. The optimal environmental conditions for livestock production include a range of temperatures and other environmental conditions for which the animal does not need to significantly alter behavior or physiological functions to maintain a relatively constant core body temperature. Optimum animal core body temperature is often maintained within a 2°C to 3°C range. For many species, deviations of core body temperature in excess of 2°C to 3°C cause disruptions of performance, production, and fertility that limit an animal's ability to produce meat, milk, or eggs. Deviations of 5°C to 7°C often result in death. For cattle that breed during spring and summer, exposure to high temperatures decreases conception rates. Livestock and dairy production may be more affected by changes in the number of days of extreme heat than by adjustments of average temperature. The combined effect of temperature and humidity affect animal response and are quantified through the thermal-humidity index.

Livestock production systems that provide partial or total shelter to mitigate thermal environmental challenges can reduce the risk and vulnerability associated with adverse weather events. Livestock such as poultry and swine are generally managed in housed systems where airflow can be controlled and housing temperature modified to minimize or buffer against adverse environmental conditions. However, management and energy costs associated with increased temperature regulation will increase for confined production enterprises. Protection of animals against exposure to high temperatures will require modification of shelter and perhaps even methods of increasing cooling.

Warmer, more humid conditions will also have indirect effects on animal health and productivity through promotion of insect growth and spread of diseases. Such effects may be substantial; however, exact relationships between climate change and vectors of animal health are not well understood. Climate affects microbial populations and distribution, the distribution of vector-borne diseases, host resistance to infections, food and water shortages, and food-borne diseases. Earlier springs and warmer winters may enable greater proliferation and survivability of pathogens and parasites. Regional warming and changes of rainfall distribution may lead to changes in the spatial or temporal distributions of diseases sensitive to temperature and moisture, such as anthrax, blackleg, hemorrhagic septicemia, as well as increased incidence of ketosis, mastitis and lameness in dairy cows.

Effects of Climate Change on Soil and Water

Climate change effects on agriculture also include the effects of changing climate conditions on resources of key importance to agricultural production, such as soil and water. Seasonal precipitation affects the potential amount of water available for crop production, but the actual amount of water available to plants also depends upon soil type, soil water-holding capacity, and infiltration rate. Healthy soils have characteristics that include appropriate levels of nutrients necessary for the production of healthy plants, moderately high levels of organic matter, a soil structure with good aggregation of the primary soil particles and macro-porosity, moderate pH levels, thickness sufficient to store adequate water for plants, a healthy microbial community, and absence of elements or compounds in concentrations toxic for plant, animal, and microbial life. Several processes act to degrade soils including, erosion, compaction,

Climate affects microbial populations and distribution, the distribution of vector-borne diseases, host resistance to infections, food and water shortages, and food-borne diseases.

acidification, salinization, toxification, and net loss of organic matter.

Several of these processes are sensitive to changing climate conditions. Changes to the rate of soil organic matter accumulation will be affected by climate through soil temperature, soil water availability, and the amount of organic matter input from plants. Erosion is of particular concern. Changing climate will contribute to the erosivity from rainfall, snowmelt, and wind. Rainfall's erosive power will increase if increases in rainfall amount are accompanied by increases of intensity. Shifts of rainfall intensity have begun to occur in the United States with more extreme events expected for the future. Although there is a general lack of knowledge about the rates of soil erosion associated with snowmelt or rain-on-thawing-soil erosion, if decreased days of snowfall translate to increased days of rainfall, erosion by storm runoff is likely to increase.

Changes in production practices can also have effects on soil erosion that may be greater than other effects of climate change. Tillage intensity, crop selection, as well as planting and harvest dates can significantly affect runoff and soil loss. Though the magnitude of these effects is still highly uncertain, studies have shown potential for significant increases of erosion loss, in part due to a reduction of projected crop biomass, which results in less overwintering residue available to protect the soil. As soil erosion changes under climate change, so does the potential for associated, off-site, non-point-source pollution. Soil conservation practices will therefore be an important element of agricultural adaptation to climate change.

Changing climate conditions over the coming decades will also significantly affect water resources, with broad implications for the U.S. crop sector. Climate change will affect surface-water resources, which account for 58% of water withdrawals for irrigated production nationally. Rising temperatures and shifting precipitation patterns will alter crop-water requirements, crop-water availability, crop productivity, and costs of water access across the agricultural landscape. Temperature and precipitation shifts are expected to alter the volume and timing of storm and snowmelt runoff to surface water bodies. Annual streamflow may increase in the northern and eastern United States, where annual precipitation is projected to increase. Precipitation declines for regions such as the Southwest and Southern Plains will result in reduced streamflow and a shift of seasonal flow volumes to the wetter winter months in areas already dominated by irrigation.

Climate change effects on snowpack have important implications for surface-water availability and stored water reserves, particularly in the West, where much of the surface-water runoff comes from mountain snowmelt. Higher temperatures will continue to restrict the snow storage season, resulting in reduced snow accumulations and earlier spring snowmelt. Stored water reserves are projected to decline in many river basins, especially during critical summer growing season months when crop-water demands are greatest. As a result, agriculture may become increasingly water constrained across the central and southern portions of the Mountain and Pacific Southwest regions, while projected precipitation increases in the Northern Rockies and Pacific Northwest could improve surface-water supplies for those areas.

The effect of precipitation changes on surface-water flows may be offset or compounded by temperature-induced shifts of potential evapotranspiration. Higher temperatures are projected to increase both evaporative losses from land and water surfaces, and transpiration losses from non-crop land cover, potentially reducing annual runoff and streamflow. The resulting shifts of water stress, crop yields, and crop competitiveness, in turn, will drive changes of cropland allocations and production systems within and across regions.

Groundwater is a primary water source for irrigation in the Plains States and an important irrigation water supply for the Eastern United States, as well as areas of the Mountain and Pacific West regions. While groundwater aquifers are generally less influenced in the short term by weather patterns, changing climate effects on precipitation, streamflow, and soil water evaporation can affect groundwater systems over time through changes in groundwater recharge.

Extreme Events

Climate change projections into the future suggest an increased variability of temperature and precipitation. Extreme climate conditions, such as dry spells, sustained drought, and heat waves can have large effects on crops and livestock. Although climate models are limited in their ability to accurately project the occurrence and timing of individual extreme events, emerging patterns project increased incidence of areas experiencing droughts and periods of more intense precipitation. The occurrence of very hot nights and the duration of very low (agriculturally insignificant) rainfall events are projected to increase by the end of the 21st century. The timing of extreme events relative to

Changes in production practices can also have effects on soil erosion that may be greater than other effects of climate change. Tillage intensity, crop selection, as well as planting and harvest dates can significantly affect runoff and soil loss.

sensitive phenological stages could affect growth and productivity.

Crops and livestock production will be affected by increased exposure to extreme temperature events and increased risk of exceeding the maximum temperature thresholds, potentially leading to catastrophic losses. Ruminants, including, goats, sheep, beef cattle and dairy cattle tend to be managed in more extensive outdoor facilities. Within limits, these animals can adapt to and cope with gradual thermal changes, though shifts in thermoregulation may result in a loss of productivity. Lack of prior conditioning to rapidly changing or adverse weather events, however, often results in catastrophic deaths of domestic livestock and losses of productivity by surviving animals.

Adaptation

U.S. agriculture has demonstrated a remarkable adaptive capacity over the last 150 years. Crop and livestock production systems expanded across a diversity of growing conditions, responded to variations in climate and other natural resources, and to dynamic changes in agricultural knowledge, technology, markets, and, most recently, public demands for sustainable production of agricultural products. This adaptive capacity has been driven largely by public sector investment in agricultural research, development, and extension activities made during a period of climatic stability and abundant technical, financial, and natural resource availability.

Climate change presents an unprecedented challenge to the adaptive capacity of U.S. agriculture. Current climate change effects are increasing the complexity and uncertainty of agricultural management. Projected climate changes over the next century may require major adjustments to production practices, particularly for production systems operating at their marginal limits of climate. Because agricultural systems are human-dominated ecosystems, the vulnerability of agriculture to climatic change is strongly dependent not just on the biophysical effects of climate change, but also on the responses taken by humans to moderate those effects within the United States and worldwide. Effective adaptive action undertaken by the multiple dimensions of the U.S. agricultural system offers potential for capitalizing on the opportunities presented by climate change, and minimizing the costs via avoidance or reduction of the severity of detrimental effects from changing climate.

Vulnerability and adaptive capacity are characteristics of human and natural systems, are dynamic and multi-dimensional, and are influenced by complex interactions among social, economic, and environmental factors. Adaptive decisions are shaped by the operating context within which decision are made (for example, existing natural resource quality and non-climate stressors, government policy and programs), access to effective adaptation options, and the individual capability to take adaptive action. Adaptation strategies in use today by U.S. farmers coping with current changes in weather variability include changing cultivar selection or timing of field operations, and increased use of pesticides to control higher pest pressures. In California's Central Valley, an adaptation plan consisting of integrated changes in crop mix, irrigation methods, fertilization practices, tillage practices, and land management was found to be the most effective approach to managing climate risk. Adaptation options for managing novel crop pest management challenges may involve increased use of pesticides, new strategies for preventing rapid evolution of pest resistance to chemical control agents, the development of new pesticide products and improved pest and disease forecasting. Adaptation options that increase the resilience of agricultural systems to increased pest pressures include crop diversification and the management of biodiversity at both field and landscape scale to suppress pest outbreaks and pathogen transmission. Given the projected effects of climate change, some U.S. agricultural systems will have to undergo more transformative changes to remain productive and profitable.

Adaptation measures such as developing drought, pest, and heat stress resistance in crops and animals, diversifying crop rotations, integrating livestock with crop production systems, improving soil quality, minimizing off-farm flow of nutrients and pesticides, and other practices typically associated with sustainable agriculture are actions that may increase the capacity of the agricultural system to minimize the effects of climate change on productivity. For example, developing drought and heat stress resistant crops will improve the ability of farmers to cope with increasing frequencies of temperature and precipitation variability. Similarly, production practices that enhance the ability of healthy soils to regulate water resource dynamics at the farm and watershed scales will be particularly critical for the maintenance of crop and livestock productivity under conditions of variable and extreme weather events. Enhancing the resilience of agriculture to climate change through adaptation strategies that promote the development of sustainable agriculture is a common multiple-benefit recommendation for agricultural adaptation planning.

National agricultural adaptation planning has only recently begun in the United States and elsewhere. Broad policy measures that may enhance the adaptive capacity of agriculture include strengthening climate-sensitive assets, integrating adaptation into all relevant government policies, and addressing non-climate stressors that degrade adaptive capacity. Because of the uncertainties associated with climate change effects on agriculture and the complexity of adaptation processes, adaptive management strategies that facilitate implementation and the continual evaluation and revision of adaptation strategies as climate learning proceeds will be necessary to ensure agricultural systems remain viable with climate change. Synergies between mitigation and adaptation planning are also possible through the use of coherent climate policy frameworks that link issues such as carbon sequestration, greenhouse gas emissions, land-use change, regional water management, and the long-term sustainability of production systems.

High adaptive capacity does not guarantee successful adaptation to climate change. Adaptation assessment and planning efforts routinely encounter conditions that limit adaptive action regardless of the adaptive capacity of the system under study. Potential constraints to adaptation can arise from ecological, social and economic conditions that are dynamic and vary greatly within and across economic sectors, communities, regions, and countries.

Adapting agricultural systems to dramatic changes in the physical environment may be limited by social factors such as values, beliefs, or world views. Those factors can be affected by access to finance, political norms and values, and culture and religious ideologies.

Other limits to adaptation include the availability of critical inputs such as land and water, and constraints to farm financing and credit availability. These constraints may be substantial, especially for agricultural enterprises with little available capital or those without the financial capacity to withstand increasing variability of production and returns, including catastrophic loss. Differential capacity for adaptation, together with the variable effects of climate change on yield, creates significant concerns about agricultural productivity and food security.

Research Needs

The research needs identified in this report are categorized within a vulnerability framework and address specific actions that would serve to improve understanding of the exposure, sensitivity, and adaptive

capacity of U.S. agriculture to climate change. Attention to these research needs will enhance the ability of the U.S. agriculture sector to anticipate and respond to the challenges presented by changing climate conditions.

Some broad research needs include the following:

- Improve projections of future climate conditions for time scales of seasons to multiple decades; enable more precise projections of the changes and durations of average and extreme temperatures, precipitation, and related variables (e.g., evapotranspiration, soil moisture).

- Evaluate and develop process-level understanding of the sensitivity of plant and animal production systems, including insect, weed, pathogen, soil and water components, to key direct, indirect and interacting effects of climate change effects.

- Develop and extend the knowledge, management strategies and tools needed by U.S. agricultural stakeholders to enhance the adaptive capacity of plant and animal production systems to climate variability and extremes. While existing management and agronomic options have demonstrated significant capacity for expanding adaptation opportunities, new adaptive management strategies, robust risk management approaches, and breeding and genetic advances offer much potential, but have yet to be evaluated.

Understanding Exposure

The vulnerability of an agricultural system to climate change is dependent in part on the character, magnitude and rate of climate variation to which a system is exposed. Effective adaptation will be enhanced by research to:

- Improve projections of future climate conditions for time scales of seasons to multiple decades, including more precise information about changes of average and extreme temperatures, precipitation, and related variables (e.g., evapotranspiration, soil moisture). Such projections are needed to better understand exposure to climate risks, and support effective assessment, planning, and decisionmaking across the multiple dimensions of the U.S. agricultural system.

- Enable projection of future climate conditions at finer temporal scales (hourly and daily versus weekly, monthly, or annual averages) and spatial scales (1-10 km, as opposed to 50-100 km). This finer-scale information would permit decision-

Adapting agricultural systems to dramatic changes in the physical environment may be limited by social factors such as values, beliefs, or world views. Those factors can be affected by access to finance, political norms and values, and culture and religious ideologies.

makers to examine the potential effects of climate change on specific crop and livestock production systems in specific regions. There is also a need to include more precise decadal-scale projections to integrate climate information into longer term planning and improved information about the probability of potential changes to effectively manage climate risks.

- Develop modeling systems that produce climate and effects projections through the use of standard socioeconomic scenarios and access to more accurate and comprehensive observations of climate change and its effects on agricultural systems. Improve process-level understanding and validate model simulations.

- Improve the accuracy and range of weather predictions (as opposed to longer term, scenario-dependent climate projections) and seasonal forecasts. Better forecasts are needed to understand near-term exposure and support tactical decision-making at all levels of the agricultural system. Improved forecasting is particularly critical given the expected increases in the variability of weather and the incidence of extreme conditions.

Understanding Sensitivity

The nature and degree of response to key climate change drivers determines the sensitivity of the agricultural system to climate change effects. Critical thresholds, feedbacks, and synergies operating at multiple temporal and spatial scales complicate efforts to assess agricultural system sensitivity to climate change. Effective adaptation to climate change will be enhanced by research to:

- Improve understanding of both direct and indirect climate change effects and their interactions on plant and animal production systems, together with new tools for exploring their dynamic interactions throughout the multiple dimensions of the U.S. agricultural sector;

- Enhance capabilities to quantify and screen plant and animal response to water and temperature extremes;

- Improve understanding of climate change effects on the natural and biological resources upon which agricultural productivity depends, particularly soil and water resources;

- Improve understanding of climate change effects on existing agricultural landscape patterns and production practices;

- Improve understanding of the economic impacts of climate change and how those impacts are distributed.

- Develop improved integrated assessment models and ecosystem manipulation sites to enable experiments that examine the impacts of simultaneous interacting multiple stresses on plant and animal production systems.

Enhancing Adaptive Capacity

Because agricultural systems are human-dominated ecosystems, the vulnerability of agriculture to climate change is strongly dependent on the responses taken by humans to adapt to climate change effects. The adaptive capacity of U.S. agriculture will be enhanced by research to:

- Improve understanding of the key determinants (social, economic, and ecological) of adaptive capacity and resilience in agricultural systems;

- Develop effective methods for the assessment of adaptive capacity;

- Identify and extend information about existing best management practices that offer "no-regrets" and "low regrets" adaptation options;

- Develop resilient crop and livestock production systems and the socio-economic and cultural/institutional structures needed to support them;

- Develop, assess, and extend adaptive management strategies and climate risk management tools to improve decisionmaking throughout the U.S. agricultural sector;

- Improve understanding of the social limits to adaptation, including the effects of cost/benefit considerations, technological feasibility, beliefs, values and attitudes, and resource constraints on adaptive response.

- Develop effective methods of adaptation planning and assessment useful to decisionmakers operating throughout the multiple dimensions of the U.S. agricultural system.

Chapter 1

U.S. Agriculture and Climate

"The dogmas of the quiet past are inadequate to the stormy present. The occasion is piled high with difficulty, and we must rise with the occasion. As our case is new, so we must think anew, and act anew."

Abraham Lincoln, 1862*

Strong scientific consensus highlights that anthropogenic effects of climate change are already occurring and will be substantial (Intergovernmental Panel on Climate Change 2007a). U.S. agriculture is a multi-billion-dollar industry that stands to be significantly influenced by the effects of climate change. This document presents an overview of the latest research available related to climate change impacts on U.S. agriculture and the potential options for adaptation in the agricultural sector. Building upon the extensive scientific literature, the impacts and risks of climate change, climate variability, and adaptation options for managed and unmanaged ecosystems and their constituent biota and processes are considered. The report also highlights changes in resource conditions that scientific studies suggest are most likely to occur in response to climate change.

Today, the United States Department of Agriculture (USDA) classifies 116 plant commodity groups as agricultural products, as well as four livestock groupings (beef cattle, dairy, poultry, and swine) and products derived from animal production, e.g., cheese or eggs. U.S. crops and livestock varieties are grown in diverse climates, regions, and soils. No matter the region, however, weather and climate characteristics such as temperature, precipitation, carbon dioxide (CO_2), and water availability directly affect the health and well-being of plants and livestock, as well as pasture and rangeland production. The distribution of crops and livestock is also determined by the climatic resources for a given region, and U.S. agriculture has benefited from optimizing the adaptive areas of crops and livestock. For any commodity, variation in yield between years is related to growing-season weather effects. These effects also influence how insects, disease, and weeds affect agricultural production.

Report Goals and Scope

Within this report, information is presented that enables framing and evaluation of existing vulnerabilities of U.S agriculture to climate change

and adaptation strategies. Timeframes for the assessments are the present, 25 years in future, and 90 years in future. This report focuses particularly on the near future, because the climate projections are relatively more certain and address more immediate planning and management needs for crops, livestock, economic needs, and risk concerns, among other considerations. However, projections and expectations are considered out to the century's end, in some cases.

This technical document builds on the 2008 report, *The Effects of Climate Change on Agriculture, Land Resources, Water Resources, and Biodiversity (CCSP 2008)*. While including up-to-date scientific analysis of the subjects included in that assessment, e.g., temperature and precipitation effects on crop and animal agriculture, this report builds on the earlier report in three important ways by covering climate change adaptation processes, looking at the economic effects of changing climate, and including new findings on the indirect (biotic) effects of climate change on U.S. agriculture.

U.S. landscapes include a mosaic of agricultural, urban, and wildland ecosystems. Within this mosaic, agricultural components play a large role in how climate change affects natural resources (water, soil, and air). Responses, whether environmental or economic, to changing climate are termed "adaptations" and play an important role in how climate change will influence agricultural landscapes and management needs. Equally important to adaptive changes are considerations related to the economics of changes in climate, e.g., how climate influences agricultural production economics, how economically driven choices influence agricultural management decisions, and how such decisions influence climate effects on the landscape. The last of the three advances on the earlier report is inclusion

---

* This quote is from Lincoln's 2nd State of the Union Address. In this address, he also announced the creation of USDA.

The National Climate Assessment

Created under the leadership of the U.S. Department of Agriculture (USDA), some information for this report may be used for the National Climate Assessment (NCA). The NCA provides a status report on climate change science and the impacts of climate on sectors of the United States to the U.S. Congress and the President of the United States. The NCA is a comprehensive compilation of information on the state of climate and its effects on U.S. ecosystems, infrastructure, and society to enhance the ability of the United States to anticipate, mitigate, and adapt to changes in the global environment.

of new scientific findings related to how changes in temperature and precipitation may affect pests, weeds, and disease, and how those changes play out in agricultural systems.

This technical assessment was driven by the following questions:

1. How does a changing climate directly influence agriculture?

2. What non-climate stresses need to be considered in interpreting climate change effects on U.S. agriculture?

3. How do economic factors respond to climate or alter the effects of climate change in agricultural systems?

4. How might agricultural decisionmakers take adaptive actions that capitalize on the opportunities and minimize or avoid the negative effects on production under changing climate conditions?

The mitigation of greenhouse gases (GHGs), either by reducing atmospheric emissions or through removal by various biological or technological means, including carbon sequestration in soils, is not within the scope of this report. A recent review of greenhouse gas mitigation can be found in Task Force Report 142, *Carbon Sequestration and Greenhouse Gas Fluxes in Agriculture: Challenges and Opportunities by the Council for Agricultural Science and Technology* (2011).

Document Organization

This document consists of seven chapters, as well as an Executive Summary and three appendices. This chapter, the *Introduction*, is intended to help orient the reader to the report. Chapter 2, *An Overview of U.S. Agriculture*, presents the context of U.S. agriculture. Chapter 3, *An Overview of the Effects of Changing Climate on U.S. Agriculture*, discuss global and national effects of climate change, and how these manifest within the national agroecosystem. Chapter 4, *Climate Change Science and Agriculture*, reviews the scientific literature on direct and indirect effects of climate change on agriculture and introduces a number of non-climate stressors that have effects on crops or livestock that change our understanding of climate change's relationship to agriculture. Chapter 5, *Climate Change Effects on U.S. Agricultural Production*, addresses the aggregate effects of climate change on specific cropping systems, livestock production, soils, and water and ecosystem services, including water resources for agriculture. Chapter 6, *Climate Change Effects on the Economics of U.S. Agriculture*, looks at the economic effects of climate change on agriculture, assesses economic risks due to changing climate, and considers economic means of adaptation under a changing climate. Chapter 7, *Adapting to Climate Change*, provides information on the adaptive capacity of agriculture. Chapter 8, *Conclusions and Research Needs*, expresses the authors' conclusions based upon the findings within the report.

Appendix A provides literature citations by chapter. Appendix B provides a glossary of terms used commonly in this report, and Appendix C lists the report authors and their affiliations.

Authors

This document was coordinated by the U.S. Department of Agriculture and composed by 60 authors from the Federal service, universities, non-governmental organizations, and private industry. Authors provide the depth of expertise required by the subject matter and the geographic diversity of the issues under consideration in this report. The lead author team includes Charles L. Walthall, Jerry Hatfield, Peter Backlund, Laura Lengnick, Elizabeth Marshall, and Margaret Walsh.

The authors wish to thank Dr. John Reilly (Massachusetts Institute of Technology), Dr. Louise Jackson (University of California - Davis), Dr. David Schimel (National Ecological Observation Network), and John Antle (Oregon State University) for providing expert review of its contents, and members of the USDA Global Change Task Force for their comments.

Chapter 2

An Overview of U.S. Agriculture

Agriculture is a major economic sector within the United States, with more than 2 million farms covering about 900 million acres and gross annual farm income between $300 and $350 billion. The farm sector – i.e., farmers, seed companies, and producers and distributors of agriculturally based products – has a long history of innovating and adapting to changing economic, environmental, regulatory, and climate conditions and has become much more productive over time. For example, in 1910, U.S. farmers cultivated 330 million acres and supplied food and fiber to a population of 92.2 million. By 2006, on the same cultivated land area, U.S. farmers supplied food and fiber to 297.5 million people.

Agriculture in the United States is a dynamic, self-adjusting system that responds to changes or fluctuations in trade, policy, markets, technology, and climate. The United States is a global supplier of agricultural products. With agricultural exports totaling slightly less than $140 billion, and agricultural product imports totaling less than $90 billion, agriculture offers a net positive to the U.S. trade balance. In addition to crops and related agricultural goods, more than 200 different products from

across the United States are produced from livestock; these account for slightly more than half of the total economic value of the agricultural sector. Common to all of these commodities is sensitivity to climate variability and change.

Since 1900, U.S. farms have grown larger, more mechanized, less labor intensive, and more specialized. While the number of farms has fallen, the total amount of land used for agricultural practices has remained fairly constant, and production and productivity have increased dramatically. Today, agriculture accounts for a declining share of employment and gross domestic product (GDP) (Dmitri et al. 2005; Hoppe et al. 2007) (Figure 2.1), with U.S. farms generating less than 1% of total U.S. GDP as of 2007 (O'Donoghue et al. 2011).

At the same time that average farm size has increased, U.S. farms have also become more specialized over time, concentrating on the production of fewer commodities per farm (Figure 2.2). Specialization offers both benefits and risks. Specialization allows the farmer to concentrate on particular areas of expertise and minimize the different types of capital investments and inputs required for production,

Fig. 2.1. Farms, land in farms, and average acres per farm in the U.S. 1850–2007. Most of the decline in farms occurred between 1935 and 1974. The break in the lines reflects an adjustment in the methods employed by the Census of Agriculture. Source: USDA ERS 2002.

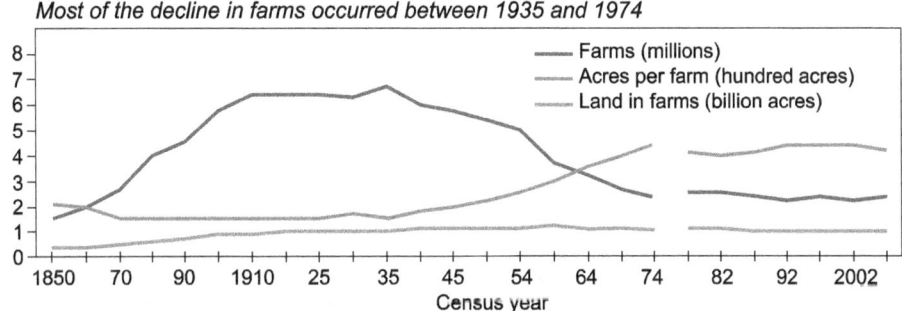

Note: The break in the lines after 1974 reflects the introduction of an adjustment to estimates of the farm count and land in farms. Beginning in 1978, the data are adjusted to compensate for undercoverage by the Census of Agriculture. For more information, see Allen (2004). Source: USDA, Economic Research Service, compiled from Census of Agriculture data.

while simultaneously making a farm more vulnerable to catastrophic loss of a particular crop due to insects, pathogens, or extreme weather events, for example (O'Donoghue et al. 2011). Though specialization trends vary by sector, the smallest farms tend to be the most specialized, while larger farms generally produce a wider variety of commodities (Hoppe et al. 2007; Melhim et al. 2009a, b).

Forces Affecting U.S. Agriculture

A diverse set of forces have sculpted U.S. agriculture, including productivity increases, integration of national and global markets, and

changing consumer demands for convenience, healthier products, and environmentally friendly production (Dmitri et al. 2005). A large part of the success of U.S. agriculture results from the dynamic, self-adjusting characteristics of the system, which responds constantly to changes or fluctuations in environmental conditions, trade, policy, markets, and technology. This capacity to react and adapt to shifting circumstances will be invaluable in the face of changing climate; however, the pace and intensity of projected climatic changes present novel challenges to U.S. agriculture.

Economic Factors and U.S. Agriculture

After remaining steady from 1982 to 2002, farm product prices since 2002 have trended upward (O'Donoghue et al. 2011). However, these averages mask significant fluctuations; six major spikes of world crop prices have occurred since 1970, and producers regularly adapt production decisions to compensate for such price variability (Figure 2.3). Over the past decade, one of the most prominent characteristics of the domestic and world food system has been rare back-to-back price swings, first in 2007-2008, and then again during 2010-2011. While many factors contributed to these price swings, both occurrences were in part attributable to short-term, weather-driven yield shortages (Trostle et al. 2011). For example, 2010-2011 saw major drought effects on Russian and Chinese wheat production that, coupled with increased demand, increased prices. Severe weather in other nations affects the U.S. agricultural system because of the global scope of agricultural production.

Fig. 2.2. As U.S. farms have become more specialized, the number of commodities produced per farm has decreased. Source: USDA ERS 2002.

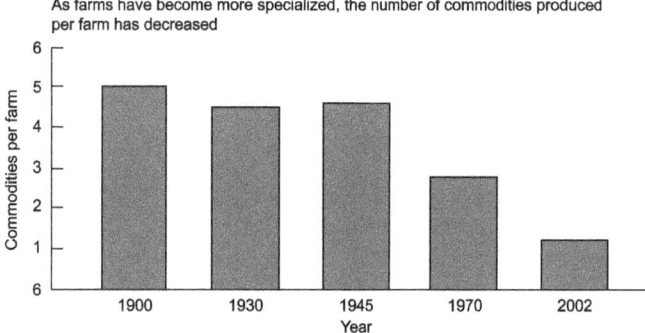

Fig. 2.3. Crop price spikes since 1970. The graph shows the weighted average of four crops (wheat, soybeans, corn and rice) based on IMF monthly export-weighted prices. Source: USDA ERS 2002.

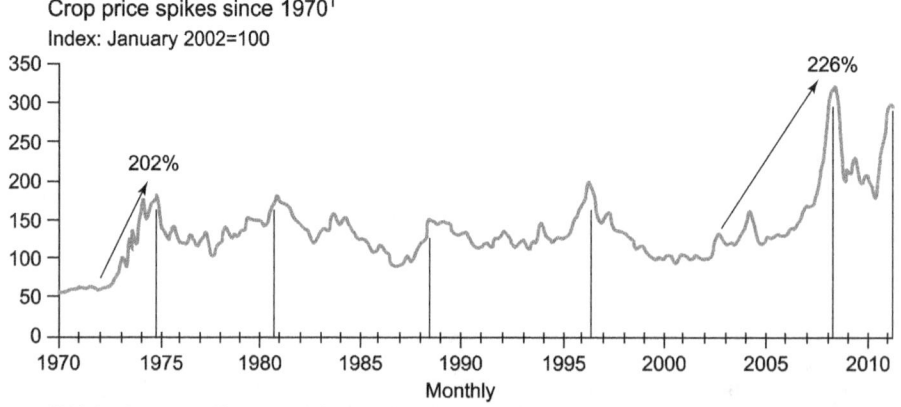

Producers employ a number of production, organizational, and marketing strategies to manage risks associated with farming (O'Donoghue et al. 2011). For example, farmers use production and marketing contracts to ensure outlets for their products and to reduce exposure to price and production shocks. While the share of field crops under contract has remained steady at 25-30%, the share of production in the livestock sector under contract increased from 33% from 1991-1993 to 50% between 2006 and 2007 (O'Donoghue et al. 2011). The share of field crops under contract may also be starting an upward trend, as one might expect due to recent volatility of field crop prices (Figure 2.4). Federal crop insurance is another increasingly important risk management tool, with the total number of insured acres up from 100 million in 1989 to more than 270 million acres in 2007 (O'Donoghue et al. 2011). Federal subsidies supporting the federal crop insurance program have increased at a similar rate, jumping from roughly $200 million in 1989 to more than $3.8 billion in 2007 (O'Donoghue et al. 2011).

The nature of other Federal Government policies providing additional farm support has changed over the past few decades, moving from price support and supply management policies to conservation and commodity payments that support farm income, reduce trade distortions, and promote environmentally sustainable production practices (Dimitri et al. 2005; O'Donoghue et al. 2011). Between 1999 and 2008, commodity payments such as direct payments, loan deficiency payments, and emergency disaster assistance represented between 74% and 93% of total farm program payments. During 2007, large-scale family farms and non-family farms received roughly 75% of such payments (Hoppe et al. 2010).

Large-scale family farms and non-family farms also received more than 60% of Federal working-lands conservation program funding, while conservation payments for land retirement went largely to small family farms (Hoppe et al. 2010). Rising prices significantly increased the value of agricultural production during 2007 relative to that of 2002, contributing to the recent reductions in farmer reliance on government commodity payments as a source of farm and farm household income (O'Donoghue et al. 2011).

The aggregate value of agricultural production remains fairly evenly split between livestock and crop production (Figure 2.5). The source of this value is widely distributed geographically across the United States (Figure 2.6), although some sectors are more regionally specialized than others. For example, corn and soybean production, which accounts for the largest share of crop value, is concentrated in the Midwest, Great Plains, and Delta States. Cattle and calf production, which dominates

Fig. 2.5. Comparison of market value of crops and livestock sold in the United States. Source: USDA-ERS data.

Market Value of Products Sold

Fig. 2.4. Production under marketing or production contracts for selected crops, 1991-2007. Source: USDA ERS 2002.

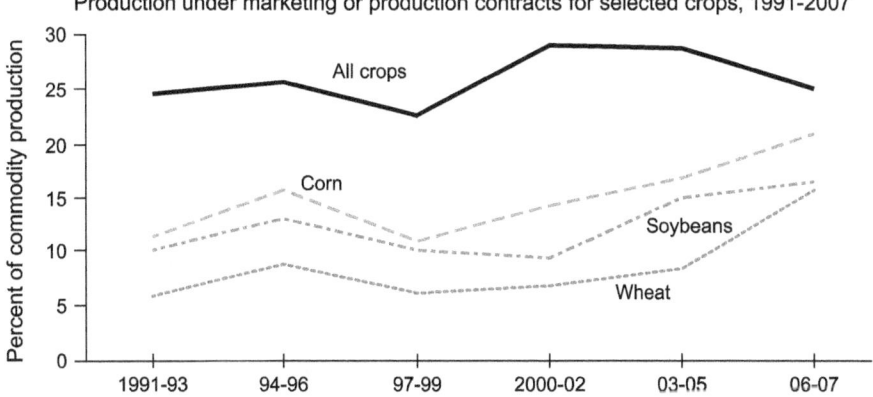

Production under marketing or production contracts for selected crops, 1991-2007

Note: Data include both marketing and production contracts.
Source: USDA, Economic Research Service and National Agricultural Statistics Service,
1991-1995 Farm Costs and Returns Survey and 1996-2007 Agricultural Resource Management Survey.

Fig. 2.6. Market value of all agricultural products sold in the United States. Source: USDA NASS 2007.

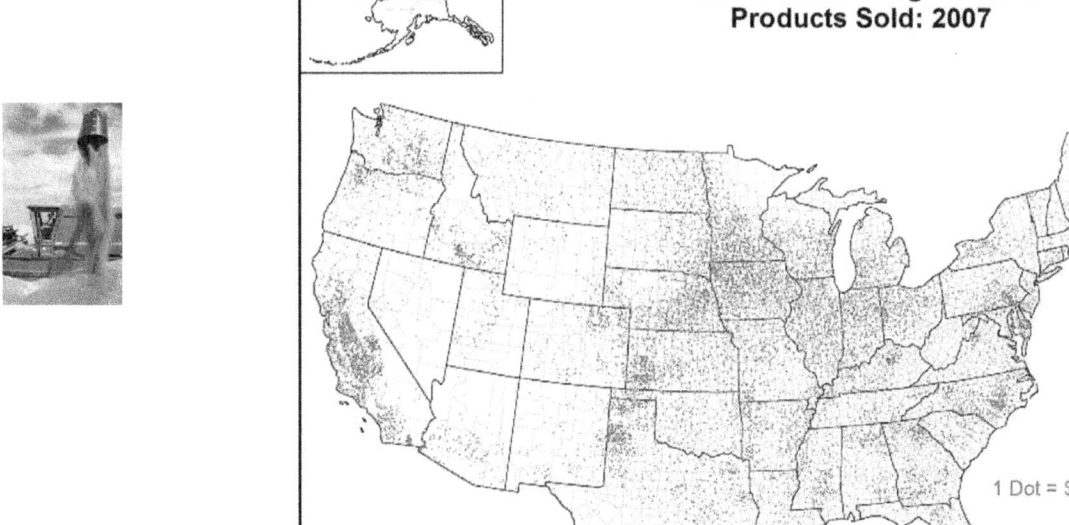

Fig. 2.7. Distribution of hog and pig production in the United States, 2002–2007. Source: USDA NASS 2007.

livestock production values, is distributed across the United States, with hog and pig production increasingly concentrated in only a few regions (Figure 2.7).

U.S. agriculture has become increasingly integrated into world markets, with both imports and

exports of agricultural products growing since 1935 up to present (Figure 2.8). U.S. exports constitute a large fraction of international markets in several export markets, with primary agricultural exports being soybeans, corn, wheat, and cotton (Table 2.1). The amount of each of these exported crops varies

Fig. 2.8. Annual fiscal year (Oct–Sept) U.S. agricultural trade, imports and exports, 1935-2011.

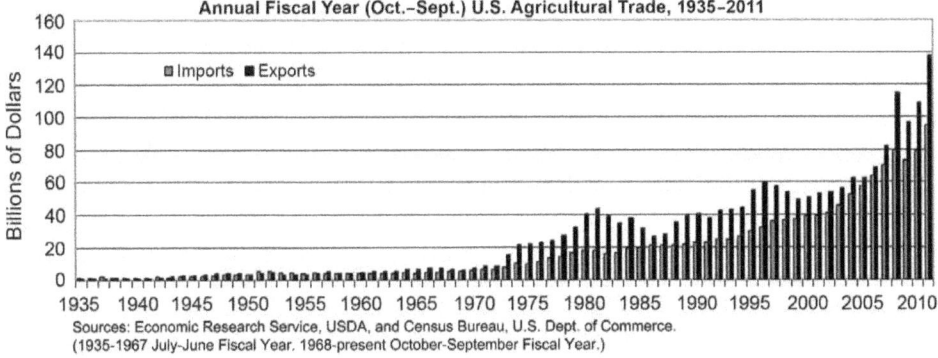

Table 2.1. Top 25 agricultural export and import commodities for fiscal year 2011 fiscal expressed in current dollars. Source: Compiled by USDA ERS from U.S. Department of Commerce data.

Exports 2011		Imports 2011	
Soybeans	20,347,317,208	Coffee Incl Prods	7,335,626,992
Corn	12,903,604,236	Wine	4,733,790,184
Wheat Unmilled	11,477,204,419	Cocoa and Prods	4,685,418,643
Cotton Ex Linters	8,861,356,654	Rubber/allied Gums Crude	4,419,760,212
Other Feeds & Fodder	5,486,053,783	Malt Beverages	3,526,937,192
Beef & Veal Fr/Froz	4,387,315,343	Beef and Veal Fr/Froz	2,747,998,291
Pork Fr/Froz	4,266,308,341	Biscuits and Wafers	2,629,883,334
Misc Hort Products	4,079,909,279	Misc Hort Products	2,562,187,257
Chickens Fr/Froz	3,348,722,320	Sugar Cane and Beet	2,534,132,361
Soybean Meal	3,341,173,322	Other Grains and Preps	2,362,366,330
Other Grain Prods	3,135,381,498	Other Beverages	2,130,744,738
Almonds	2,670,069,077	Essential Oils	2,090,178,270
Rice-Paddy Milled	2,096,410,318	Tomatoes Fresh	2,066,804,158
Soybean Oil	1,732,970,324	Bananas/Plantains Fr/Froz	1,969,880,327
Other Veg Oils/Waxes	1,732,735,395	Rapeseed Oil	1,760,653,987
Related Sugar Prod	1,599,701,798	Other Fruits Prep/Pres	1,666,332,390
Essential Oils	1,479,155,219	Cattle and Calves	1,450,996,963
Nonfat Dry Milk	1,451,990,267	Drugs Crude Natural	1,392,815,374
Seeds Field/Garden	1,354,074,918	Confectionery Prods	1,387,491,573
Other Dairy Prods	1,335,177,127	Feeds/Fodders EX Oilcake	1,364,875,618
Other Veg Prep/Pres	1,288,893,350	Cheese	1,061,226,335
Wine	1,229,833,343	Palm Oil	1,060,916,683
Beverages Ex Juice	1,228,014,393	Other Dairy Products	1,060,022,264
Chocolate & Prep	1,152,045,508	Grapes Fresh	988,555,117
Bovine Hides Whole	1,089,536,141	Berries EX Strawberries	964,566,293

widely, ranging from 11% for corn to 78% for cotton, as does the percentage of the global market share that each of these crops comprises. U.S. prominence as a corn supplier – supplying about 50% of global exports – makes world markets highly sensitive to U.S. supply and demand relationships, as well as to the weather conditions in the Corn Belt (Figure 2.9).

In contrast, a wide diversity of exporters in wheat (Figure 2.10) and soybean markets helps protect against variability in supply occurring due to yield shortages or excesses in any one region. The share of the world soybean market represented by U.S. exports, which is currently about 40%, has been falling steadily since the mid-1970s, in large part due to increasing production and export by Brazil and Argentina. The situation is similar for wheat, where the United States produces about 20% of the world's supply; other major suppliers include Argentina, Australia, Canada, Europe, and the Former Soviet Union. Of note, the U.S. position as a relatively major exporter but minor consumer of cotton makes cotton farmers in the United States more dependent on the world market, and poten-

Fig. 2.9. World's leading corn exporters. Source: USDA ERS 2011.

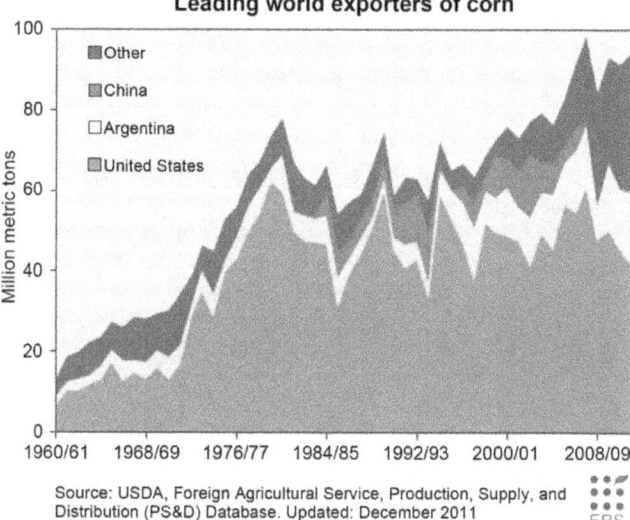

Source: USDA, Foreign Agricultural Service, Production, Supply, and Distribution (PS&D) Database. Updated: December 2011

tially more vulnerable to the trade policies of major importers than is the case for the Nation's other major export commodities.

U.S. engagement in global livestock markets has also increased over the past few decades. U.S. pork producers have increased exports from 2% of production in 1990 to 22% of production in 2011, while poultry exports have increased from 6% to 18%

Fig. 2.10. Market share of major wheat producers.

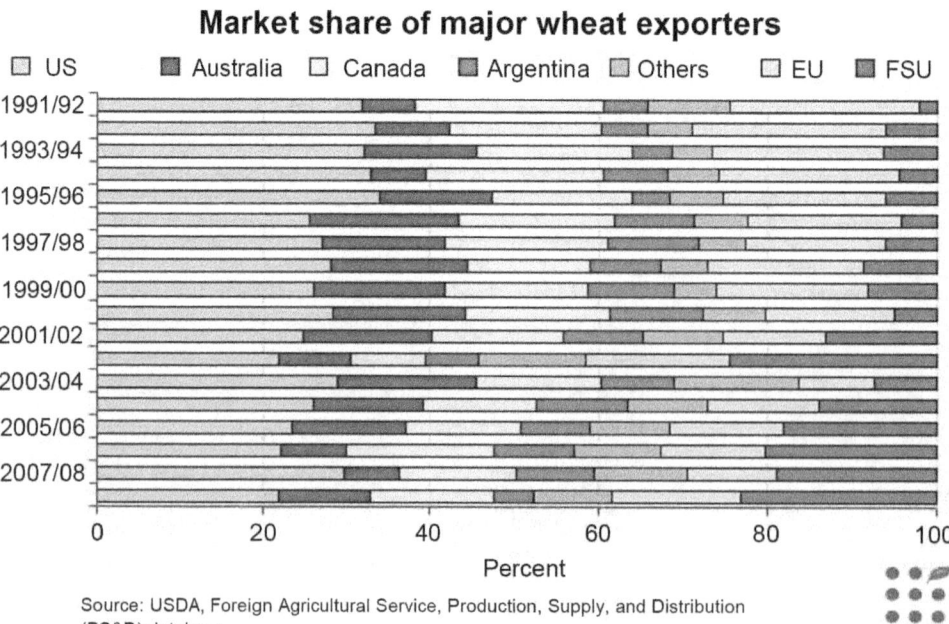

Source: USDA, Foreign Agricultural Service, Production, Supply, and Distribution (PS&D) database

of production over the same period. The United States is the largest producer of high-quality, grain-fed beef in the world. Though interrupted in 2004 by concerns of bovine spongiform encephalopathy (BSE), U.S. beef exports have been steadily increasing since that time (Figure 2.11). In part because imports and exports generally represent different grades of beef, imports to the United States have also trended higher over the past few decades, but since 2010 the United States has been a net exporter of beef. Strong global demand for dairy products has increased U.S. exports in recent years. As a percent of production, U.S. dairy exports have reached record levels (17% of production in 2011 on a skim-solids basis). However, greater participation in world markets has resulted in increased exposure across livestock sectors to global safety and disease concerns, trade policies, and demand shifts driven by changing consumer preferences.

Effects of Technology on U.S. Agriculture

Agricultural production has steadily increased since the 1940s, with the introduction of improved genetics, inorganic fertilizers, and crop protection chemicals, and cultural management practices. Yields of corn, wheat, soybean, and rice for the United States have shown increases over the period from 1940 to present, with corn showing the largest annual increase (Figure 2.12a) and wheat (Figure 2.12b) showing an increase in yield but at a much slower pace than corn. Increases in rice yields (Figure 2.12c) have been similar to corn in terms of the annual increase. Soybean yields (Figure 2.12d) have shown

a steady increase since 1940 with annual variations throughout the period. All of these crops show a common feature with variation in production among years; however, there are differences in the years that show decreases in yield because of the effect of weather differences in the areas in which these crops are grown and crop growing season.

Crop and livestock distribution across the United States largely exists in locations where individual commodities are best suited for growth due to some combination of climate, soil, and/or economic return on production. For example, winter wheat can be found in areas across the United States in which winter temperatures support crop survival, yet provide adequate exposure to the chilling temperatures needed for vernalization that lead to crop flowering and grain production. Grain crops, fiber crops, vegetable crops, horticultural crops, and fruit trees are distributed across the United States in areas where production is optimized. Production of each of these commodities is reported in the Census of Agriculture (USDA-NASS 2007). Also notable, however, are the expanded regions of production which have occurred due to changes in technology, climate, and economics; corn production in North Dakota and South Dakota provides one example of such expansion from the Midwest.

Climate Effects on U.S. Agriculture

Agricultural systems are primarily defined by prevailing spatial and temporal distributions of climatic and edaphic (soil-related) conditions. As

Fig. 2.11. U.S. Beef Trade, 1980–2008. The drop in demand associated with fears of Bovine Spongiform Encephalopathy (BSE) shows clearly in 2004. Source: USDA-ERS.

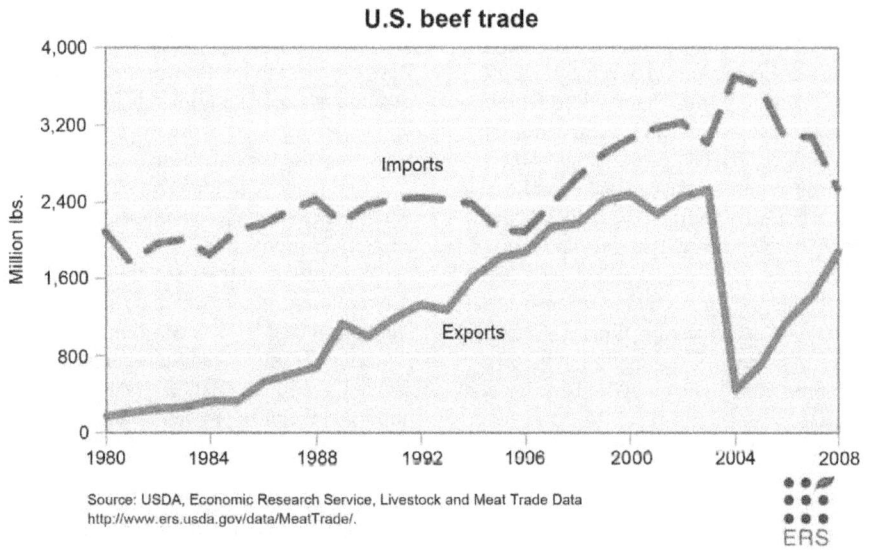

Fig. 2.12. U.S crop yields 1940 to 2010; corn(a), wheat (b), rice (c), and soybean (d). Source: USDA-NASS.

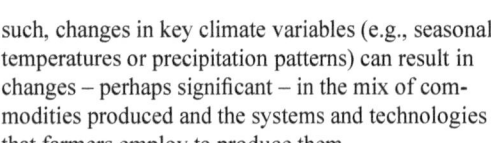

such, changes in key climate variables (e.g., seasonal temperatures or precipitation patterns) can result in changes – perhaps significant – in the mix of commodities produced and the systems and technologies that farmers employ to produce them.

Climate change presents a novel challenge to U.S. agriculture because of the sensitivity of agricultural system response to climatic variability and the complexity of interactions between agriculture and the global climate system. Interactions within the agricultural social-ecological system can result in synergistic effects that dampen or amplify the system response to climate change and complicate development of effective mitigation and adaptation options for U.S. agriculture (McLeman and Smit 2006; Reidsma et al. 2010; Smith and Olesen 2010). Developing the knowledge needed to manage agricultural production in a changing climate is a critical challenge to sustaining U.S. agriculture in the 21st century (Robertson and Swinton 2005; Howden et al. 2007; NRC 2010).

While the U.S. agricultural system has the ability to respond to changes or fluctuations in markets, technology, and the environment to a great degree,

individual agricultural products differ in their ability to adapt to changing climate conditions. For example, crops have different cardinal temperatures – the critical temperature range for ideal lifecycle development. These vary by species and between vegetative and reproductive growth stages. Basic temperature responses by crops range from a base-temperature requirement, i.e., the point at which growth begins, and a temperature maximum where growth ceases. Between these extremes exists an optimum temperature where plant growth is fastest. In general, optimum temperatures are lower for the reproductive stage than the vegetative stage, i.e., plants are less able to tolerate high temperatures during the reproductive stage. Increasing temperature generally accelerates progression of a crop through its lifecycle (phenological) phases, up to the species-dependent optimum, above which development (node and leaf appearance rate) slows. Temperature increases projected for the United States under high and low scenarios of future GHG emissions are therefore an important factor in projecting future U.S. agricultural productivity.

However, increasing air temperature is only one factor to consider under current and future climate change scenarios; local management practices such as irrigation will also influence effects on agriculture. For example, amply irrigated plants growing under arid conditions create microenvironments that are 10°C cooler than ambient air temperature due to evapotranspiration cooling. Variables such as solar and reflected long-wave radiation, wind speed, air humidity, and plant stomatal conductance also affect to what degree temperature will influence crop growth and development. Many climatic factors affect agricultural performance, and a complete understanding of climate change effects on U.S. agriculture requires an understanding of these variables and how they interact.

Like temperature, precipitation has a direct influence on agriculture. In many areas of the Nation, precipitation is projected to increase, particularly in northern regions, but the incidence of drought is also expected to increase in some areas, and changes in timing and rain/snow mix may increase the management challenge of delivering water to crops at the right time through irrigation systems and practices. The intensity of precipitation events is also expected to increase. Excess precipitation, both in the form of short bursts or through increased amounts over longer episodes, can be just as damaging as too little precipitation, leading to increased erosion and decreased soil quality. Increased evapotranspiration due to warmer temperatures can result in less available water – even with increased precipitation – especially in soils with limited soil water holding capacity. Corn is susceptible to excess water in the early growth stages, which can result in reduced growth or even plant death, while deficit soil water leads to less growth and yield if the stress occurs during the grain filling period of growth (Hatfield and Prueger 2011).

In addition to their direct effects on plants, changes in temperature and precipitation also affect the amount of water in the atmosphere. With increases in water vapor, cloud cover is expected to increase, leading to a decrease in incoming solar radiation. This effect has already been observed in the solar radiation record around the world. Stanhill and Cohen (2001) observed a 2.7% reduction per decade during the past 50 years, with the current solar radiation totals reduced by 20 W m^{-2}. Changes in solar radiation will directly affect crop water balance and evapotranspiration and have less effect on crop productivity due to other factors limiting productivity (e.g., water and temperature) (Hatfield et al. 2011). In a later, U.S.-centered study, Stanhill and Cohen (2005) evaluated data from across the United States for sunshine duration and global irradiance (solar radiation), finding that after 1950 there has been a

decrease in solar duration, with sites in the Northeast, West, and Southwest showing notable decreases. They suggested that more detailed solar radiation records will be required to quantify temporal changes in solar radiation related to cloudiness and aerosols. Reduction in solar radiation in agricultural areas in the last 60 years as revealed by models (Qian et al. 2007) is projected to continue (Pan et al. 2004) due to increased concentrations of atmospheric GHGs, which may partially offset acceleration of plant growth. A study on solar radiation by Medvigy and Beaulieu (2011) examined the variability in solar radiation around the world. They concluded there was an increase in solar radiation variability that was correlated with increases in precipitation variability and deep convective cloud amounts that may affect solar energy production and terrestrial ecosystem photosynthesis. Any change in solar radiation resources under climate change will affect the agricultural system.

Finally, changes in CO_2, temperature, precipitation, and radiation over the next century will be accompanied by other changes in atmospheric chemistry that have implications for agriculture. One of the most significant of these is expected changes in concentrations of ground level ozone. The number and complexity of these biophysical interactions demonstrates the necessity of systemic analyses of potential climate effects on agriculture. All of the factors mentioned above will affect U.S. agriculture over the coming century, but their ultimate effect will also depend on social and economic feedbacks.

Agriculture: A Complex Social-Ecological System (SES)

Agriculture in the United States is a dynamic social-ecological system (SES) of plant and animal production that is dependent on a complex flow of resources regulated by the internal processes and interactions between ecological and social elements of the system that exist and function at multiple scales of space, time, and social organization (Figure 2.13). Recognition of the interactions within and across scales is fundamentally important in understanding the behavior of the SES at any particular focal scale (Gunderson and Holling 2002). For example, consider the linkages to management decisions at the enterprise focal scale: producer decisionmaking is driven by perceptions of risk and other personal considerations and preferences, knowledge of the production capacity of the enterprise, and multiple external drivers (for example, consumer preference, market demand, government policies, and climatic variability, etc.). Thus the process of plant and

animal production is a dynamic interplay between society-driven demand and natural resource supply capacity operating within the context of policy structures and producer knowledge.

The U.S. agricultural SES interacts with the global climate system through multiple linkages that are increasingly responsive to climatic change (Figure 2.14); these interactions challenge the sustainability of U.S. agriculture in the 21st century (NRC 2010). Agricultural adaptation to climate change will be particularly challenging because of the sensitivity of agricultural SES response to climatic variability and extremes, the complexity of interactions between the agricultural SES and the global climate system, the uncertainties associated with how the climate will change and the resulting effects at different temporal and spatial scales, and the increasing pace of climatic change (Howden et al. 2007). Multiple stresses such as the limited availability of water, the loss of biodiversity, and reduced soil, water, and air quality interact to increase the sensitivity of the agricultural SES to climatic change (Easterling et al. 2007; NRC 2010).

The complexity of the SES response to climatic change is a critical challenge to research and development efforts seeking to sustain U.S. agriculture in the 21st century (Robertson and Swinton 2005; Howden et al. 2007; NRC 2010). Understanding the vulnerabilities of key components of the U.S. agricultural SES and the linkages across spatial and temporal scales within it, under multiple uncertainties (climate, economic, policy, etc.), is critical to the development of effective adaptation strategies.

New Research for a Novel Challenge

New research, development, management and governance strategies utilizing complex-systems science to address the complicated interactions and multidimensional nature of agricultural SES response to climate change are needed (e.g. Easterling et al. 2007; Howden et al. 2007; Jackson et al. 2010; NRC 2010; Hatfield et al. 2011; Lin 2011; Newton et al. 2011; Tomich et al. 2011). The climate change challenge requires an innovative framework to facilitate holistic systems thinking across the multiple dimensions of the agricultural SES, e.g., farm (Rivington et

Fig. 2.13. The United States agricultural social-ecological system can be viewed as a dynamic system of interacting social and ecological components and processes linked to global scale biophysical systems such as climate system and the nitrogen cycle and global scale social systems such as international trade and governance (Humphrey 2011).

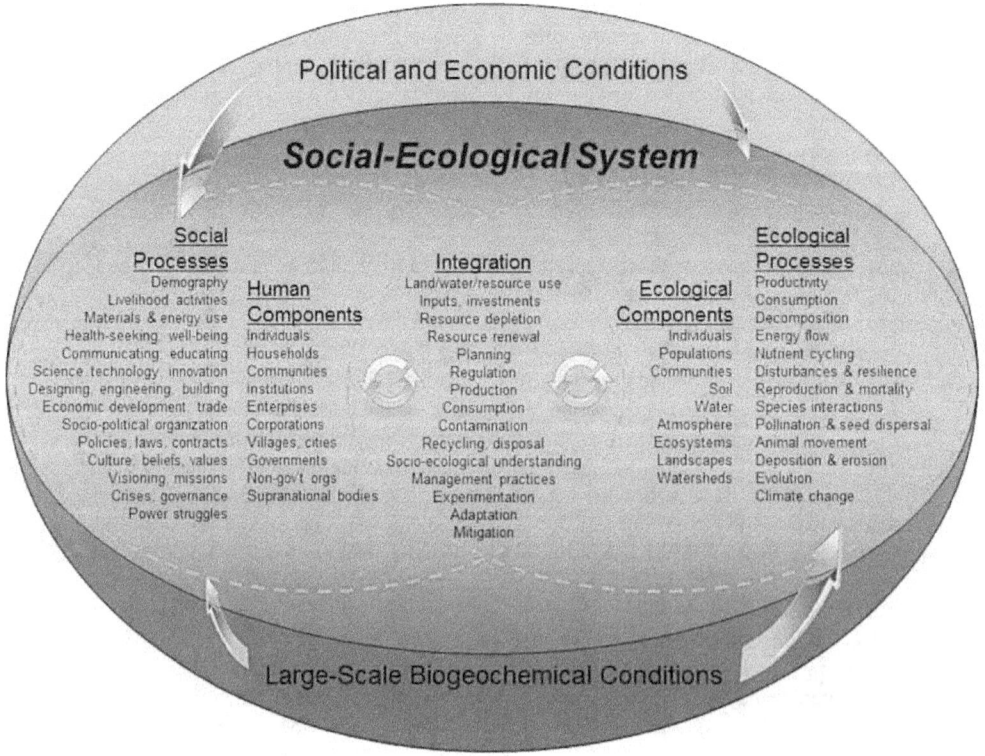

al. 2007), region (Reidsma et al. 2010), institutional and political structures (Romero and Agrawal 2011), trade globalization (Young et al. 2006), and multiple actors and the way they are represented (Rounsevell et al. 2012). A specific feature of the SES approach is that it places the resource manager (or producer or consumer) within and as part of the system, rather than external to it (Walker et al. 2002; Janssen et al. 2006). The SES approach incorporates the benefits of both focused disciplinary research to tackle specific problems, and yet offers the breadth and flexibility required to enable integration of multiple disciplines (including the social sciences) within the overall analysis and development of holistic adaptation strategies. Without such a common framework to organize findings, isolated knowledge does not accumulate and policies may be disjointed or even contradictory (Ostrom 2009).

Also important in the analysis of SES dynamics is the rates of change (fast or slow) and the spatial scales at which each occurs, the durability of each

scale (i.e., how resilient a particular component such as a farm system is), and the consequences for the response of the whole SES (Young et al. 2006). The SES conceptual framework thus presents an approach that can integrate across research disciplines and practitioner differences to develop appropriate policy and management responses to build resilient systems (Walker et al. 2002; Young et al. 2006; Nelson et al. 2007; Ostrom 2009). The risk of not using an integrated systems approach such as the SES is that a greater probability exists of relying overly on technological packages that lack site specificity, on developing policies with conflicting objectives, and failing to recognize the social or ecological thresholds that limit local adaptive capacity (Nelson et al. 2007) or wider planetary boundaries (Rockstrom et al. 2009).

Past adaptation in the U.S. agricultural SES has focused on objectives of production and profitability operating within limited ranges of uncertainty (e.g., fuel costs, demand, and market prices, etc.). Climate

Fig. 2.14. A schematic framework representing key linkages between the anthropogenic drivers of climate change and the global climate system (IPCC 2007, p 26, Figure I.1). An assessment of the interactions between key components of this system may inform the development of adaptation options to reduce future climate change impacts on the United States agricultural SES.

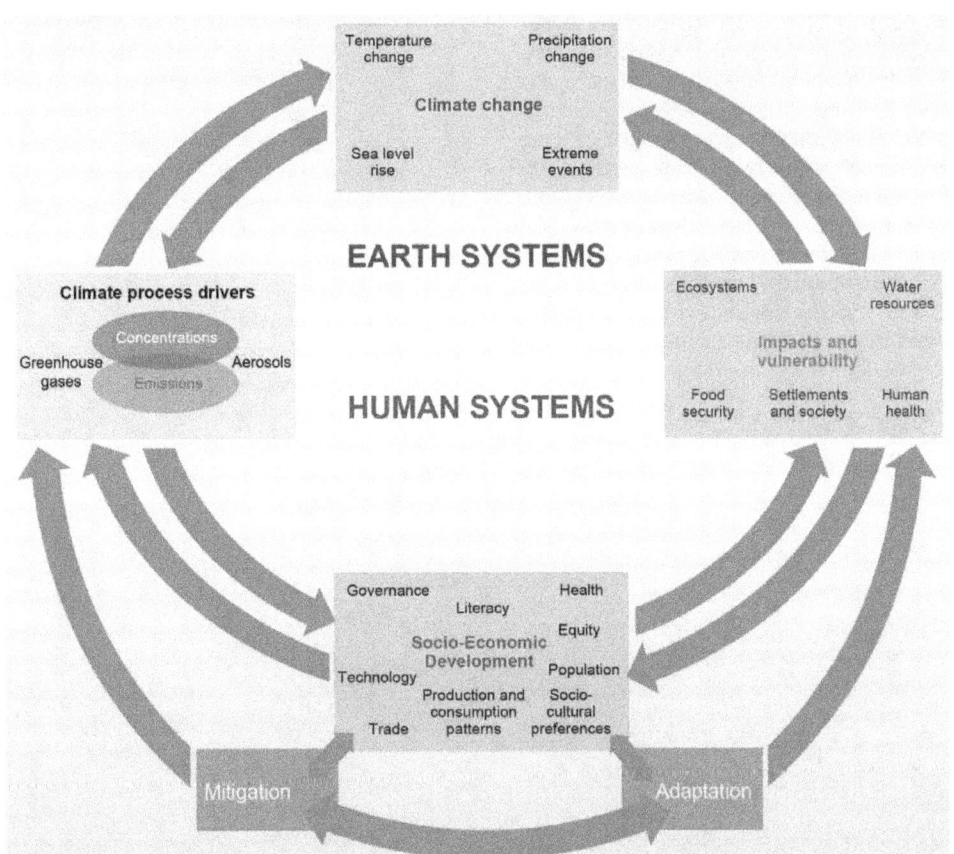

change presents a novel adaptation driver, one that involves greater uncertainty in future projections of risk that is further complicated by tradeoffs between mitigation and adaptation. A conceptual framework such as the SES approach may aid in the management of the uncertainties faced by stakeholders operating across multiple dimensions of the U.S. agricultural SES under climatic change. For example, Dessai et al. (2009) argue that a need exists for robust adaptation strategies that are flexible enough to account for the range of such uncertainties and multiple objectives. The use of probabilistic climate projections (based on multi-model ensembles) provides information about the likely range of climate change effects and the development of risk-based assessments (New et al. 2007), while adaptive management can provide a buffer against uncertainty (Howden et al. 2003; Littell et al. 2011). This implies that adaptation planning needs to exist within a conceptual structure that facilitates the consideration of risks across multiple components of the system and responses.

SES research strategies draw on the knowledge and methodological approaches of agroecology, transdisciplinary problem-solving, integrated analysis, adaptive management, and resilience science in an effort to integrate the biophysical, economic, and social dimensions of agricultural production. Experience from the development of the ecosystem management approach (Liu et al. 2011) and the integration of social and ecological objectives taking an SES perspective may be instructive in guiding strategies for U.S. agriculture adaptation. Taking an ecosystems-management perspective facilitates incorporation of multiple considerations and objectives, and potential for attainment of multiple benefits (Munang et al. 2011). The novel analyses made possible with these strategies also support efforts to identify and explore

the critical thresholds and dynamic cross-scale interactions that drive agricultural SES response to climatic change. These new strategies are being increasingly employed to understand agricultural resilience to climate change effects across a variety of focal scales, including at the farm level (Hendrickson et al. 2008; Moriondo et al. 2010; Reidsma et al. 2010; van Apeldoorn et al. 2011), in rural communities (Atwell et al. 2008; Nelson et al. 2010a; Nelson et al. 2010b; Arbuckle 2011), and across regions (Allison and Hobbs 2004; Wolfe et al. 2008; Easterling 2009; Jackson et al. 2011).

Chapter 3

An Overview of the Changing Climate

Evidence of Changing Climate Across the Globe

The United States and the U.S. agricultural system are part of a changing world. There is broad scientific agreement that the climate conditions affecting agriculture are being changed on a global scale by human activities. Burning of fossil fuels, deforestation, and a variety of agricultural practices and industrial processes are rapidly increasing the atmospheric concentrations of CO_2 and other greenhouse gases (GHGs) (IPCC 2007a, pg. 10). These changes in atmospheric composition are increasing temperatures, altering the timing and distribution of precipitation, and affecting terrestrial and marine ecosystems (IPCC AR4 WGI and WGII SPM's; IPCC 2007a, b). Scientific evaluation of the effects of global climate change done as part of the Intergovernmental Panel on Climate Change (IPCC) Fourth Assessment Report (AR4), new studies in the peer-reviewed scientific literature (e.g., Allison et al. 2009), and assessments by the U.S. Global Change Research Program, the U.S. National Research Council, and other scientific bodies provide strong

evidence of ongoing changes in the Earth climate system. Among the findings:

- Global-average surface temperature has increased by about 0.74°C (0.56-0.92°C) over the 20th century (IPCC 2007a, pg. 10).

- Long-term temperature records from ice sheets, glaciers, lake sediments, corals, tree rings, and historical documents demonstrate that every decade in the late 20th century has been warmer than the preceding decades (NOAA NCDC 2011; Hansen et al. 2012; Jones et al. 2012).

- The most recent 50 years likely have been the warmest worldwide in at least the last 1,300 years (IPCC 2007a, pg. 9), and 10 of the 11 warmest years on record have occurred since 2001 (NOAA NCDC 2011; Hansen et al. 2012).

- Observations since 1961 show that at depths of at least 3,000 meters, the average temperature of the global ocean has increased; this deep storage

Rainfall Intensities

Rainfall intensities have also increased in many parts of the world over the last few decades, including in the United States. Karl and Knight (1998) found that more than half of the observed increases in total annual precipitation for the United States between 1910 and 1996 were due to increases in frequency of large events, defined as occurring in the upper 10 percentile of measured daily values. Analyses using data from 1910 through 1999 (Soil and Water Conservation Society, 2003) showed that the proportion of precipitation coming in the form of heavy (>95th percentile), very heavy (>99th percentile), and extreme (>99.9th percentile) daily precipitation events increased by 1.7%, 2.5%, and 3.3% per decade, respectively, on average across the United States. The number of large events is on the increase, and the increase has been greatest for the most extreme of events. Groisman et al. (2005) looked at measured daily rainfall data from over half the land area of the world and found trends of increased probability of extreme events for many regions outside of the tropics. The IPPC 4th Assessment Report (Meehl et al., 2007) also projects general increases in precipitation intensities across much of the Earth.

of heat together with the higher heat capacity of water is causing the ocean surface to warm more slowly than the land surface (IPCC 2007a)

- Global sea level has increased about 12-22 centimeters (cm) during the 20th century, but satellite records confirm that the rate of sea level rise has now almost doubled to about 3.4 millimeters (mm) per year (IPCC 2007a; Allison et al. 2009).

- Precipitation is highly variable and trends are more difficult to isolate, but overall precipitation and heavy precipitation events have increased in most regions; at the same time the occurrence of drought has also been on the rise, particularly since 1970 (IPCC 2007a; Allison et al. 2009).

- Mountain glaciers and ice caps, as well as snow cover, are receding in most areas of the world. Both the Greenland and Antarctic ice sheets are now losing mass at increasing rates. The extent and thickness (volume) of Arctic sea ice is declining, and lakes and rivers freeze later in the fall and melt earlier in the spring (IPCC 2007a; Allison et al. 2009).

- Winter temperatures have increased more rapidly than summer temperatures, and nighttime minimum temperatures have warmed more than the daytime maxima. Across the United States (and elsewhere), the observed number of record high temperatures is about three times higher than the number of record cold events (IPCC 2007a; Meehl et al. 2009).

Projections of Future Global Climate

Human influences will continue to alter Earth's climate throughout the 21st century. Our current scientific understanding, supported by a large body of theoretical, observational, and modeling results (e.g., the IPCC AR4), indicates that continued changes in atmospheric composition will result in further increases in global average temperature, rising sea level, and continued declines in snow cover, land ice, and sea ice extent. The IPCC AR4 contains projections of the temperature increases that would result from many different emissions scenarios (Nakicenovic et al. 2000). For this report, we concentrate on low and high emissions alternatives defined by the IPCC. The characteristics and ways in which these alternative scenarios might be achieved are described below:

- A low emissions scenario for the 21st century (IPCC *Special Report on Emissions Scenarios* (SRES) B1) could be achieved by continued

improvements in technology, low or no growth in population, and effective action by individuals, corporations, and governments to limit emissions. In such a scenario, atmospheric concentration of CO_2 would increase to about 550 parts per million by volume (ppm), which would increase global average surface temperature by about 1.1°C to 2.9°C in 2100 relative to 1980-1999.

- A high emissions scenario for the 21st century (IPCC SRES A2) would result from a slowing in technological improvement, significant population growth, and less effective actions taken by individuals, corporations, and governments to limit emissions. In this scenario, atmospheric concentration of CO_2 would increase to about 800 parts per million (ppm), which would increase global average surface temperature by about 2.0°C to 5.4°C by 2100 relative to 1980-1999.

It is important to note that the average surface temperature for each of the above scenarios would vary by region (see Figure 3.1). Polar areas will warm more than lower latitude areas, land more than oceans, and continental interiors more than coastal areas.

Climate change in the 21st century will be driven predominantly by overall emissions of GHGs and aerosols, as well as by the strength of feedbacks in the climate system. The lower the emissions during the next 100 years, the lower the climate change experienced over this time and beyond. It is important to note, however, that the climate differences between high- and low-emissions scenarios will mainly occur in the latter half of the 21st century due to the inertia of the climate system. The climate changes being experienced today are mainly the consequence of past emissions, and today's emissions will continue to cause climate change into the future. Even if atmospheric concentrations of greenhouse gases are stabilized (which would require large decreases from current emissions levels), land surface temperatures will continue to rise for decades, while ocean temperatures and sea level will continue to rise for centuries (IPCC 2007a; Solomon et al. 2009).

Changing Climate Across the United States: The Last 100 Years

The United States is a large country with complex topography and thus has a considerable variety of climate across its different regions. Alaska has high annual precipitation and relatively cool average temperatures due to very cold winters, while Florida has

Fig.3.1. Projected global temperature changes for the 2020s (left side) and 2090s (right side) compared to 1980-1999 for low emission (B1) and a high emission (A2) scenarios. The differences between scenarios get wider as time progresses. Source: IPCC 2007.

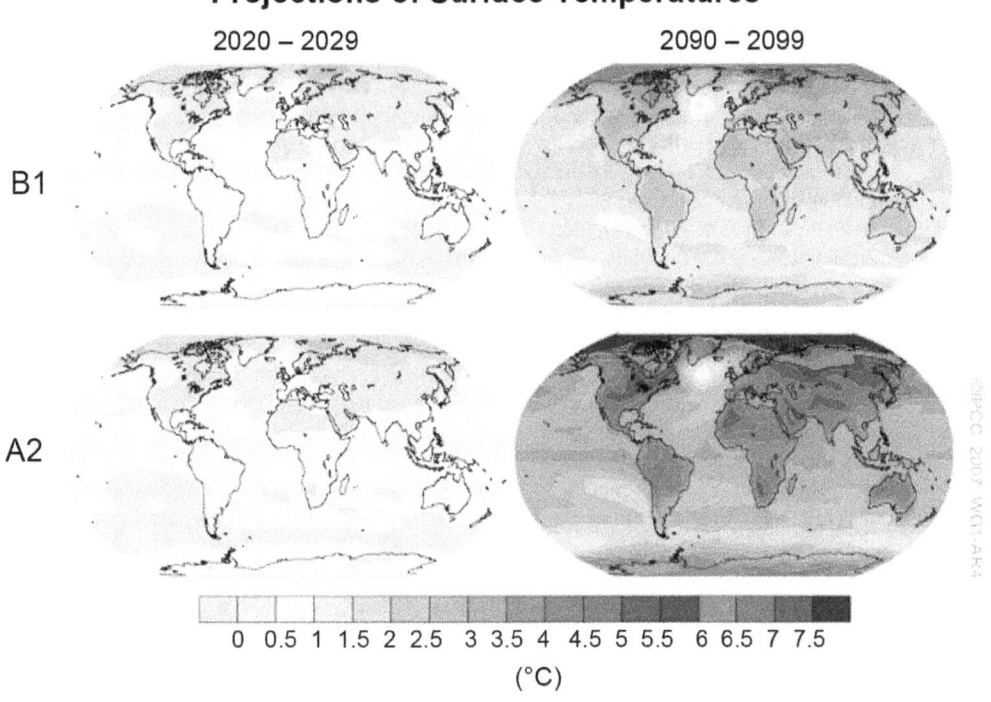

Climate Models and Climate Research

Scientists often rely on computer models to better understand Earth's climate system because they cannot conduct large-scale experiments on the atmosphere itself. Climate studies are largely based on general circulation models, which consist of mathematical representations of physical, chemical, and biological processes that drive the Earth's climate. Climate models, like weather models, use a three-dimensional mesh that reaches high into the atmosphere and into the oceans. At regularly spaced intervals, or grid points, the models apply laws of physics to compute atmospheric and environmental variables, simulating the exchanges among gases, particles, and energy across the atmosphere. To investigate possible future changes in climate, different scenarios of future greenhouse gas concentrations are used as inputs for the model calculations that produce simulations of climate for the next century and beyond. The primary focus in climate simulations is on large regional to global scale interactions of the various components of the climate system rather than local scales. This approach enables researchers to simulate global climate over years, decades, and millennia. Most current-generation global models use grid points that are about 100-200 km apart. Scientists then often use these global model results to drive finer scale, regional models with grid spacing ranging from 2-50 km (similar to weather prediction models) for "small regional" and local-scale studies. There are also a number of statistical methods that downscale the global models based on available high-resolution observations to estimate finer scale change. Most recently, a small number of climate modeling centers are experimenting with very high resolution global simulations, however such experiments require very large and expensive amounts of supercomputing time and produce very large data sets that are still challenging to analyze.

high annual precipitation and relatively warm average temperatures throughout the year. The Southwest experiences warm summers and low average precipitation, while the Northeast has warm summers and relatively high average annual precipitation. U.S. regional climates have been very different in the distant past due to large-scale natural climate fluctuations – e.g., 18,000 years ago much of New England was under a thick layer of ice – but it has been relatively stable during the last 1,000 years as Europeans explored and migrated to North America, and since the founding of the United States and its development into a modern nation. However, there have been significant inter-annual variations within U.S. regions during this period. For instance, year-to-year variations in the El Niño-Southern Oscillation, fluctuations in the North Pacific and tropical Atlantic, as well as other large-scale patterns of natural variability have been responsible for extended droughts or temperature changes and shifts in the timing and distribution of precipitation in some areas, but these have been relatively short-lived (i.e., seasonal to decadal) anomalies followed by return to more typical regional conditions.

The observational records for the last century clearly show that these natural year-to-year fluctuations are superimposed on long-term changes in temperature and precipitation (NOAA NCDC 2011; Hansen et al. 2012; Jones et al. 2012. See also Figure 3.2). This trend over the past century is consistent with observations of long-term climate change in many other areas around the globe, which, as described in

the previous section, are almost certainly the consequence of human-induced changes in the Earth's atmosphere.

Temperature

In most regions of our country, annual mean temperatures have increased significantly, though considerable variability exists across regions. While Alaska has experienced the largest changes, the northern Midwest and the Southwest have also warmed significantly throughout, with autumn temperatures increasing the least. The only large region of the United States to experience a linear cooling trend over the last century is the Southeast. This region warmed during the early part of the 20th century, but then cooled markedly during the middle part of the century and is now warming again (Figure 3.2, right upper pullout).

Changes in temperature have varied by season as well as by region (Figure 3.3). As noted above, during the most recent decades, the cooling of the Southeast has slowed and then reversed, particularly in the cold seasons. Summer has warmed in most areas, but not as pronounced as winter. Spring is also warmer in most regions, likely related to more rapid melting of snow. In much of the United States, the century-long linear trend for autumn is still largely dominated by the warming in the 1930s and 1940s, and therefore the long-term trends remain small, with the Southwest a notable exception. This overall warming is reflected in a lengthening of the growing

Fig. 3.2. Observed linear 20th century (1901-2006) temperature trends for North America based on stations with complete, consistent, and high quality records. The spatial resolution is high in the contiguous U.S., but lower at high latitudes, where interpolation was applied to achieve a 0.5-degree (about 50 km) resolution. The two pullouts illustrate that linear trends don't always represent the underlying variability well. Data source: University of Delaware, Matsuura and Willmott 2009, Version 2.01, based on augmented Global Historical Climatology Network, Version 2, http://climate.geog.udel.edu/~climate/html_pages/download.html.

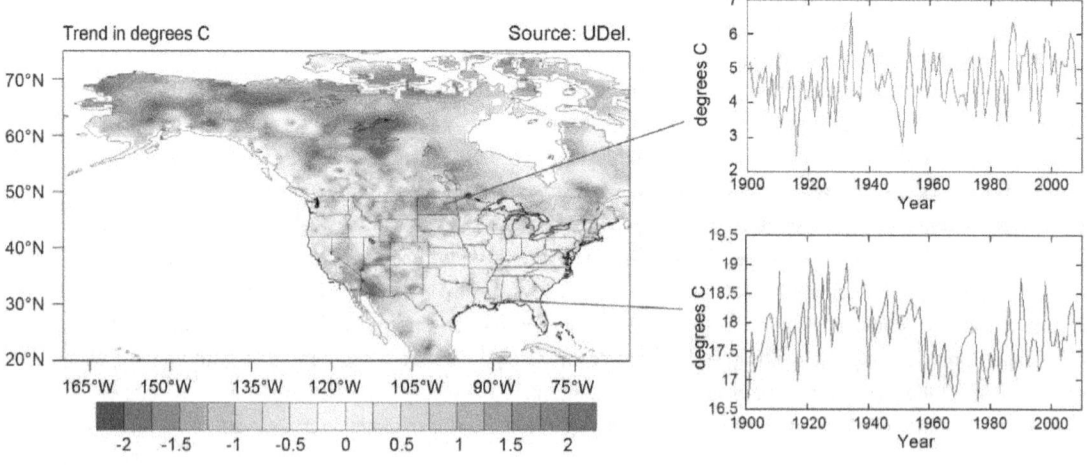

season in the Northern Hemisphere by about 4 to 16 days since 1970 (i.e., 1 to 4 days per decade) (US EPA 2010).

Overall, new record warm temperatures are becoming more common than record cold throughout the year; across the United States (and elsewhere), the observed number of record high temperatures is now about three times higher than the number of record cold events (IPCC 2007a; Meehl et al. 2009).

Precipitation

In contrast to temperature, precipitation is often a very small-scale process and thus has greater

heterogeneity than is the case with temperature across the continent. Much of the Northwest, Central, and Southern United States now receive more precipitation than 100 years ago, while other areas, such as parts of the Eastern Seaboard and the Rocky Mountains and much of the Southwest, receive less (Figure 3.4, lower panel).

As can be seen in Figure 3.5, century-long trends are not continuous through time or across seasons. Natural variability has led to decadal fluctuations with distinct periods of both drought (e.g., the 1930s Dust Bowl, and droughts in western regions) and wet intervals. It is important to recognize that analyses of average precipitation trends across years and

> Overall, new record warm temperatures are becoming more common than record cold throughout the year.

Fig. 3.3 Variation of 20th Century U.S. temperatures by season: Linear trends of observed surface temperatures over North America (left panels), and time evolution for spatial averages by season for three selected regions (right panels) relative to 1901-1930. Data source: Univ. of Delaware, Matsuura and Willmott 2009.

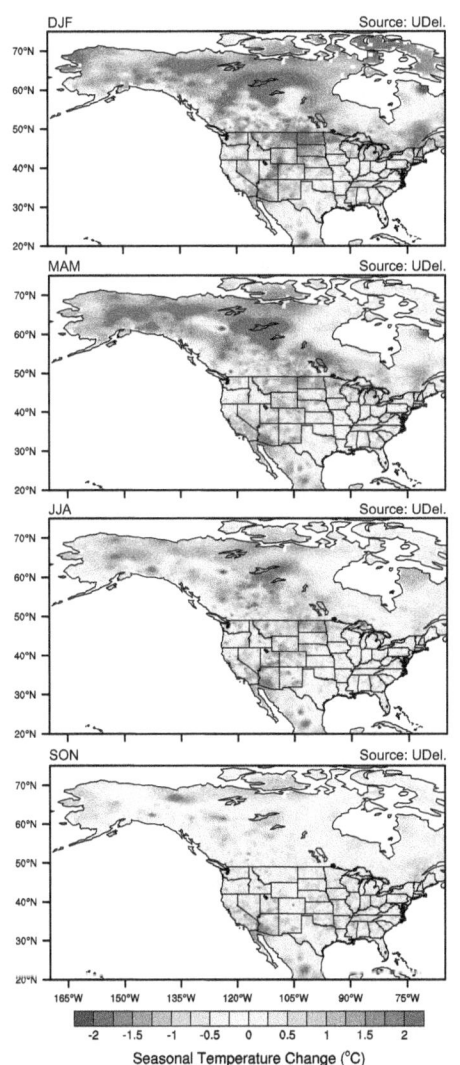

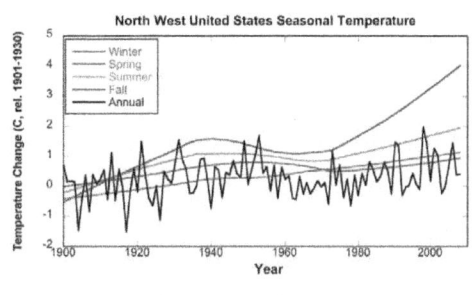

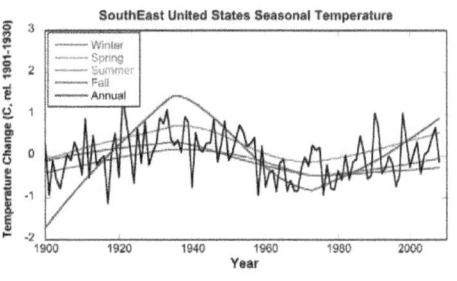

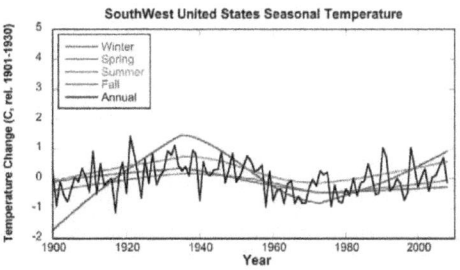

Fig. 3.4. Same as 3.2, but for precipitation: Observed linear 20th century (1901-2006) precipitation trends for North America. Data source: University of Delaware, Matsuura and Willmott 2009.

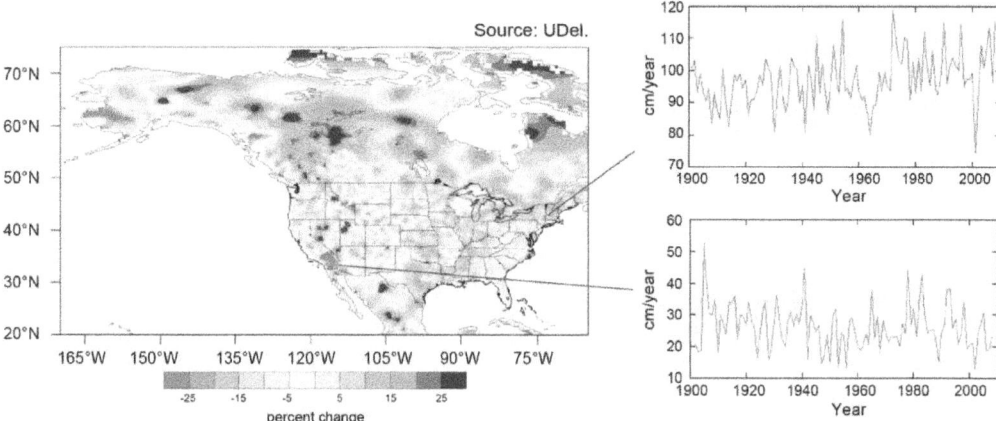

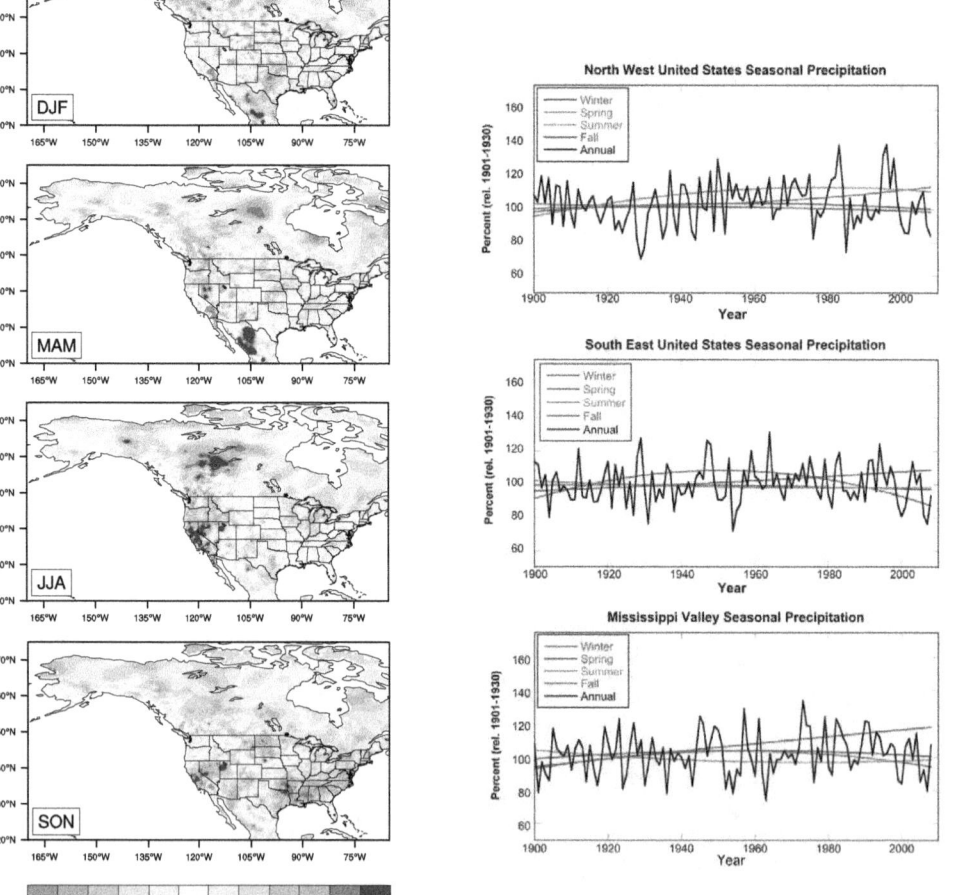

Fig. 3.5 Same as 3.3, but for precipitation. Variation of 20th Century U.S. precipitation by season relative to 1901-1930. Data source: Univ. of Delaware, Matsuura and Willmott 2009.

large regions can blur the variability that occurs over smaller regions or at specific periods in time (see Figures 3.4 and 3.5).

The intensity of precipitation has also increased in most areas, even if some regions get less water overall. This trend is consistent with an overall warming, since the atmosphere's water-vapor carrying capacity increases due to higher average temperature. However, this does not mean that all of the additional moisture is available for agriculture and other biological and ecological processes. More intense rain leads to faster surface runoff, and higher temperatures enhance evapotranspiration losses to the atmosphere, both resulting in less available moisture in soils.

Projections of Future U.S. Climate Change

It is very likely that U.S. climate conditions will continue to change throughout the 21st century. For the purposes of this document, we have chosen to show projections for low and high emissions scenarios for the coming decades (centered around the 2040s) and the end to the 21st century (centered around the 2080s) to illustrate how different levels of global GHG emissions could affect future U.S.

climate conditions. The differences between high and low scenarios of future GHG emissions are much more noticeable near the end of the century than they are in coming decades, similar to the global analysis described above. The results shown here (Figures 3.6–3.11) are based on multi-model ensemble averages produced for the IPCC AR4 that have been downscaled to 12-km horizontal resolution and bias-corrected to provide as much detail as possible about the projected regional changes (see Maurer 2007).

Temperature

The entire United States is likely to warm substantially over the next 40 years, with an increase of 1°C to 2°C over much of the country (Figures 3.6 and 3.7). This is a substantially greater rate of change than that observed over the last century, reflecting the accelerated rate of increase in GHG concentrations and temperatures observed during the last few decades.

Much of the interior United States is likely to see increases of 2°C to 3°C, while the southeastern and western coastal areas will experience about 1°C to 2°C degrees of warming. The cooling in the Southeast during the middle of the 20th century is projected to become warming in throughout the 21st

Fig. 3.6. Projections of U.S. summer surface temperature from a 16-model ensemble for a low emissions scenario (SRES-B1, top panels) and a high emissions scenario (SRES-A2, bottom panels) relative to 1970-1999. The near-term differences between scenarios (left panels showing the 2040s) are much smaller than the long-term differences (right panels showing the 2080s). Data source: CMIP3.

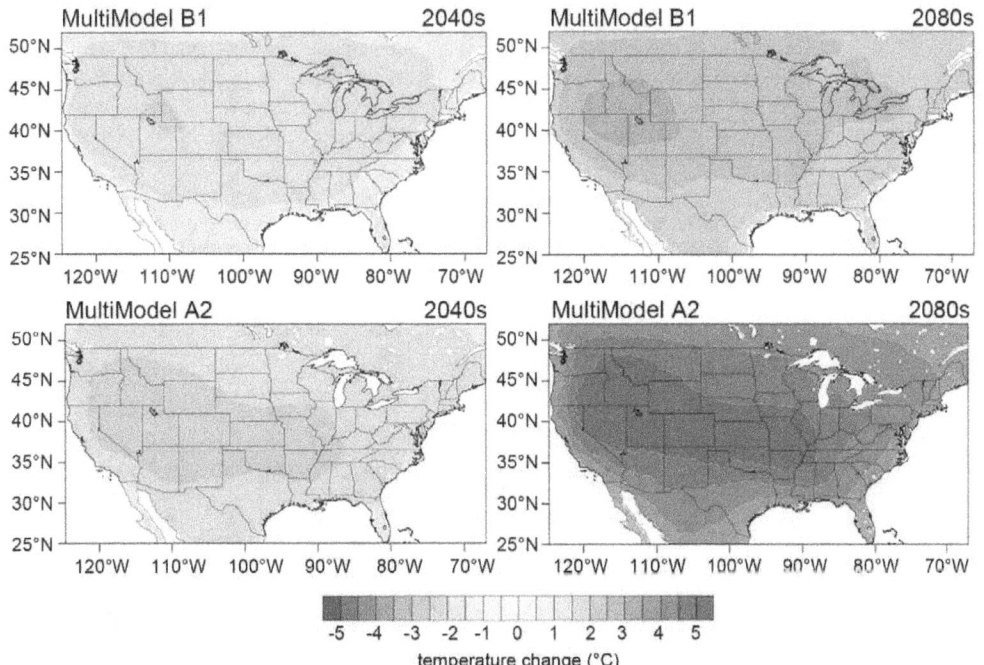

Fig. 3.7. Same as Fig. 3.6, but during winter. Source: CMIP3.

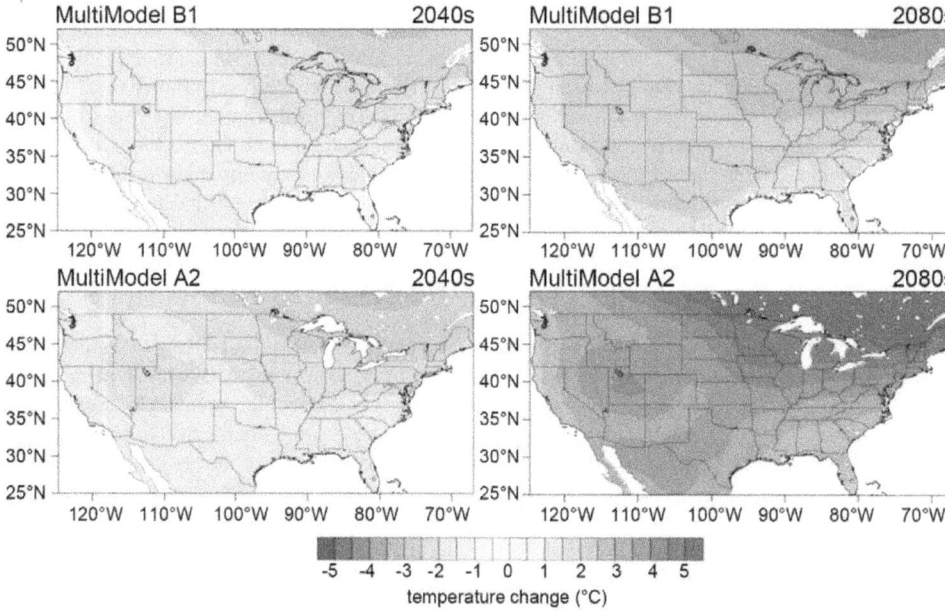

century, continuing the warming seen during the last few decades in that region.

Looking ahead to 2100, a low-emissions scenario is likely to produce a summer-time warming of 3°C to 4°C in much of the Interior West, with warming of 2°C to 3°C almost everywhere else. A high-emissions scenario is likely to result in warming of 5°C to 6°C in much of the Interior West and Midwest, with warming of 3°C to 5°C in the Southeast and far western regions. These changes in mean temperatures will very likely be accompanied by significant increases in hot nights (Figure 3.8, left panel). This widespread warming will lead to a further shift in the length of the growing season, reaching the scale of

a month or two. Occurrence of frost days will also change significantly, particularly in the West (Figure 3.8, right panel).

Precipitation

Projected changes in precipitation are more uncertain because they are sensitive to local conditions as well as shifts in the large-scale atmospheric circulation; these uncertainties are probably larger in summer than in winter. Figure 3.9 shows projections of change in summer precipitation. Over the next 30 to 40 years, models agree that the Northwest is likely to become noticeably drier, with reductions of 15-25% in summertime precipitation. Much of the central

Fig. 3.8. In a high emissions scenario, the U.S. growing season will lengthen by as much as 20-40 days by the end of the century (left panel). The number of frost days (days with minimum temperatures below freezing) will be reduced by 20-60 days in much of the United States. Both panels produced from multi-model ensemble projections based on simulation results from CMIP-3.

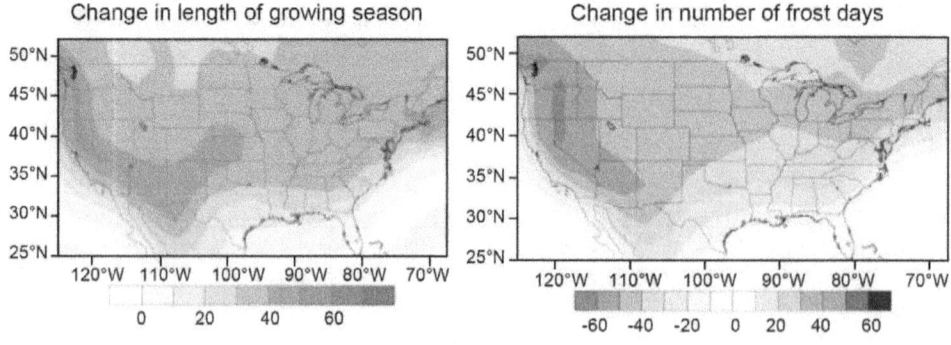

South will likely sees decreases of about 5%, while some northern central and eastern U.S. regions are projected to experience increases of 5-15%.

Interestingly, the simulations for the low-emissions scenario indicate that summer precipitation might remain largely stabilized during the second half of the 21st century after substantial change during the first part of the century. In the higher emissions scenario in which emissions continue to increase, however, the emerging summer precipitation pattern shows a substantially drier Northwest and South, while the increasing moisture input along the Eastern Seaboard and Northeast is likely to strengthen even further.

Snow availability and timing of snowmelt runoff is an important seasonal concern, particularly in western regions which are dependent on snow accumulation and gradual release of water stored in snowpack throughout the spring and summer. Figure 3.10 shows projected changes in U.S. winter precipitation.

Most regions of the northern and central U.S. are projected to see an increase of 5% to 15% in winter precipitation over the next 30-40 years, but areas along the southern U.S. border will likely see much less, with decreases of 5% to 10% possible; southern Texas will likely experience the largest decreases, which could be as much as 15-20%. By 2100, the low-emission scenario produces smaller further changes in climate, particularly in the southern regions where drying is expected. At higher latitudes,

however, the increase in winter precipitation is projected to continue quite substantially, particularly in the Northwest and Northeast. For the high-emissions scenario, the models produce a similar precipitation pattern but with substantially larger enhancement of the near-term trends. Throughout the far South, particularly in Texas and Florida, reductions in precipitation may reach 20% to 25%, with strong precipitation increases in the North of 20% or more (Figure 3.10).

Although precipitation increases are anticipated for large areas of the United States in both the low- and high-emission scenarios, it is again important to note that this does not necessarily translate into substantially more available moisture for agriculture at the time when water is needed. Higher temperatures lead to both earlier melt and runoff of water stored in snow cover, and to increased evapotranspiration losses to the atmosphere. In addition, more precipitation is projected to fall in shorter, more intense storms, leading to more rapid runoff. These factors may offset the projected increase in mean precipitation amounts in the United States and thus lead to less available moisture in soils and less surface water for organisms or ecosystems.

Extreme Conditions

Average temperature and precipitation are not the only factors that affect agricultural systems. Extreme climate conditions, such as dry spells, sustained droughts, and heat waves can have large effects

Fig. 3.9. Summer precipitation projections for a low emissions scenario (SRES-B1, upper panels) and a high emissions scenario (SRES-A2, lower panels) relative to 1970-1999. Data source: CMIP-3.

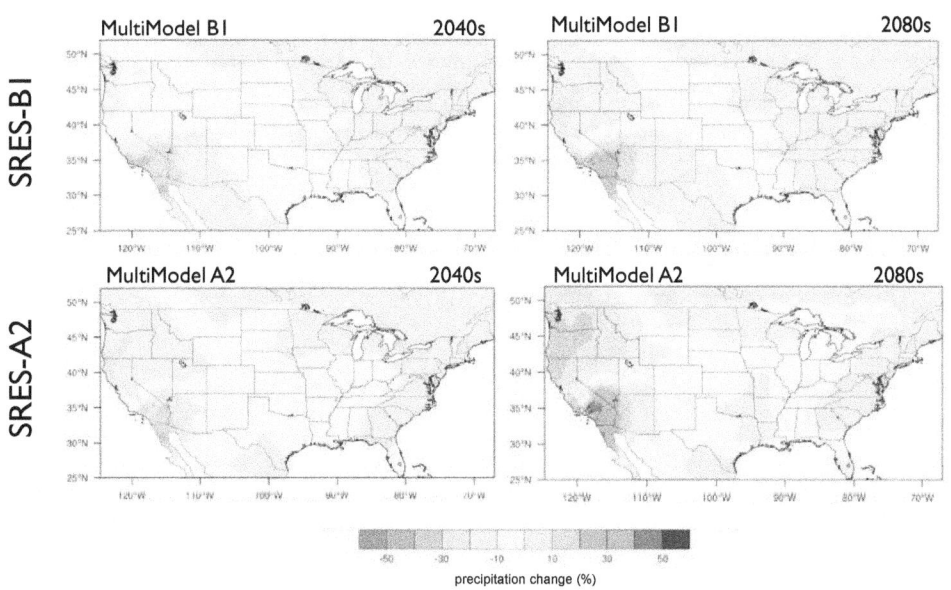

Fig. 3.10. Same as 3.9 but for winter precipitation projections. Data source: CMIP-3.

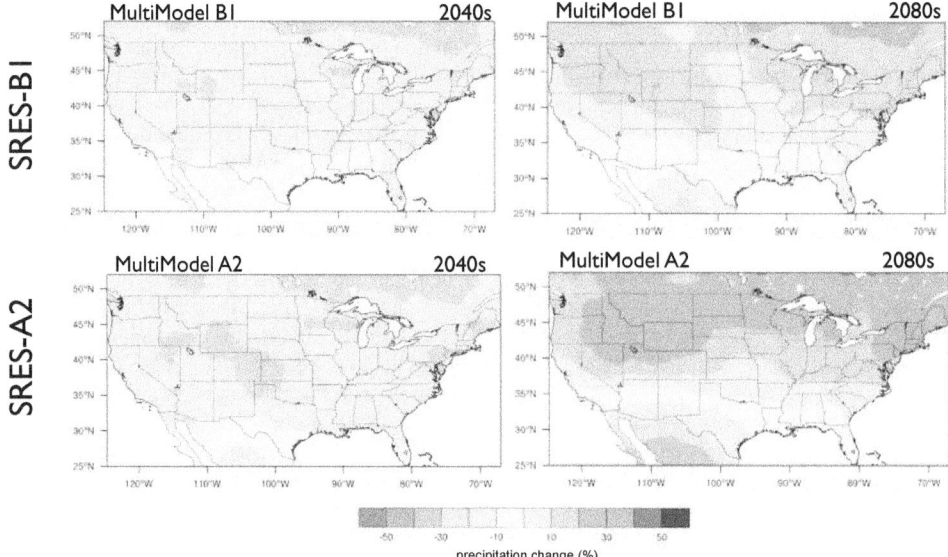

on crops and livestock. Changes in the incidence of these extreme events could thus have major effects on U.S. agricultural productivity and profitability. Although models are limited in their ability to accurately project the occurrence and timing of individual extreme events, observations indicate an emerging signal that is consistent with projections of an increase in areas experiencing droughts and the occurrence of more intense precipitation events (Alexander et al. 2006; IPCC 2007a; Zhang et al. 2007). Figure 3.11 shows how the number of hot nights and the duration of very low (agriculturally insignificant) rainfall periods are projected to change by the end of the 21st century under a high-emissions scenario.

Changes in Tropospheric Ozone

Current ground-level ozone concentrations are considerably higher in the Northern Hemisphere than the Southern Hemisphere, with background monthly mean ozone concentrations in the Northern

Fig. 3.11. The left panel shows projected changes in duration of dry spells (consecutive number of days with less than 2 mm of precipitation). Areas in the West and Southwest can expect increase in dry intervals by more than 12 days. In some parts of the Northwest and south-central Texas, this increase could be as much as 2-3 weeks, mostly concentrated in the summer season. Some North Central, as well as the East and Southeast regions, are expected to experience little change. The right panel shows increases in the number of hot nights (defined as nights with a minimum temperature warmer than 90% of the minimums between 1971 and 1990) across the United States projected for the high emissions scenario by the end of the 21st century. By 2100 many parts of the United States could experience 30-40 additional hot nights. Data Source: CMIP-3.

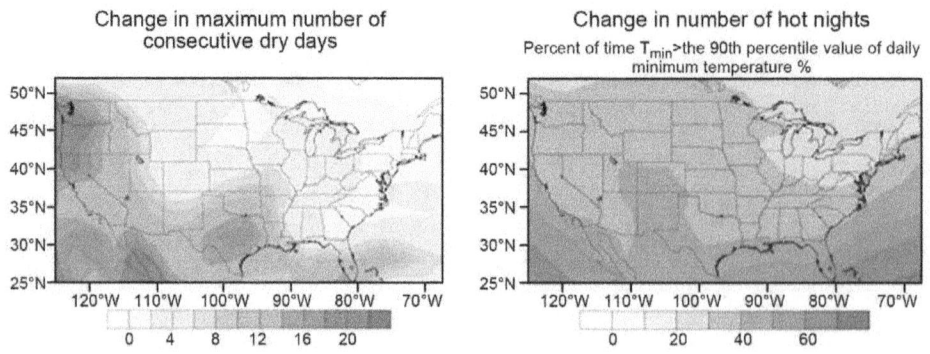

Ozone

Ozone is formed photochemically in the troposphere when its precursors (nitrogen oxides and carbon monoxide) generated mainly from fossil fuel combustion (e.g., from the energy-generating and transportation sectors), methane emissions, large fires and industrial processes, react with volatile organic compounds (some from natural vegetation) and oxygen in the presence of sunlight. Ozone and its precursors can be transported hundreds of kilometers into rural areas where agricultural activities occur and native and managed forests exist. There is evidence that ozone, along with it precursors, is increasingly transported from Asia over the Pacific Ocean to North America. Climate change will alter the dynamics, temperature, and humidity of the troposphere affecting the occurrence of stagnation episodes that lead to high ozone conditions. Efforts to reduce pollution emissions have mostly succeeded in lowering peak ozone concentrations, however mean ozone levels in many areas remain high enough to impact crops and forests (Booker et al. 2009). Concern exists that future reductions in local ozone formation may be offset by rising background levels as global industrialization increases.

Hemisphere ranging from 35 to 50 parts per billion by volume (ppb) (Emmons et al. 2010). In North America and Europe, higher ozone concentrations occur in the summer with peak daily concentrations occurring in the late afternoon. Future ground-level ozone concentrations have been explored using a variety of possible emissions scenarios (Dentener et al. 2006; Stevenson et al. 2006; Dentener et al. 2010; Lamarque et al. 2011).[1] While multi-model studies using the IPCC SRES scenarios showed increases in ozone of 2-7 ppb in the Northern Hemisphere between 2000 and 2030 (Prather et al. 2003), more recent studies have indicated smaller changes or even reductions in ozone if current air quality legislation is implemented (Dentener et al. 2006). Changes in temperature and water vapor will also affect future ozone concentrations. Increased temperatures on the order of 0.7°C and associated changes in water vapor are expected to decrease surface ozone in cleaner regions, but tend to have the opposite effect in more polluted areas (Dentener et al. 2006; Lamarque et al. 2011). A larger influx of stratospheric ozone under climate change conditions and an increasing contribution of imported ozone from intercontinental transport could also lead to changes in ground level ozone in the future (Dentener et al. 2010).

Conclusions

The climate of the U.S. has changed during the last 100 years, and the rate of climate change has increased during the last several decades. A large number of observations and simulation experiments clearly show that this long-term change is distinct from the natural variability of climate that the U.S. has always experienced. In most areas of the United States, temperatures have increased. Precipitation changes have been more variable; while some regions have experienced increases, others have seen decreases. The growing season has become longer all across the U.S., and the number of frost days has decreased.

U.S. climate will continue to change during the next century. It is very likely that the amount of change will be significantly greater and the rate of change more rapid than that experienced during the last 100 years. There will be more warm nights and longer periods of extreme heat, and the incidence of both drought and very heavy precipitation events is expected to increase. Continued increases in greenhouse gas emissions will increase the amount of climate change the United States will experience in the next 100 years. Limiting the increase in greenhouse gas emissions will reduce the rate and amount of climate change during this period.

There is still some mismatch between the typical spatial scales of climate science and agricultural science experimentation. The agricultural community requires information on local to regional scales (tens of kilometers or less) to support studies of climate effects and adaptive capacity. Higher resolution simulations and projections of change, accompanied by improved documentation of probabilities and model biases, is a critical overarching research need for study of the effects of climate on agriculture.

---

[1] The scenarios for ground-level scenarios do not take into account the rapid increase in the recent rapid development of natural gas using unconventional methods (e.g., hydraulic fracturing); such techniques may increase the amount of ground-level ozone.

The current body of scientific literature on climate change effects on agriculture clearly shows that availability of water is one of the most important elements of adaptive capacity.

Time scale is another important issue. The 100-year projections that have been the main focus of climate change forecasting are of minimal use in agricultural planning. The decadal to multi-decadal projections that are being undertaken by many climate modeling groups are more relevant but still not well suited for informing agricultural decisions. Improved seasonal to yearly forecasts would be a major step forward in providing information useful for production decisions and near-term planning.

The current body of scientific literature on climate change effects on agriculture clearly shows that availability of water is one of the most important elements of adaptive capacity. Yet representation of precipitation and other elements of the water cycle is one of the most difficult challenges in climate and weather modeling. Simulation and prediction of precipitation is less robust than simulation and prediction of temperature. Improving our ability to accurately predict changes in the timing and amount of precipitation is also a high priority research need for agriculture and climate.

There is also a profound need for design and development of more sophisticated and complete modeling systems and simulation experiments that include the simultaneous interacting effects of multiple stresses on plants and animals, such as increased temperatures, increased GHG levels, decreased water availability, and increased pest populations. Better integration of biological, ecological, economic and climate models is needed to develop a more complete picture of climate change vulnerability, adaptive capacity, and effects. More accurate representation of the complexity of change will result in the production of more accurate and usable projections.

Chapter 4

Climate Change Science and Agriculture

The preceding two chapters have provided the backdrop for the next series of chapters, describing the U.S. agroecosystem and some of the social, physical, and economic components, as well as the past, present, and likely future state of the global and national climate. The next series of chapters will explore many of these aspects in greater detail. In this chapter, focus will be on the complex and myriad interactions between climate and the U.S. agroecosystem, highlighting both the direct and indirect effects of current and future climate change.

Often assessments of the effects of climate change on agricultural focus on rising air temperatures, changing precipitation patterns, and increasing atmospheric CO_2 concentrations. All of these are critical factors on growth of crops, forage, livestock, and other agricultural products. But, as mentioned in the previous chapter, another effect of climate change with important consequences for U.S. agriculture is the incidence of air stagnation, which can lead to episodes of higher ozone concentration in agricultural regions. Together, these characteristics comprise direct (abiotic or physical) effects of climate change.

Equally important to consider are indirect effects of climate change. Included in this mix are effects of changing air temperature and precipitation on non-crop species found in agroecosystems, such as insects, weeds, pathogens, and invasive species. As is the case with direct effects, these indirect (biotic or biological) effects of climate change can have considerable influence on the vitality of U.S. agriculture.

The complexity of the system and how direct and indirect effects of climate change influence both the system as a whole and individual species within the system makes projecting the net outcome of changes to climate change challenging. In the sections below, some of the current science describing these direct and indirect climate effects on agriculture is presented.

Direct Climate Change Effects

Air Temperature

Average air temperatures in the contiguous United States are expected to increase during the next 30 years (Karl et al. 2009). Such temperature increase will almost inevitably affect agricultural products, as all plants have minimum, maximum, and optimum temperatures that define their response to temperature. The minimum and maximum temperatures are the boundaries for growth; between these extremes is an optimum temperature that allows greatest growth. Beyond a certain point, higher air temperatures adversely affect plant growth, pollination, and reproductive processes (Klein et al. 2007; Sacks and Kucharik 2011). However, as air temperatures rise beyond the optimum, instead of falling at a rate commensurate with the temperature increase, crop yield losses accelerate. For example, an analysis by Schlenker and Roberts (2009) indicates that yield growth for corn, soybean, and cotton gradually increases with temperatures up to 29°C to 32°C and then decreases sharply as temperature increases beyond this point.

Temperature minimum, maximum, and optimum have been summarized by Hatfield et al. (2011) for a number of different species, providing thresholds to use when assessing the potential effects of increasing temperature on crop growth. This information was used in crop simulation models to show that continued increases of temperature will lead to yield declines between 2.5% and 10% across a number of agronomic species throughout the 21st century. Other evaluations of temperature on crop yield have had varying outcomes: Lobell et al. (2011) showed estimates of yield decline between 3.8% and 5%; Schlenker and Roberts (2009) used a statistical approach to produce estimates of wheat, corn, and cotton yield declines of 36% to 40% under a low-emissions scenario, and declines between 63% to 70% for a higher emissions scenario. Note that these simulation exercises did not incorporate effects

The complexity of the system and how direct and indirect effects of climate change influence both the system as a whole and individual species within the system makes projecting the net outcome of changes to climate change challenging.

of rising atmospheric CO_2 on crop growth, yield reductions due to pests, crop genetic variability, or management innovations such as new fertilizers, rotations, tillage, or irrigation. Experiments are currently underway to update simulation models to account for interacting effects of temperature, CO_2, and moisture on crop growth, genetic variability, and production effects.

Research tends to focus on the effects of average air temperature changes on crops; however, minimum air temperature changes may be of greater importance (Knowles et al. 2006) because minimum temperatures are more likely to be increased by climate change over broad geographic scales (Knowles et al. 2006). Minimum air temperatures affect nighttime plant respiration rate and can reduce biomass accumulation and crop yield (Hatfield et al. 2011). Even as climate warms and minimum average temperatures increase, years with low maximum temperatures may more frequently be closer to achieving the temperature optimum, which will result in higher yields than is the case today during years when average temperatures are below the optimum. Welch et al. (2010) found this to be the case for a historical analysis of rice in Asia – higher minimum temperatures reduced yields, while higher maximum temperature raised yields; notably, the maximum temperature seldom reached the critical optimum temperature for rice. As future temperatures increase, the authors found that the maximum temperatures could decrease yields if they rise substantially above the critical zone.

Maximum temperatures are affected by local conditions, especially soil water content and evaporative heat loss as soil water evaporates (Alfaro et al. 2006). Hence, in areas where changing climate is expected to cause increased rainfall or where irrigation is predominant, large increases of maximum temperatures are less likely to occur than will be the case in regions where drought is prevalent.

Increasing air temperature can enable earlier planting during the spring if suitable moisture and soil temperature conditions exist, resulting in a longer growing season. A longer growing season creates more time to accumulate photosynthetic products for greater biomass and harvestable yields as long as the temperatures do not exceed optimum values. However, increasing temperatures will also increase crop water demand and larger plants will use more soil water as part of the growth process (Betts et al. 2007). The positive effects of temperature could be offset by increased variation of precipitation and soil water availability to the crop. At the same time, a longer growing season can affect water availability

(Betts et al. 2007), as well as weed and insect interactions with crops.

In addition to effects on crops, increasing air temperatures will affect livestock production through increases in animal stress, with such stress further amplified when higher air temperature is coupled with higher relative humidity. Animal stress, as evidenced by reduced pregnancy rates, longer time needed for the animals to reach market weight, and reduced milk production, can result in livestock production declines. Of note, minimum temperature is the environmental variable having the closest relationship with beef cattle pregnancy rate, with 12.6°C to 14.9°C being the optimum temperature for supporting beef pregnancy.

Water

Precipitation has a direct influence on agriculture and is projected to increase for some areas of the United States and decrease for others. Changes of the timing, intensity, and amount of rain/snow mix for a location are expected to increase the management challenge of delivering water to crops at the right time through irrigation systems and practices. Excess precipitation can be as damaging as receipt of too little precipitation due to the increase in flooding events, greater erosion, and decreased soil quality. Increases in evapotranspiration can result in less available water even in cases when precipitation amounts increase, particularly in soils with limited soil water holding capacity. For example, excess water during corn's early growth stages may cause a reduction in growth or even death, while soil water deficit may lead to less growth and lower yields if the stress occurs during the grain filling period of growth (Hatfield and Prueger 2011).

Water requirements for agriculture are expected to increase due to rising temperatures. An example of the regional effect of changing temperature, the U.S. West will experience declining snow accumulation and early, faster snow-melt rates due to earlier spring-time warming and higher average winter-time air temperatures; this region depends on snowpack runoff both for early-season crop growth and irrigation needs later in the growing season (Knowles et al. 2006).

Atmospheric Carbon Dioxide

Carbon dioxide concentrations of the well-mixed atmosphere, as sampled at the summit of Mauna Loa, Hawaii, have increased rapidly since measurements began in 1958. Because enhanced atmospheric CO_2 concentrations stimulate photosynthesis and plant growth, much work has been focused on determining the responses of crops and weeds to elevated CO_2, often in single-variable experiments. To fully appreciate the implications of CO_2 for weeds, it is necessary to understand the nature of individual species versus crop population responses. Although higher CO_2 levels typically increase growth, the response varies by species. Part of this variability is related to photosynthetic biochemistry. For example, plants with the C_3 photosynthetic pathway (about 95% of all plant species) are likely to respond more strongly than plants possessing the C_4 photosynthetic pathway (for which photosynthetic rates are saturated at current, ambient CO_2).

Most experiments have used one or two elevated CO_2 concentrations, most often near 550 or 700 μmol mol^{-1} (i.e., 550 to 700 ppm[2]), rather than

The Difference Between C_3 and C_4 Plants

Most plant life on Earth can be broken into two categories based on the way they assimilate carbon dioxide (CO_2) from the atmosphere into different physiological components. More than 95% of the world's plant species fall into the C_3 category. As CO_2 is taken up from the air by C_3 plants the first component formed is a three-carbon compound as the first stable product of carbon fixation, while C_4 plants make a four-carbon compound during the initial stages of photosynthesis. The most recognizable C_4 plants, include sugarcane and corn. One of the most important differences between C_3 and C_4 species for rising CO_2 levels is that C_3 species continue to increase photosynthesis with rising CO_2, while C_4 species do not. Another important difference between C_3 and C_4 plants is evident in stomatal conductance and water use by plants. There is a decrease in the stomatal conductance of the leaves (Wand et al 1999; Ainsworth et al 2002) as the atmospheric concentration of CO_2 increases. The result of this decrease in conductance is a reduction in the rate of water use and an increase in water use efficiency (amount of biomass produced per unit of water transpired).

concentrations based on particular future target dates and emission scenarios; the "ambient" control concentration has increased gradually over the years of this research, which complicates comparisons of responses. Hurdles involved in moving similar studies forward is that debate exists about the most realistic experimental techniques to expose crops to simulated future CO_2 concentrations (Holtum and Winter 2003; Long et al. 2005; Long et al. 2006; Tubiello et al. 2007; Ziska and Bunce 2007).

Yields of wheat, rice, and soybeans under field conditions increased approximately 12% to 15% under 550 ppm compared with 370 ppm CO_2 concentrations, with the percentage increases about 1.6 times those for elevated CO_2 concentrations of approximately 700 ppm. As compared with most other annual crop species, cotton had an exceptional 43% yield increase under increased CO_2 concentrations, but it should be noted that some varieties of rice and soybean also had yield increases as large as cotton. Corn had negligible yield increases. Within C_3 species, we might expect differences in CO_2 responsiveness between sexual and vegetative commodities (e.g., seed crops versus pasture species), and between root and shoot crops. However, given the variation in response among varieties within species, these expected differences in response have not been substantiated. Also, response differences may exist between annual and perennial species because the stimulation of growth by perennial species grown with little competition may be cumulative over years.

Elevated atmospheric CO_2 can modify responses of crops to environmental stresses. Some modifications tend to reduce effects of stress, such as elevated CO_2 causing partial stomatal closure and reducing penetration of ozone into leaves, which in turn lowers yield losses due to ozone (Fiscus et al. 1997; Booker and Fiscus 2005). Partial stomatal closure at elevated CO_2 also reduces crop water loss (Jarvis and McNaughton 1986; Wilson et al. 1999; Bunce 2004). However, elevated CO_2 increases crop tissue temperatures, which may exacerbate damage to reproductive processes caused by high air temperatures.

Rising atmospheric CO_2 concentrations over the last 150 years have likely increased productivity of pastures (Polley et al. 2003; Izaurralde et al. 2011). Based on simulation studies, it is expected that the productivity of Great Plains native grasslands will continue to increase over the next 30 years as air temperature and atmospheric CO_2 concentrations

---

[2] One μmol mol^{-1} is equivalent to one ppm by volume; ppm will be the units used throughout the rest of this report.

increase (Parton et al. 2007; Izaurralde et al. 2011). Rangeland species encompass a wide variety of types of plants and include both C_3 and C_4 species; elevated CO_2 can increase the proportion of C_3 relative to C_4 species (Owensby et al. 1999). Rangeland species' responses to increased temperature and CO_2 are similar to those of the major crops, though interactions among species are more important as rangelands consist of a mixture of species.

As is the case for rangelands, the mixed nature of pasture crops has important implications for the response to water and nutrients under elevated temperatures and CO_2. In Texas, average pastureland biomass increased with CO_2 concentration, with increases ranging from 120 to 160 g m^{-2} per 100 ppm increase in CO_2 (Polley et al. 2003). Rangeland species will grow faster with higher temperatures and experience a longer growing season.

Beneficial to growth of woody plants, encroachment of such species into pastures may reduce the available nutrients for livestock and will, as a result, require management changes to address. An analysis of cattle fecal chemistry over the past 14 years suggests that changes in pasture makeup and effects of increased temperature and decreased rainfall have resulted in a general decline in forage quality (Craine et al. 2010). This includes a decrease in dietary crude protein and digestible organic matter. Consequently, it seems likely that the livestock industry will have to provide increased supplemental feeds to pasture-raised cattle in the future to prevent decreased cattle growth (Craine et al. 2010).

In addition to production quantity, the quality of agricultural products may be altered by elevated CO_2. For example, some non-nitrogen-fixing plants grown at elevated CO_2 have reduced nitrogen (N) content (Ainsworth and Long 2005). Nitrogen is a critical agricultural crop nutrient. The mechanism for this is unclear. One implication may be that changes of N application-practices may be useful in dealing with climate change effects, both for the economic gains by producers, and to reduce the environmental effects of elevated residual soil N. For instance, non N-fixing cereal and forage crops grown at elevated CO_2 often have lower protein contents (Erbs et al. 2010), which will affect human and animal nutrition, and could also affect the behavior of pests. More subtle product quality responses, especially to temperature and water stress, may also be very important economically.

In addition to production quantity, the quality of agricultural products may be altered by elevated CO_2.

Tropospheric Ozone

Recently, a number of innovative studies have advanced understanding of plant responses to ozone, refining researchers' knowledge of ozone-agriculture interactions under changing climate conditions. This enhanced understanding of ozone's effects on U.S. agriculture is increasingly important because ozone and its precursors are transported hundreds of miles into rural areas affecting native and managed forests, including culturally important Class I Wilderness Areas, as well as farm, pasture, and other regions of agricultural activity (Mickley et al. 2004; Dentener et al. 2006).

Ozone, after uptake through the leaf's stomata, interacts with plants' cellular processes, inhibiting photosynthesis, growth, and yield. Gene expression and proteomic studies show that detrimental ozone effects are likely caused by a combination of chemical toxicity and plant-mediated responses that either amplify or inhibit injury (Cho et al. 2011). Already, studies indicate that current ambient ozone levels are suppressing yields of crops such as alfalfa, bean, clover, cotton, peanut, potato, rice, soybean, sugar cane, and wheat in many regions of the United States and worldwide (Booker et al. 2009; Grantz and Vu 2009). In addition, changes in leaf chemistry due to elevated ozone exposure in common grassland species have reduced nutritional quality of the land used to support grazing animals. This loss of food quality may be more significant than biomass losses in the assessment of ozone's effect on forages (Muntifering et al. 2006). Additionally, ozone may offset potential elevated CO_2 aerial fertilization effects in some plants.

Elevated CO_2 and ozone pollution interact to affect crop yields, suggesting that projected benefits of rising CO_2 in the atmosphere may be overly optimistic because they are based on models that do not include many important confounding factors in the environment. However, in some cases, increases of atmospheric CO_2 may lessen ozone injury (Ainsworth and Long 2005; Fiscus et al. 2005), although the interaction becomes less effective as ozone concentrations increase.

At the agroecosystem level, ozone effects on soil carbon (C) and nitrogen dynamics have not been well characterized. Changes in below-ground crop processes are hypothesized to result mainly from ozone effects on plant C allocation and biomass production. Both are generally reduced by ozone, especially in plant roots (Andersen 2003; Grantz et al. 2006); for example, decreased N_2 fixation with elevated ozone has been observed (Tu et al. 2009). Expression of genes and proteins involved in photosynthesis is suppressed, while carbohydrate catabolism, oxidative signaling, antioxidant, and defense pathways are stimulated by ozone (Ahsan et al. 2010; Booker et al. 2011; Cho et al. 2011). The responses are hypothesized to be due in part to a need for energy and a reduction in the plant's power to detoxify and repair damage caused by oxidative molecules (Ahsan et al. 2010).

In a 6-year, no-till, soybean-wheat study conducted in open-top field chambers, there was no effect of ozone on soil microbial activity, biomass, community composition, or nitrogen mineralization, in contrast to elevated CO_2, which increased these processes (Cheng et al. 2011). Plant residue input in the open-top chamber experiment was less in the added ozone treatment, but seemed to have no effect on soil nitrogen (Cheng et al. 2011). A Soybean Free Air Concentration Enrichment (SoyFACE) study showed that with 20% higher ozone, bulk soil nitrogen and carbon concentrations were 12% and 15% higher, respectively, than in soil from plots exposed to ambient air (Pujol Pereira et al. 2011). It was suggested that decomposition processes may have been slower under elevated ozone due to the lower amount of plant material input. Soil NH_4^+ concentration was decreased by ozone, possibly related to decreased residue input and lower symbiotic N_2 fixation. Denitrifying bacteria increased with soil organic carbon at the SoyFACE site (Pujol Pereira et al. 2011). Decreased N_2 fixation with elevated ozone has also been observed in peanut (Tu et al. 2009). There is no evidence that soil carbon and nitrogen dynamics in highly managed agroecosystems are significantly influenced by ambient ozone in the United States, although it is possible that substantial reductions in biomass production due to higher future ozone levels could influence soil nutrient cycling processes. It should be noted, however, that the potential influence of high ambient ozone levels on soil nutrient cycling has not been evaluated in the agriculturally productive regions of southern California, eastern China, or northern India, for example, where ozone effects on crops are evident (Booker et al. 2009).

Indirect Climate Change Effects

As is the case with crops and livestock, climate change affects weeds, pests, and pathogens. Changes in temperature and precipitation patterns, coupled with increasing atmospheric CO_2 create new conditions that change weed-infestation intensity, insect population levels, the incidence of pathogens, and the geographic distribution of many of these pests. Such changes on non-crop species found in agroecosystems

are indirect effects of climate change. For agriculture, such effects can alter production yields and quality, and may necessitate changes to management practices. These indirect effects may also increase farming costs, as additional inputs may be required to manage the influence of weeds, invasive species, insects, and other pests. Weeds cause the highest crop losses globally (34%), with insect pests and pathogens showing losses of 18% and 16%, respectively (Oerke 2006). In the following sections, some of the indirect effects of climate change on weeds, pests, and pathogens and their respective effects on U.S. agriculture will be sketched out.

Weeds and Invasive Plant Species

Agronomic Weeds
Cropland agriculture, in its simplest arrangement, can be characterized as a managed plant community that is composed of a desired plant species (the crop) and a set of undesired plant species (weeds). Agronomic weeds reduce food production through competition for light, nutrients, and water, and by reducing production quality, increasing harvest interference, and acting as hosts for other pest vectors. By altering the environment (e.g., temperature) or increasing a resource (e.g., CO_2), we change not only the growth of an individual, but also the interactions among species, and the growth patterns of the entire plant community.

Temperature and Precipitation Effects on Agronomic Weeds
Weed scientists have long recognized that temperature controls weed species success (Woodward and Williams 1987). Thus, warming will affect the dissemination of weeds with subsequent effects on their growth, reproduction, and distribution. Many of the most troublesome weeds in agriculture – both warm-season (C_3) and cool-season (C_4) species – are confined to tropical or subtropical areas (Holm et al. 1997); the lower temperature extremes that occur at higher latitudes are inhospitable to many weeds. High-latitude temperature limits of tropical species are set by accumulated degree days (Patterson et al. 1999), while low-latitude limits are determined, in part, by competitive ability to survive at lower temperatures (Woodward 1988). However, because many weeds associated with warm season crops originate in tropical or warm temperature areas, northward expansion of these weeds may accelerate with warming (Patterson 1993; Rahman and Wardle 1990).

For maize and soybean crops within the United States, there is a clear latitudinal distinction between the Great Lakes (Michigan, Minnesota, Wisconsin) and Gulf States (Alabama, Louisiana, Mississippi)

Agronomic weeds reduce food production through competition for light, nutrients, and water, and by reducing production quality, increasing harvest interference, and acting as hosts for other pest vectors.

with respect to weed limitations (Bunce and Ziska 2000). The greater soybean and corn losses in the southern Gulf States are associated with a number of very aggressive weed species found in tropical or subtropical areas (e.g., prickly sida and Johnson grass). Warmer temperatures, in particular an increase in the number of frost-free days, may allow a northward expansion of these aggressive weeds into other areas of the Midwest, with subsequent effects on maize and soybean production. An analysis of such changes, using a "damage niche" hypothesis, and a "business as usual" climate scenario (IPCC 2007) showed significant changes in the range of two weed species affecting corn in the northern and southern United States (velvetleaf and Johnson grass, C_3 and C_4 weeds, respectively) (Mcdonald et al. 2009). Based on these initial evaluations, velvetleaf, a cold-tolerant annual weed, is likely to become less problematic in the Corn Belt; whereas Johnson grass, a warm-season perennial, may become more common, advancing northward by 200 to 600 km by midcentury.

Given their similar life histories and growth rates, crops and weeds are likely to have similar responses to drought; consequently, the overall effect of weeds may be reduced because of decreased growth of both crops and weeds in response to water availability (Patterson 1995). However, effects of drought are likely to vary widely among crops and weeds. In corn, for example, drought has been found to

both decrease interference from weed communities dominated by foxtail (*Setaria*) species (McGiffen et al. 1997) and increase competitive ability of Johnson grass (Leguizamon 2011).

Effects of Carbon Dioxide on Agronomic Weeds

There are only a handful of field studies that have quantified changes in crop yields with weedy competition as a function of rising atmospheric CO_2 (Ziska 2000; 2003a; 2010). These outcomes were consistent with the known kinetics of the photosynthetic pathway; i.e., plants with the C_4 photosynthetic pathway performed poorly relative to plants with the C_3 photosynthetic pathway as atmospheric CO_2 increased. For example, soybean yield losses from pigweed, a C_4 weed, were reduced from 45% to 30% with rising CO_2 (Ziska 2003a). Conversely, for dwarf sorghum (C_4 crop) and velvetleaf (a C_3 weed), yields further reduced as CO_2 increased.

However, the interaction of rising CO_2 on crop-weed competition must also consider weed-crop associations where both plant species have the same photosynthetic pathway, a situation that often occurs since agronomic practices tend to select, over time, for weeds with similar morphological and phenological characteristics to the crop. An assessment of these weed crop interactions (Table 4.1) demonstrates that agronomic weeds consistently respond more than crops to elevated CO_2.

Table 4.1. Summary of studies examining whether weed or crops grown in competition were "favored" as a function of elevated concentrations of CO_2. "Favored" indicates whether elevated [CO_2] produced significantly more crop or weed biomass.

Crop	Weed	Increasing [CO_2] Favors	Environment	Reference
A. C_4 Crops / C_4 Weeds				
Sorghum	*Amaranthus retroflexus*	Weed	Field	Ziska (2003)
B. C_4 Crops / C_3 Weeds				
Sorghum	*Xanthium strumarium*	Weed	Glasshouse	Ziska (2001)
Sorghum	*Albutilon theophrasti*	Weed	Field	Ziska (2003)
C. C_3 Crops / C_3 Weeds				
Soybean	*Cirsium arvense*	Weed	Field	Ziska (2010)
Soybean	*Chenopodium album*	Weed	Field	Ziska (2000)
Lucerne	*Taraxacum officinale*	Weed	Field	Bunce (1995)
Pasture	*Taraxacum and Plantago*	Weed	Field	Potvin & Vasseur (1997)
Pasture	*Plantago lanceolate*	Weed	Chamber	Newton et al. (1996)
D. C_3 Crops / C_4 Weeds				
Fescue	*Sorghum halapense*	Crop	Glasshouse	Carter & Peterson (1983)
Soybean	*Sorghum halapense*	Crop	Chamber	Patterson et al. 1984
Rice	*Echinochloa glabrescens*	Crop	Glasshouse	Alberto et al. (1996)
Soybean	*A. retroflexus*	Crop	Field	Ziska (2000)

Interactive Effects of Global Changes on Agronomic Weeds

To date, only one study has evaluated the interaction between temperature, CO_2, and crop/weed competition (Alberto et al. 1996). This study found increased CO_2 to favor rice, a C_3 crop, over a C_4 weed at 27/21°C; however, concurrent increases in CO_2 and temperature favored the weed, due to increased seed yield loss for rice relative to the weed (Alberto et al. 1996). Hypothetically, there are a number of additional potential interactive effects related to temperature, CO_2, and weed/crop competition. For example, growth of tropical weeds is strongly stimulated by small air temperature changes (Flint et al. 1984; Flint and Patterson 1983), but it is unknown if a greater synergy with rising CO_2 would be anticipated for these weeds relative to tropical crops. Still, the Alberto study emphasizes that effects of climate change on simple competitive outcomes will be difficult to predict based simply on a C_3 crop/C_4 weed model.

Few studies have examined interactions between drought, rising CO_2, and weed/crop competition. Although competition was not determined directly, the proportion of weed biomass increased with CO_2 to a similar extent in wet and dry treatments in a pasture mixture (Newton et al. 1996). In a study of tomato (C_3 crop) and redroot pigweed (C_4 weed), well-watered conditions resulted in reduced competition from the weedy species; however, if drought and high CO_2 occurred concurrently, redroot pigweed was a better competitor (Valerio et al. 2011). Overall, if C_4 weeds utilize less water with increasing CO_2 than do C_3 crops, C_4 weeds could potentially outcompete C_3 crops in high CO_2/drought conditions (Knapp et al. 1993).

Similarly, little information is available regarding weed/crop competition, CO_2, and nutrient availability. Under extreme nutrient limitations, stimulation of biomass with additional CO_2 may be minimal; however, under moderate nutrient limitations more relevant to agricultural situations, the increase in biomass may be reduced, but still occur (e.g., Rogers et al. 1993; Seneweera et al. 1994). In the only published study to examine competition between a C_3 crop (rice) and a C_4 weed (barnyard grass) (Zhu et al. 2008), the proportion of rice biomass increased relative to barnyard grass with a 200 ppm increase in atmospheric CO_2, but only if nitrogen was adequate. If N was low, elevated CO_2 reduced the competitive abilities of rice relative to the C_4 weed, presumably by reducing carbon sinks (e.g., tiller formation) in rice. These data indicate that in rice cropping systems with limited N, rising CO_2 could still exacerbate competitive losses, even from C_4 weeds.

Invasive weeds

Invasive weeds compete with desired plants in rangelands, pastures and other perennial agricultural systems in the United States, reducing both food production and biological diversity (DiTomaso 2000). A key difference between agronomic weeds and invasive plants, with respect to global change, is that global changes that influence plant resources (water, N, light, C) influence invasive weeds particularly strongly (Bradley et al. 2010a).

Temperature and Precipitation Effects on Invasive Weeds

Both warming and precipitation change can alter plant resources and invasion. Experimental warming has been found to favor invasion in relatively wet European grassland (Verlinden and Nijs 2010), but to have little effect on, or to inhibit invasive species in drier California and New Zealand grasslands, perhaps because it increases evapotranspiration and therefore water limitation (Williams et al. 2007; Verlinden and Nijs 2010; Dukes et al. 2011). As with agronomic weeds, warming may be most likely to favor C_4 invaders competing with C_3 species (Bijoor et al. 2008) and inhibit C_3 invaders competing with C_4 species (Williams et al. 2007). The few experiments examining how changing precipitation might influence invasion suggest that effects depend on seasonality. Increases in winter precipitation favored invasive species in mixed-grass prairie (Figure 4.1, Blumenthal et al. 2008), while increases in spring

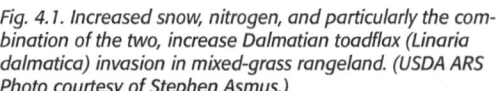

Fig. 4.1. Increased snow, nitrogen, and particularly the combination of the two, increase Dalmatian toadflax (Linaria dalmatica) invasion in mixed-grass rangeland. (USDA ARS Photo courtesy of Stephen Asmus.)

precipitation favored invasive species in California and Utah grasslands (Miller et al. 2006; Thomsen and D'Antonio 2007). Across studies and ecosystems, invasive species tend to use more water than natives (Cavaleri and Sack 2010), suggesting that invasive species may often be favored by increased water availability during their growing season (Bradley 2009). Therefore, the large sections of the United States that are expected to receive higher precipitation may need to engage more actively in invasive weed management.

In addition to altering the success of invasive species within plant communities, changes in climate are also likely to alter the distributions of those species (McDonald et al. 2009; Watt et al. 2009; Ibanez et al. 2009; Bradley 2009; Bradley et al. 2010b; Jarnevich and Reynolds 2011.) Biogeographical modeling, which uses current spatial distribution to identify suitable habitat under future climate conditions, suggests that rising temperatures and altered precipitation may not consistently increase invasive species' ranges (Bradley et al. 2010a). For some species, projected changes in climate primarily cause an expansion of invasion risk (e.g., Jarnevich and Stohlgren 2009; McDonald et al. 2009; Bradley et al. 2010b), particularly near the cooler margins of their range (poleward and upward in elevation). For other species, climate change may reduce invasion risk in portions of the invaded range (e.g., Parker-Allie et al. 2007; Beaumont et al. 2009; Bradley 2009). For example, a model of spotted knapweed risk suggests that the species' potential range will be substantially

reduced with climate change, while cheatgrass' potential range shifts, expanding into currently wetter areas and contracting from currently drier areas (Figure 4.2).

Extreme climatic events such as drought, flooding, and strong storms, which are predicted to become more frequent with climate change, can also influence weed invasion (Jimenez et al. 2011; Diez et al. 2012). While decreasing precipitation might be expected to inhibit invasion, severe or extended droughts can act as disturbances, decreasing biotic resistance from native species, and providing opportunities for invasive species once precipitation returns. For example, in Arizona rangeland, severe drought in 2004 and 2005 led to the death of many native shrubs and grasses, followed by rapid invasion and dominance by Lehmann lovegrass (*Eragrostis lehmanniana*) (Scott et al. 2010). Similarly, hurricanes in Florida and Louisiana have damaged forests and increased cover of invasive vines (Horvitz et al. 1998; Brown et al. 2011).

Effects of Carbon Dioxide on Invasive Weeds

Many plants grow faster with elevated CO_2. Inherently fast-growing plants, including many invasive plants, can respond particularly strongly (Poorter and Navas 2003; Ziska 2003b; Ziska et al. 2005; Song et al. 2009). In controlled-environment studies, these differences have not translated into consistently stronger CO_2 responses in invasive than non-invasive plants (Dukes 2000). However, in field studies that incorporate competition with native plants, elevated

Fig. 4.2. *Biogeographical models project range shifts in invasive plant distribution, creating both areas of increased and decreased risk. Colors show future climatically suitable regions for invasive plant species according to climate projections for the year 2100 from 10 Atmosphere-Ocean General Circulation Models (AOGCMs) under the IPCC A1B future climate scenario. Warmer colors represent greater overlap of AOGCM projections, increasing confidence in future risk. Hatched areas show regions that are currently unsuitable, but become suitable in at least one projection. A) Spotted knapweed distribution is affected mainly by temperature, and is projected to expand upwards in elevation, but to contract at lower elevations (dark blue areas). B) Cheatgrass distribution is affected mainly by precipitation, and is projected to expand into wetter areas, but to contract from drier areas as overall water availability decreases in the West. Source: Reprinted from Bradley et al. 2009.*

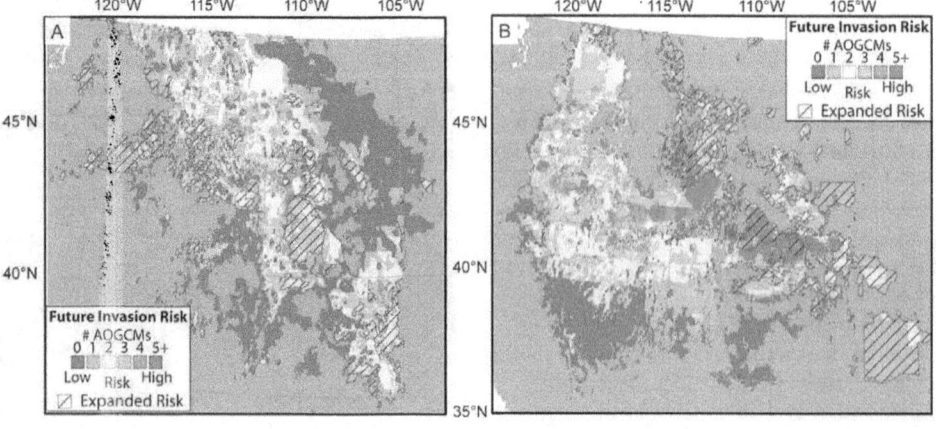

CO_2 has been found to increase invasion in grasslands (Dukes et al. 2011; but see Williams et al. 2007), desert (Smith et al. 2000), and forests (Hattenschwiler and Korner 2003; Belote et al. 2004) in some years. Carbon dioxide also increases plant water-use efficiency, and may be most likely to favor invasive species in water-limited ecosystems (Dukes 2002), as observed in the Nevada desert (Smith et al. 2000). Carbon dioxide can exacerbate nitrogen limitation, however (Luo et al. 2004), and may be least likely to favor invasive species in environments with low available nitrogen.

Interactive Effects of Global Changes on Invasive Weeds

The combined effects of multiple global changes on invasion are difficult to predict, but could have serious consequences for perennial agricultural systems. For example, in mixed-grass prairie, the combination of increased winter precipitation and simulated N deposition increased invasion much more than the sum of the two individual changes (Blumenthal et al. 2008). In contrast, while elevated CO_2 and N increased yellow starthistle (*Centaurea solstitialis*) biomass 6-fold and 3-fold, respectively, in California grassland, their combined effects were additive (Dukes et al. 2011). Multiple global changes may also influence invasion through interactions with fire. Both elevated CO_2 and severe droughts can favor fire-promoting invasive grasses in western U.S. rangelands (Smith et al. 2000; Brooks 2003; Ziska et al. 2005; Scott et al. 2010; Mazzola et al. 2011). At the same time, warmer temperatures and earlier cessation of cool-season precipitation are expected to increase the number and intensity of fires (Abatzoglou and Kolden 2011). The likely result is further transformation of diverse native rangelands into near-monocultures of invasive grasses (Bradley 2009; Abatzoglou and Kolden 2011).

Adaptation

Managing agronomic and invasive weeds under climate change requires attention to: (1) changes in the distribution and diversity of weed threats; (2) changes in the vulnerability of crop production to weed limitations under the range of weed management practices currently in use; and (3) risks posed by new weeds, including those not yet introduced to the United States.

The first step in adapting weed management to climate change is to determine which weeds will threaten agricultural production in the future, and where they will be most problematic. Although temperature and water have often been used to delineate vegetative zones, weed habitat is rarely included in those mapping efforts. To rectify this, innovative

researchers are utilizing biogeographical models to assess future weed threats related to climate change. For example, warming temperatures are predicted to lead to considerable turnover in the set of damaging weed species for any given agricultural field (McDonald et al. 2009). Although a number of studies have projected species-specific range shifts for invasive plants (for examples, see Bradley et al. 2010a), projections for agronomic species are rare. Furthermore, predictions of weed distribution rarely include weed effects, a problem that could be addressed by increasing the use of the abundant data provided by biogeographical modeling (Leibold 1995; McDonald et al. 2009; Kulhanek et al. 2011). For now, as a general rule of thumb, managers can look to neighboring States in the South for insight on what the damaging weeds of the future will be (McDonald et al. 2009). In the future, models may allow for species- and location-specific predictions of the effects of weeds.

Weed management includes the identification and implementation of cultural, mechanical, chemical, and biological options to prevent or maintain weed populations at acceptable levels. Effects of CO_2 and/or climate change on herbicide efficacy have only been examined in a handful of studies, but are likely to depend on the mode of action, the weed species, and on competitive interactions (Archambault et al. 2001). For example, although elevated CO_2 had no effect on the sensitivity of redroot pigweed, a weed that in large doses may prove toxic to animals grazing on it, to the most commonly used herbicide in the United States (glyphosate), sensitivity of lambsquarters, a commonly occurring weed, to glyphosate was reduced, such that the full, recommended dosage suppressed, but did not eliminate, growth (Ziska et al. 1999). Similarly, elevated CO_2 reduced the efficacy of glyphosate against Canada thistle (*Cirsium arvense*), quackgrass (*Elytrigia repens*) (Ziska and Teasdale 2000), and a number of exotic C_4 grasses (Manea et al. 2011), and the efficacy of glufosinate (a cell membrane disruptor) against Canada thistle (Ziska and Teasdale 2000).

Experimental data assessing the effects of climate and CO_2 change on mechanical, biological and cultural weed control are almost non-existent. Yet management strategies may also change in efficacy with changing climate and CO_2 concentrations. For example, tillage could be affected by rising CO_2, with a faster time to vegetative cover, but less time for field operations. Rising CO_2 levels could also increase asexual reproduction (e.g., Rogers et al. 1992; Ziska et al. 2004), further limiting mechanical control. Precipitation extremes of drought or flood could also hinder field operations. The efficacy of biocontrol agents (e.g., insects) is dependent on synchrony between various aspects of the plant community. While global changes may disrupt relationships between invasive weeds and biocontrol agents, it may be possible to anticipate such changes by matching prospective agents to predicted future distributions as well as present distributions of invasive species (Thomson et al. 2010). Cultural weed management may also interact with changes in climate and/or changes in CO_2. For example, flooding is used for weed control in rice. Climate change is anticipated to affect water supply through its influence on glacial runoff, snowpack, or drought severity (IPCC 2007; Kerr 2007).

It will also be important to account for climate change in policies that limit introduction and movement of new, potentially invasive species. The combination of changing environments and changing patterns of trade is likely to increase both the risks posed by and the supply of species adapted to warm, dry environments (Bradley et al. 2012). For example, as limited water supplies in the western United States increase demand for drought-tolerant horticultural species, many of which are exotic, risks of introducing species capable of invading this relatively dry region increase. By incorporating such predictions into risk assessments, and associated policies, it may be possible to reduce the number of invasive species that need to be managed in the future.

Insect Pests

The geographic ranges of insect pests are limited by the presence of the plants upon which they feed, and the ability of the insects to survive winter temperatures. However, through local dispersal and long-distance migration, some insects may reinvade colder regions annually. Spring emergence is generally defined by temperature, whereas winter dormancy is cued by photoperiod or a combination of photoperiod and temperature. Insects are capable of withstanding all but the most extreme precipitation events, thus rainfall affects growth and survival principally through increased cloud cover, which can reduce activity, and changes in the nutritional quality of the plants upon which insects feed. Insects, especially small ones and those with aquatic life stages, will desiccate and die without ready access to water. Humidity influences the prevalence of insect diseases, as well as plant diseases that insects carry. Although food quality is important to their growth, survival of many insects is dependent upon predation in natural ecosystems with chemical, biological, and microbial controls used to suppress pests, and sometimes their predators, below their natural level in agroecosystems.

....as is the case for crops, insects have optimal temperatures under which they thrive, so not all insect populations will increase with increasing temperature.

Air Temperature Effects on Insect Pests

Generally, increasing air temperature is beneficial to insect pests. As long as upper critical limits are not exceeded, rising temperatures accelerate every aspect of an insect's life cycle, and warmer winters reduce winter mortality. Although increased summer temperatures also favor growth of insect populations, extension of the growing season has a proportionately greater effect on the damage insects inflict on their host plants (Bradshaw and Holzapfel 2010). Moreover, pests' greater nutrient demands in early spring and autumn coincide with the planting and fruiting stages – stages that are particularly vulnerable for many crops and critically important for successful production.

Increasing air temperature has resulted in reduced cold stress without substantial increase in heat stress (Bradshaw and Holzapfel 2006), although decreased soil temperatures in areas with reduced frequency of snow cover can result in greater winter insect mortality (Bale and Hayward 2010) because arousal from winter dormancy is generally dependent on accumulated temperature (growing-degree days). Research shows examples of insect phenology advancing faster than previously experienced within a season (Gordo and Sanz 2006; Harrington et al. 2007; Gregory et al. 2009; Bale and Hayward 2010). Some insects spawning multiple generations per season have responded to longer growing seasons by producing more generations per year (Tobin et al. 2008; Altermatt 2010), which, in addition to adding more insects to the environment, can lead to pests developing greater resistance to insecticides (May and Dobson 1986).

The overall positive influence of increasing air temperature on expansion of insect geographical ranges is well documented in natural systems, although some insects' ranges have shifted and others have contracted (Walther et al. 2002; Parmesan and Yohe 2003; Parmesan 2006; Walther 2010). Earlier migration and maturation result in successful colonization of habitats that were formerly outside an insect population's range (Bale and Hayward 2010). However, as is the case for crops, insects have optimal temperatures under which they thrive, so not all insect populations will increase with increasing temperature.

Increased winter survival in newly colonized habitats also contributes to successful expansion (Crozier 2004). Less work has been done in agroecosystems, but Diffenbaugh et al. (2008) projected range expansion of the corn earworm, European corn borer, and the Northern and Western corn rootworms in the United States based upon tolerance to minimum absolute temperature, number of hours below -10°C, and the required growing-degree days in the first half of the year. Models project that geographic ranges will expand for all four species by 2100, indicating that insects from diverse life styles may be affected similarly by recent and future temperature changes (Diffenbaugh et al. 2008).

Projected increases of extreme precipitation events could make pest population outbreaks and crashes more common (Hawkins and Holyoak 1998; Srygley et al. 2010). Pest outbreaks are often associated with dry years (White 1984), although extreme drought is unfavorable to insects (Hawkins and Holyoak 1998). Extremely wet years are also unfavorable (Fuhrer 2003). Under changing climate, environmental thresholds currently keeping some pests in check may be exceeded because of increased variability, making pest outbreaks likely to become more common as a result of increased climate variability. Phenological shifts and geographical range shifts in interacting species can be synchronous or asynchronous, and as a result may have important ramifications on pest population (Hance et al. 2007; Memmott et al. 2007; Hegland et al. 2009). For example, as a result of warming over the last century, the larch budmoth's range has shifted to the distributional limit of its host, dampening a millennium-long cycle in outbreaks of the moth (Johnson and McNicol 2010). As another example, the northward expansion of crop ranges may have altered aphid community composition. In Europe, autumn sowing of winter wheat, barley, and rape provides a substrate for parthenogenic, non-diapausing aphids to survive the winter (Roos et al. 2011). In Poland, changes in winter survival of parthenogenic aphids may have resulted in a shift in species composition to fewer aphid species with sexual forms in their life cycles.

Blood-sucking and tissue-feeding insects and ticks on livestock may also be affected by climate change. One clear example is the recent emergence of bluetongue virus in Europe (Wittmann and Beylis 2000). The geographic range of bluetongue's principle Old World vector, the midge *Culicoides imicola*, has expanded northward into southern Europe; with the midge came several introductions of bluetongue. Once European midges picked up the virus, they transmitted it beyond its traditional range, increasing its range up to 800 km further north than was the case prior to 1998.

Effects of Enhanced Atmospheric Carbon Dioxide on Insect/Pests

The effects of increased atmospheric CO_2 on insect pests is much more complex than that of increasing temperature because insect performance is

highly dependent on the response of the host plant to increased CO_2. This indirect action of CO_2 makes for quite variable interactions between plants and insect pest. Generally, increasing C to N (C:N) ratios in plants under increased enhanced atmospheric CO_2 makes nutritionally poorer forage for insects. However, compensatory feeding can offset an insect's N needs (DeLucia et al. 2008; Johnson and McNicol 2010), and addition of N to the soil can also moderate the influence of CO_2 on insect performance by restoring the C:N ratio that is observed in plants under present-day conditions (Tylianakis et al. 2008).

Nitrogen limitations can cause plants to produce fewer of the secondary metabolites that are involved in developing resistance to insect pests (Zavala et al. 2008), while enhanced CO_2 fixation can increase C-based defenses that reduce the digestibility of a given crop for insects (Stiling and Cornelissen 2007). For example, enhanced CO_2 fixation by soybeans increases leaf toughness, but there is a coincident decrease of a plant's production of N-based compounds such as cysteine proteinase inhibitors – proteins that defend soybean from beetles (Gregory et al. 2009). Although most insects would find a plant with decreased N-based defenses more appealing, some specialized insects that cue on those specific secondary compounds to stimulate feeding will feed less.

Evidence also exists that micronutrients are less available with increasing CO_2 (Loladze 2002), which can reduce the quality of plants used for forage. Ultimately the effect of increased CO_2 on insects is quite variable, with some insects growing more slowly and maturing at smaller sizes, and others growing more quickly and becoming larger (Stiling and Cornelissen 2007). A review of the net effects of elevated CO_2 on crops and forages with insect herbivores (Table 4.2) suggests that beetles and aphids generally perform better, to the detriment of the plants, while moths often eat more and achieve similar weights.

Projections of insect distributions and abundance with climate change (Table 4.3) have different underlying assumptions. In some projections, an insect's existing geographic range is used to estimate critical temperatures that define its habitat boundaries. Then the change in mean global air temperature projected to accompany a doubling of CO_2 is used to investigate how the range might change in the future. To estimate effects of gradual changes in temperature on range or abundance, a series of step increases in ambient temperatures are also applied to the critical thermal parameters that define the insect's range. These models have advantages in their simplicity, but one critical assumption is that the host plants will show a similar change and may be available in newly colonized geographic range areas. Other significant assumptions of many model inputs are that there will be no effects on plant tissue composition that might affect the tissues' nutritional value, or that the secondary metabolites involved in attracting beneficial insects or defending the plant from harmful insects will be modified with temperature changes.

Agronomists and modelers are aware that the large uncertainties in precipitation can expand the modeled outcomes (Olfert and Weiss 2006). Modeling additional trophic levels, such as the response of host plants or parasitoids to temperature, make the assumptions more realistic (Gutierrez et al. 2008). Some projections have applied predictions from

Table 4.2. Net effect of pest herbivory on crop, forage, and invasive plants in elevated CO_2. Beetles and aphids generally perform better to the detriment of the plants; caterpillars typically eat more but enhanced plant growth results in little net effect. Host Species Effect Codes: (-) plant likely to be harmed by increased pest performance. (+) plant performance is likely to increase because insect performance decreases. (Ø) equivocal or neutral results, i.e., insect and plant performances increase more or less equally. Sources : [1]Heagle et al. 1994, [2]Karban and Thaler 1999, [3]Joutei et al. 2000, [4]O'Neill et al. 2008, [5]Dermody et al. 2008, [6]Johnson and McNicol 2010, [7]Salt et al. 1995, [8]Smith and Jones 1998, [9]Butler et al. 1986, [10]Sun et al. 2011, [11]Li et al. 2011, [12]Awmack et al. 1997, [13]Chen et al. 2004, [14]Bezemer et al. 1998, [15]Himanen et al. 2008, [16]O'Neill et al. 2011, [17]Lincoln et al. 1984, [18]Osbrink et al. 1987, [19]Lincoln and Couvet 1989, [20]Marks and Lincoln 1996, [21]Akey et al. 1988, [22]Chen et al. 2005, [23]Wu et al. 2006, [24]Karowe and Migliaccio 2011, [25]Johnson and Lincoln 1990, [26]Johnson and Lincoln 1991, and [27]Heagle 2003.

Order	Herbivore Species	Host Species	Effect on Host
Acarina	Tetranychus urticae (red spider mite)	Trifolium repens[1] (white clover) Gossypium hirsutum[2] (upland cotton) Phaseolus vulgaris[3] (kidney bean)	- - +
Coleoptera	Popillia japonica (Japanese beetle)	Glycine max[4] (soybean)	-
	Diabrotica virgifera (western corn rootworm)	Glycine max[4] (soybean)	-
	Sitona lepidus (clover root weevil)	Trifolium repens[6] (white clover)	-
Diptera	Pegomya nigritarsis (leaf-mining fly)	Rumex crispus (invasive dock) R. obtusifolius[7] (invasive dock)	- -
	Chromatomyla syngenesiae (leaf-mining fly)	Sonchus oleraceus[8] (invasive sow thistle)	+
	Bemisia tabaci (sweet potato white fly)	Gossypium[9-11] (cotton)	Ø
Hemiptera	Aulacorthum solani (glasshouse potato aphid)	Vicla faba[12] (broad bean)	-
	Stiobion avenae (grain aphid)	Triticum aestivum[13] (spring wheat)	-
	Myzus persicae (green peach aphid)	Poa annua[14] (grass) Brassica napus[15] (oilseed rape)	- +
	Brevicoryne brassicae (cabbage aphid)	Brassica napus[15] (oilseed rape)	Ø
	Aphis glycines (soybean aphid)	Glycine max[5,16] (soybean)	-
Hymenoptera	Aphidius matricariae (green peach aphid parasitoid)	Poa annua[14] (grass)	Ø
Lepidoptera	Pseudoplusia includens (soybean looper)	Glycine max[17] (soybean)	-
	Trichoplusia ni (cabbage looper)	Phaseolus lunata[18] (lima bean)	Ø
	Spodoptera eridania (southern armyworm)	Mentha piperita[19] (peppermint)	Ø
	Spodoptera frugiperda (fall armyworm)	Festuca arundinaceae[20] (tall fescue)	Ø
	Pectinophora gossypiella (pink bollworm)	Gossypium hirsutum[21] (upland cotton)	Ø
	Helicoverpa armigera (cotton bollworm)	Gossypium[22] (cotton) Triticum aestivum[23] (spring wheat)	+ +
	Colias philodice (clouded sulfur butterfly)	Trifolium repens[24] (white clover) Medicago sativa[24] (alfalfa) Lotus corniculatus[24] (birdsfoot trefoil)	Ø Ø Ø
Orthoptera	Melanoplus sanguinipes (migratory grasshopper)	Artemisia tridentata[25] (sage)	Ø
	Melanoplus differentialis (differential grasshopper)	Artemisia tridentata[26] (sage)	-
Thysanoptera	Frankliniella occidentalis (western flower thrip)	Trifolium repens[27] (white clover)	+

two or more global circulation models reduced to the region of interest to compare their effects on projected insect responses to climate change (Newman 2006; Trnka et al. 2007).

The most common model projections for pest insects show an expansion or shift in range with increasing

temperature (Table 4.3). For example, corn earworm, European corn borer, and the Northern and Western corn rootworms are expected to expand their ranges northward in the United States into what is currently unsuitable habitat (Diffenbaugh et al. 2008). Swede midge is projected to expand into the northern Midwest and central Canada (Mika et al. 2008). Pink

Table 4.3. Projected effects of climate change on agricultural pest insects. [1]Jeffree and Jeffree 1996, [2]Kocmankova et al. 2011, [3]Diffenbaugh et al. 2008, [4]Olfert and Weiss 2006a, [5]Stephens et al. 2007, [6]Gutierrez et al. 2009, [7]Mika and Newman 2010, [8]Mika et al. 2008, [9]Musolin et al. 2010, [10]Gutierrez et al. 2008, [11]Harrington et al. 2007, [12]Newman 2006, [13]Newman 2005, [14]Porter et al. 1991, [15]Trnka et al. 2007, [16]Gutierrez et al. 2006, [17]Olfert and Weiss 2006b, [18]Bergant et al. 2005. Note that small body size generally results in lower fecundity and higher mortality. Larger body size results in higher fecundity and higher survival.

Herbivore Species	Simulation	Region	Response
Coleoptera			
Leptinotarsa decemlineata (Colorado beetle)	2x CO_2	Europe	range expansion[1]
	2021-2100	central Europe	northward expansion of additional generations per year[2]
Diabrotica barberi (Northern corn rootworm)	2071-2099	U.S.A.	range expansion[3]
Diabrotica virgifera (Western corn rootworm)			range expansion[3]
Ceutorhynchus obstrictus (cabbage seedpod weevil)	+1-7°C		range expansion and greater abundance, inhibited by increased precipitation[4]
Meligethes viridescens (rape blossom beetle)	±60% of precipitation	Canada	range expansion and greater abundance[4]
Oulema melanopus (cereal leaf beetle)			range expansion and greater abundance, sensitive to precipitation[4]
Diptera			
Batrocera dorsalis (oriental fruitfly)	2080	U.S.A.	range expansion following invasion[5]
Batrocera oleae (olive fly)	+1-3°C	California	shift northward and to coastal areas, contraction in Central Valley and deserts[6]
Liriomyza huldobrensis (pea leaf-miner)	2020-80s	North America	range GCM dependent[7]
Contarinia nasturtii (swede midge)	2020-80s	Canada & U.S.A.	westward shift to central Canada and Northern Great Plains; optimal range GCM dependent[8]
Hemiptera			
Nezara viridula (Southern green stinkbug)	+2-2.5°C	Japan	additional generation, smaller summer and larger autumn body size, greater winter survival[9]
Planococcus ficus (vine mealybug)	+2-3°C	California	increases in abundance across extant range due to reduced biological control[10]
Aphis, Brevicoryne, Myzus, etc. (aphids)	2050	Europe	8-day advance in first flight[1]
Rhopalosiphum padi, etc. (cereal aphids)	2080s	Canada	increased abundance in northern or coastal regions, less abundant in southern or central regions, depending on climate model[12]
Rhopalosiphum padi, etc. (cereal aphids)	2080s	Great Britain	5-92% decline in abundance in southern England[13]
Lepidoptera			
Ostrinia nubilalis (European corn borer)	+1°C	Europe	range expansion 165-665 km north[14]
	2x CO_2	Europe	range expansion 1220 km north; additional generation annually[14]
	2021-2100	central Europe	northward expansion of additional generations per year[2]
	2025-2050	central Europe	spring advancement, additional generation, northward range expansion[15]
	2071-2099	U.S.A.	range expansion to occupy all of lower 48 except Rocky Mountains[3]
Heliothis zea (corn earworm)	2071-2099	U.S.A.	northward range expansion to upper Midwest[3]
Pectinophora gossypiella (pink bollworm)	+0.5-2.5°C	California	little change at or below 1 C; at 1.5 C and above, expansion into Central Valley and population increases in extant range[16]
Orthoptera			
Melanoplus sanguinipes	+1-7°C	Canada	increased range and abundance[17]
Thysanoptera			
Thrips tabaci (onion thrips)	2021-2050	Slovakia	0.5-4 additional generations per year[18]
	2051-2080		0.9-6.9 additional generations per year[18]

bollworm is predicted to expand its range into the Central Valley of California (Gutierrez et al. 2006), and olive fly is projected to shift its range from the Central Valley and desert regions northward and westward to coastal habitats (Gutierrez et al. 2009). In addition to shifting range, additional generations being born during a single season have been projected for insects such as the Colorado beetle, the European corn borer, and onion thrips in central Europe, and the southern green stinkbug in Japan (Table 4.3). Consequently, insect abundances are also projected to increase. The diversity of insects modeled lends credence to the suggestion that these modeled predictions may extend to many insects capable of more than one generation per year in U.S. regions with similar projected increases of temperature.

Projections are generally made with the assumption that the traits defining the insect's range, phenology, and abundance will not evolve. This is true in so far as there is no evidence that novel climatic tolerances have evolved on an ecological time scale that allows a species to inhabit a previously hostile environment (Parmesan 2006). However, even if the species as a whole does not evolve, population-level genetic changes could have large local consequences. Researchers have documented rapid shifts in a population's critical photoperiod so that the insects migrate or diapause (i.e., go into a dormant state) later in autumn in accordance with an extended growing season (Bradshaw and Holzapfel 2001; Gomi et al. 2007). At the edge of their range, insects' abilities to disperse may also change, as made evident by newly colonized habitats having insects with larger flight muscles and more active metabolic enzymes (Haag et al. 2005; Hill et al. 2011).

Insect Vectors of Pathogens

Aphids are important vectors of plant pathogens. Their short generation times make them likely to gain from global warming with a high risk of damage to crops. Range expansion of both the aphids and the pathogens they transmit will also result in increased genotypic diversity, making resistance to control efforts more likely to evolve (May and Dobson 1986). For example, green peach aphid populations are becoming more genetically variable in Scotland in association with warmer winters and earlier dispersal (Malloch et al. 2006). Projected changes in cereal aphid abundance in Canada in 2080 were temperature dependent, with increases in aphid populations predicted in more northerly latitudes or coastal regions, whereas southern or central regions had projected decreases, depending on the climate model (Newman 2006). Note that these projections are very different from the uniform decrease in cereal aphid abundance projected for southern Great Britain due

to interactions of increased CO_2 and limited N in a region that will experience greater drought (Newman 2005); projections of the response of an aphid parasitoid to climate change in Great Britain did not qualitatively alter the projections for the effects of climate change on the parasitoid's cereal aphid hosts (Hoover and Newman 2004).

Adaptation

With more pests shifting northward, generation times decreasing, and abundances increasing in the future, management costs are expected to increase due to more frequent application of pesticides. For example, pesticide applications to control lepidopteran pests (e.g., moths) on sweet corn decrease with increase in latitude from 15 to 32 times per year in Florida, four to eight times per year in Delaware, and zero to five times per year in New York (Hatfield et al. 2011). It can also be expected that resistance to chemical control agents will evolve more rapidly because of the increased genotypic diversity that comes with pest insects' range expansion and greater numbers of generations of particular pests undergoing selection for resistant forms each year (May and Dobson 1986). Crop diversification and landscape management for natural pest control can result in greater suppression of pest outbreaks and pathogen transmission in a changing climate (Lin 2011). It is also likely that some biological control agents will become less effective due to mismatched sensitivity between agent and effects on pests due to changes in the environment that increase pest resistance. For example, with increases in temperature, the vine mealybug is projected to find refuge from parasitoids introduced for its control in California vineyards, as the

Insects and Trade

There are some very specific Animal and Plant Health Inspection Service (APHIS) regulations that allow import/export of a commodity that might have a quarantine pest on it to specific locations, during specific times of the year when the weather is considered so unfavorable for the pest that if the pest were present and if it escaped into the environment, it would have a zero chance of surviving and reproducing. If climate changes and the receiving location becomes habitable for the pest, then some of the regulations may need to change.

parasitoids cannot survive under increased temperature (Gutierrez et al. 2008). Thus, the performance of candidate biological control agents under changing climate conditions will need to be assessed prior to selection and use.

Pathogens

Plant Pathogens

With non-vector-borne pathogens, plant pathogen responses to climate change must be considered within the context of a "disease triangle" that involves the pathogen, the host, and the environment; together these component parts determine whether a disease, itself a process, will occur (Agrios 2005). With vector-borne pathogens, the vector must be included in the disease triangle, with the microbial pathogen, the host, and the vector all interacting separately with the environment (e.g., Thresh 1983). In addition to having the basic components – pathogen, host and vector – as the required drivers of plant disease, plant pathogens and their vectors are influenced by other factors that complicate our ability to predict pathogen movement, incidence, severity, and evolution (Van der Putten et al. 2010).

Under current climate conditions, even with efforts to manage disease in place, crop losses to pathogens are estimated to be approximately 11% of overall worldwide production (Oerke 2006). Pathogen growth and reproduction can be evaluated independently with regard to the epidemiological parameters necessary for disease development (i.e., cardinal temperatures and responses to individual atmospheric influences).

These effects have been determined for some pathogenic viruses, fungi, and bacteria, leading to weather-based decision-support models designed to address seasonal production issues and disease management protocols (Jones et al. 2010; Savary et al. 2011). One of the first comprehensive reviews of the potential effects of climate change on plant disease recognized that it would most certainly affect plant disease at many levels of complexity, although generalizations would be difficult to make (Coakley et al. 1999). More than 10 years later, this remains true, in spite of significant progress in defining parameters potentially driving plant disease processes in a changing climate.

Yield and quality losses caused by diseases are influenced by 1) the direct consequences of climate change, e.g., increased temperatures, elevated CO_2 concentrations, altered rainfall patterns, drought and greater wind speeds; 2) regional alterations in areas cropped and ranges of crops grown, and 3) changes in vector ranges and activity. These factors alter the geographic ranges and relative abundance of pathogens, their rates of spread, the effectiveness of host resistances, the physiology of host-pathogen interactions, rates of pathogen evolution and host adaptation, and the effectiveness of control measures (Jones 2009). Effects of such changes on the frequency and duration of epidemics will vary depending on the pathosystem involved and geographic location, as well as continued environmental conditions that are conducive to the pathogen's survival and thriving (e.g., moisture and temperature conditions) (Garrett et al. 2006).

Role of Scale in Disease/Pests

Pests and diseases offer particular challenges for predicting and adapting to climate change effects because of the strong temporal and spatial correlation produced by their spread (Garrett et al., 2011; Shaw and Osborne, 2011). Greater pathogen or pest reproduction in one place and time will have important effects on risk in other places and times. New invasive species may have impacts greater than the impacts of climate change. These two forms of global change need to be considered together (Anderson et al., 2004; Coakley et al., 1999). Adaptation strategies will need to consider how regional patterns of cropping areas may change in response to climate change, and how this may change risks for transmission (Coakley et al., 1999; Margosian et al., 2009). Disease and pests are thus also an example of how management may need to adapt at multiple scales to changes in risk. Farmers must be prepared with strategies for addressing new types or degrees of problems, crop insurance programs may need to adapt, pesticide manufacturers may need to modify production, plant breeders may need to change their breeding priorities, decision-support systems and other management support systems may need modification, extension services may need to more frequently update their training programs, and policies related to management may need to be altered.

Extreme weather events projected with climate change include episodes of torrential rain with strong winds, in addition to heat waves and droughts, all of which influence plant pathogen epidemics (Jones 2009). Also, the rate of spread of contact-transmitted viruses will be accelerated through greater plant wounding arising from intense storms that feature torrential rainfall, or hail and high winds.

Drought and heat stress may affect the expression of crop resistance genes that would normally provide protection from pathogens, but even this can be variable within a given host, depending on the resistance genes present.

It was hypothesized that elevated ozone (resulting in increased plant tissue necrosis) would lead to increased disease by fungal necrotrophs, whereas elevated CO_2 was predicted to favor infection by fungal biotrophs. In some cases these hypotheses have proven true, but there are also many examples where the opposite effects have been observed (Eastburn et al. 2011). It is evident that these types of effects are difficult to predict, due in part to the non-linear biological responses of pathogens to increased CO_2, ozone, temperature, and humidity (Garrett et al. 2011). The majority of the studies that measure the effects of these parameters on pathogen growth are not able to incorporate all of the potential changes that may take place, particularly where including the host in conjunction with the pathogen or vector is logistically challenging.

In a thorough review, Garrett et al. (2006) provide a framework for considering climate change effects across multiple changing variables, with individual plant responses to single factors such as increased CO_2 or temperature well characterized for many crop plants. Generally, if host-plant survival can be linked to a single factor that overrides all others, then pathogen survival can likewise be linked to this overriding factor. For example, increased plant growth associated with elevated CO_2 can result in a canopy that is more conducive to fungal foliar diseases due to higher humidity occurring at the microclimate level (Pangga et al. 2011).

Increasing temperature may cause plant stress or may decrease plant stress depending on whether a crop is being grown in its optimal range or near a heat-tolerance threshold. Unfortunately, rarely does a single plant-growth or -health factor change as a result of climate change. When a combination of changes exist that result in temperatures, for example, that are no longer ideal for the crop host, this effect can be compounded when the change coincidently favors increased growth, formation of spores, earlier initial infection, shorter latent periods, or increased rates of disease progress (Campbell and Madden 1990).

More recently, studies involving pathogens in Free Air CO_2 Enrichment (FACE) facilities have combined variations in CO_2 and ozone concentrations while measuring effects on multiple pathogens. Work by Eastburn et al. (2010) at a FACE facility evaluated the effects of increased CO_2 and ozone on downy mildew, brown spot, and sudden death syndrome (SDS) in soy crops. Elevated CO_2 alone or in combination with increased ozone reduced downy mildew, increased brown spot severity (associated with changes in soybean canopy structure), and had no effect on SDS.

In addition to field studies, understanding host-pathogen interactions related to climate change has dramatically improved as a result of new molecular research methods. For example, elevated CO_2 has been shown to induce non-specific plant defense responses effective against Potato virus Y (Matros et al. 2006).

Information on the influence of changing climate on crop development, physical structure, and biochemistry is critical for determining pathogen response. For example, pathogens that require entry via plant stomata are likely to encounter conditions of increased cuticular wax and higher stomatal resistance (Eastburn et al. 2011). Changes in wax composition will also likely affect plant-pathogen biochemical interactions that influence infection processes (Eastburn et al. 2011).

Drought and heat stress may affect the expression of crop resistance genes that would normally provide protection from pathogens, but even this can be variable within a given host, depending on the resistance genes present. The effectiveness of some plant genes for resistance to virus diseases is known to be temperature sensitive; for instance, the gene for Tobacco mosaic virus (TMV) resistance is markedly reduced in efficacy above 28°C (Samuel 1931). This same temperature effect has been observed in transgenic tomato plants containing the same gene (Witham et al. 1996). Transient expression of the resistance genes for TMV (N gene) or Potato virus X (potato Rx gene) in a model system further demonstrated the reduced efficacy of these genes at high temperatures (Wang et al. 2009). Conversely, some resistance genes have been found to be more effective at higher temperature. One example of this is the wheat gene Yr36, which confers resistance to many races of the wheat stripe rust at temperatures between 25°C and 35°C, but loses the resistance at lower temperatures (Uauy et al. 2005). Similarly, the bacterial blight resistance gene Xa7 restricts disease more effectively at high temperatures than at low temperatures, although the crop and the pathogen are both present during cool and warm production seasons (Webb et al. 2010).

Increased temperature decreases efficacy of plant antiviral resistance mechanisms based on gene silencing, a process by which a plant gene is "turned off" so that it does not respond to the presence of a virus (Webb et al. 2010).

In the face of climate-related change, cultural control measures are likely to be less reliable in suppressing virus epidemics. Such techniques include planting upwind of virus sources when prevailing wind patterns vary, planting early maturing cultivars or harvesting early to avoid exposure of crops at peak insect vector flight times, and manipulation of sowing date to avoid coincidence of peak times for insect vector flights with vulnerable early crop growth (reviewed by Jones 2009).

Changes in individual host-plant structure and shifts in range that affect whole crop populations result in significant alterations in microclimate, pathogen dynamics, and multi-trophic interactions (Pangga et al. 2011); these interactions have far-reaching consequences. Range expansion has been predicted for many pathogens, based on models that incorporate changes in crop distribution and requirements for pathogen survival and reproduction (Savary et al. 2011).

Other interactions will also contribute to potential outcomes. Most economically important plant viruses, for example, are vectored by insects (predominantly aphids, whiteflies, or thrips), mites, nematodes, or soil fungi. Plant viruses are responsible for more emerging diseases (due to increasing host numbers and/or expanded geographic ranges; reviewed in Fargette et al. 2006) worldwide than any other pathogen group (Anderson et al. 2004, reviewed in Jones, 2009). International movement of plant material that may be infected with virus(es) or infested with viruliferous vectors is a key route of human involvement in the emergence of virus diseases (e.g., Jones 2009; Chellemi et al. 2011; Navas-Castillo et al. 2011).

Climate change is also likely to affect the emergence of virus diseases in new encounter scenarios when vulnerable, newly introduced crops or weeds are grown next to indigenous vegetation infected with viruses the new crops had not been exposed to previously. Although such circumstances have been relatively little studied (Jones 2009; Navas-Castillo et al. 2011), it is well known that viruses with wide host ranges adapt to new plant hosts better than viruses with narrow host ranges (Jones 2009).

Additionally, many viruses and associated vectors and some pathogens (see Asian soybean rust and the invasive weed kudzu, Eastburn et al. 2011) have non-crop (often weed) reservoirs that provide bridges between cropping periods (e.g., Adkins et al. 2011). Climate change is likely to indirectly affect virus diseases by altering the geographic range of both vectors and non-crop reservoirs, and the feeding habits of vectors (Canto et al. 2009; Jones 2009; Navas-Castillo et al. 2011). Projected climate changes are similarly predicted to alter populations and distributions of other insect, mite, nematode and soil fungi vectors of viruses and non-crop reservoirs, and thus the viruses transmitted or hosted, although effects are likely to vary by geographic region (reviewed in Jones, 2009). Examples include:

- Increased temperatures in temperate regions, which result in earlier appearance of spring aphids (and hence earlier appearance of aphid-vectored viruses). In Mediterranean-type, subtropical and tropical regions, summer aphids may not survive the warmer conditions, thus reducing incidence of aphid-vectored viruses (Jones, 2009).

- Increased temperatures and altered rainfall result in more favorable conditions for whitefly population, which can lead to a wider distribution of whitefly vectors (Morales and Jones 2004), and thus whitefly-transmitted viruses.

- Like insect pests, there is also an increased potential for rapid changes in composition of pathogen communities due to greater numbers of reproductive cycles occurring under intensified crop management. This can lead to more rapid evolution of new races, which may compromise crop resistance strategies and result in resistance to currently used pesticides (Juroszek and von Tiedemann 2011)

Livestock Pathogens

Climate change may indirectly affect animal production by altering the frequency, intensity, or distribution of animal pathogens and parasites. Climate affects microbial density and distribution, the distribution of vector-borne diseases, host resistance to infections, food and water shortages, or food-borne diseases (Baylis and Githeko 2006; Gaughan et al. 2009; Thornton et al. 2009). Earlier springs and warmer winters may allow for greater proliferation and survivability of pathogens and parasites. For example, bluetongue was recently reported in Europe for the first time in 20 years (Baylis and Githeko, 2006). Regional warming and changes in rainfall distribution may lead to changes in the spatial or temporal distributions of diseases sensitive to moisture, such as anthrax, blackleg, hemorrhagic septicemia, and vector-borne diseases (Baylis and

Githeko, 2006). Climate change also may influence the abundance and/or distribution of the competitors, predators, and parasites of vectors themselves (Thornton et al. 2009). Hotter weather may increase the incidence of ketosis, mastitis, and lameness in dairy cows and enhance growth of mycotoxin-producing fungi, particularly if moisture conditions are favorable (Gaughan et al. 2009). However, no consistent evidence exists that heat stress negatively affects overall immune function in cattle, chickens, or pigs.

Conclusions

Climate and climate change affect agriculture directly through the immediate effects of temperature, precipitation, and CO_2. The growth and development of crops, rangelands, and livestock are also influenced indirectly by climate change, through its actions upon weeds, insects, and disease. These variables interact with one another to further influence agricultural outcomes. The complexities of the crop-climate-environment interactions make projecting the net outcome of climate change difficult. Agricultural responses to climate change depend on the specific environmental and agroecosystem conditions, in combination with the characteristics of a given agricultural product. Some of these complexities will be further explored in the subsequent chapters, with information specific to particular agricultural systems found in Chapter 5 of this report.

Chapter 5

Climate Change Effects on U.S. Agricultural Production

Aggregate Effects

Agriculture is a complex system linked closely to climate through the direct effects of temperature, precipitation, solar radiation, and atmospheric composition on plant growth and yield, as well as livestock production. The soil and water resources of agricultural landscapes are linked with the same environmental factors. As the effects of climate change on soil, water, and environmental goods and services are examined, it becomes apparent that aggregate effects of climate transcend effects on individual agroecosystem components. For example, precipitation affects the potential amount of water available, however the actual amount of available water depends upon soil type, soil water holding capacity, and infiltration rate, such that the aggregate effect is not directly determined by precipitation amount. Actual climate change effects will thus depend on the cumulative effects of climate change factors on resources that are of key importance to agriculture, such as soil and water. Many of these effects are described by the following sections.

Agricultural Soil Resources

Soils provide ecosystem services that are necessary to society and the survival of life on the planet, including our own species. The roles soils play in delivering ecosystem services include nutrient cycling and the delivery of nutrients needed by growing plants. Soils act as a water filter and reservoir, purifying water as it passes through the soil substrate, and oftentimes providing water storage for later plant uptake. Soils also provide a structure for supporting plants and animals. They regulate climate through processes of carbon sequestration and uptake of other greenhouse gases. They contribute to conservation of ecosystem biodiversity and provide a direct source of human resources such as important minerals, peat, and clay (Dominati et al. 2010).

A few of the many important ecosystem services provided by soils include provision of food, wood, fiber, and raw materials; flood mitigation; recycling

of wastes; biological control of pests; provision of the physical support for roads and buildings, as well as cultural services, which include both general aesthetics and a sense of place (Dominati et al. 2010). Healthy soils have characteristics that include the appropriate levels of nutrients required for production of healthy plants, moderately high levels of organic matter, a structure that has a good aggregation of primary soil particles and macro-porosity, moderate pH levels, thickness sufficient to store adequate water for plants, a healthy microbial community, and absence of toxicity.

It may be possible to draw inferences about the effects of climate change on agroecosystem services from observations about soil erosion and herbicide and nutrient movement from the edge of fields into adjacent areas. Erosion is a primary source for soil particles and agrochemicals transported from agricultural fields to streams and other water bodies. Under changing climate, some regions will experience greater drying, while other areas will have more intensive rainstorms or increased rate of snow melt – each of these factors may increase soil erosion. Movement of chemicals and soil material will affect the quality of water and will be affected by changes in the intensity of meteorological events. As soil erosion changes under climate change, so does the potential for associated offsite, non-point source pollution. Riparian buffers and wetlands often serve as sinks for pollutants moving from upland fields (Hill 1996; Mayer et al. 2007; Vidon 2010), thus making them important components in possible conservation practices for climate change adaptation in cases where offsite, non-point pollution is a concern.

Soil Degradation and Soil Erosion

Several processes, both natural and anthropogenic, act to degrade soils. These processes include erosion, compaction, salinization, toxification, and net loss of organic matter. Of these, soil erosion is the effect most directly affected by climate change and also the most pervasive. Soil erosion is a natural process and occurs regardless of human activity; however, human activities, including intensive agriculture, have caused accelerated erosion across many regions of the planet, including the United States (Montgomery 2007). Excessive erosion rates decrease soil productivity, increase loss of soil organic carbon and other essential nutrients, and reduce soil fertility (Quine and Zhang 2002; Cruse and Herndl 2009). The major factors affecting soil erosion are: (1) erosive effects of rainfall, irrigation, snowmelt, and wind; (2) plants, cropping, and management; (3) soil erodibility; (4) conservation practices; and (5) topography. Of these, climate change will most likely have the greatest effects on the first three, however strategies for adaptation to climate change effects generally are related to conservation practices (Delgado et al. 2011).

The most direct effect of climate change on rainfall-driven erosion is related to rainfall's erosive power.

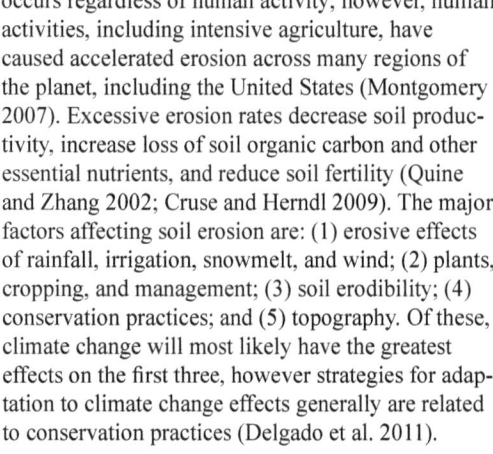

Rainfall

The most direct effect of climate change on rainfall-driven erosion is related to rainfall's erosive power (Favis-Mortlock and Savabi 1996; Williams et al. 1996; Favis-Mortlock and Guerra 1999; Nearing 2001; Pruski and Nearing 2002a, 2002b). The power or ability of a storm or series of precipitation events to cause soil erosion, or rainfall erosivity, is highly correlated with the interaction effect of storm energy and maximum prolonged precipitation intensity (Wischmeier 1959; Wischmeier and Smith 1965; Nearing et al. 1990; Nearing et al. 2005). With regard to erosivity, the dominant variable is rainfall *intensity*, which is the amount of rainfall reaching the soil surface per unit time, rather than total rainfall *amount* (Nearing et al. 2005). If both rainfall amount and intensity were to change together in a statistically representative manner, assuming temporally stationary relationships between amounts and intensities, the predicted erosion rate would increase on the order of 1.7% for every 1% increase in total rainfall (Pruski and Nearing 2002b) .

Effects of changing climate on plant biomass will also affect rainfall-driven erosion. The mechanisms by which climate change affects biomass, and by which biomass changes affect runoff and erosion are complex (Williams et al. 1996; Favis-Mortlock and Guerra 1999; Pruski and Nearing 2002a). As an example, increases of atmospheric CO_2 concentrations increase plant production rates for some species, which could translate into increased soil surface canopy cover and, more importantly, biological ground cover (Rosenzweig and Hillel 1998). Biological material, which includes materials such as ground cover and crop residue, comes in direct contact with the soil surface, and therefore such materials have a

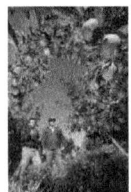

greater impact on effects of rain-driven erosion than plant canopy cover. Biological materials protect the soil from raindrop splash and substantively increase surface water flow roughness, which reduces flow velocities and the ability of water to move sediment. Conversely, increases in soil and air temperatures will trend toward faster rates of residue decomposition via increased microbial activity – the higher the temperature, the faster the microbes work. The rate of microbial activity is moderated by the amount of soil moisture, and, as is the case for all other organisms, beyond a critical temperature threshold the microbes die. Climate change may affect biomass production through changes in temperature and precipitation. Increased precipitation, for instance, could increase biomass production because of the removal of the water limitations on plant growth, which may in turn increase the amount of leaf litter on the ground and reduce effects of erosion.

Temperature changes also affect biomass production levels and rates. Corn biomass production, for example, may increase with increasing temperature, particularly if the growing season is extended; however, biomass may decrease due to temperature stresses as temperatures become too high (Rosenzweig and Hillel, 1998). Studies have also shown that even in areas where the overall amount of precipitation may remain constant or decrease, erosion will likely still increase because of increased event intensities (Pruski and Nearing 2002a; Zhang et al. 2012).

Irrigation

To date, no large-scale studies or reviews exist that investigate the anticipated effects of climate change on future irrigation erosion rates. Only limited data have been published on irrigation-induced erosion (Reckendorf 1995; Sojka et al. 2007), however, existing data suggest that both sprinkler irrigation and surface irrigation (particularly in furrows) are susceptible to irrigation-induced erosion. No generally recognized erosion problems are associated with drip, sub-irrigated, or flood irrigation. Changes in irrigation erosion under climate change will occur as a function of the complex interactions between the increasingly greater stresses being placed on water resources, increased food demand, changes in rainfall, and the ability to adopt improved irrigation practices for greater water-use efficiency. Climate-change-related stresses may lead to improvements in irrigation technology, including enhanced water-use efficiency, which may work in concert with soil conservation gains, a case in point being the use of drip over furrow or sprinkler irrigation.

Snow and Winter Processes

For parts of the Northern United States, including 4 million hectares of cropland in the northwestern wheat region, soil erosion is linked to snowfall amounts, snowmelt, and thawing soils (Van Klaveren and McCool 2010). Snow-associated erosion rates may be particularly high when snowmelt or rainfall occurs on thawed soil overlying a frozen layer of soil underneath (Zuzel et al. 1982; Schillinger 2001). Recently thawed soil is highly erodible because of the freezing effect on soil structure and aggregation, which increases soil erodibility, but equally or more importantly because of high moisture content and low soil water suction (Van Klaveren and McCool 2010). Although some process-based and plot-scale research has taken place, there is a general lack of knowledge about the rates of soil erosion associated with snowmelt or rain-on-thawing-soil erosion on a regional or national basis. A potential effect of climate change is associated with a change from snowfall to rainfall. If decreased days of snowfall translate correspondingly to increased days of rainfall, erosion by storm runoff is likely to increase. The potential trends of snow-induced erosion and the effects of snow-melt on thawing soils have not been assessed.

Wind

Wind erosion rate is a function of the wind velocity, soil moisture content, soil surface roughness, soil structure, field length, and vegetation characteristics (Chepil and Woodruff 1954; Skidmore 1965; Skidmore et al. 1970; Ravi et al. 2011). The primary region of concern for wind erosion on U.S. croplands stretches across the Great Plains, from Texas north to Montana, North Dakota, and western Minnesota (USDA 2010). Additional areas of concern include the Northwestern United States (Washington and Idaho) and scattered areas of the Intermountain West. Areas of high wind erosion also occur on grazing lands in the arid and semi-arid regions of the Western United States. Munson et al. (2011) have suggested that wind erosion will increase on grazing lands of the Southwestern United States because of increased aridity and associated reductions of vegetation cover. Major changes of wind erosion rates driven by climate change would likely be associated with local or regional changes in vegetation and soil moisture, however there are no published studies available that estimate the potential increases in future wind speeds.

Increased wind is also likely to increase wildfire incidence, which in turn will increase wind and water erosion rates due to the drastic reductions in ground cover associated with burns (Sankey et al. 2012). There have been declining trends in near-surface wind speed over the last several decades (Pryor et al. 2009), and model projections indicate that these trends of decreasing wind speed will continue in the future (Segal et al. 2001). This may lead to a decrease in evapotranspiration in cropping regions and also reduce the potential for wind erosion.

Changing Agricultural Production and the Effects on Soil Erosion

Agricultural producers, in response to climate change, will change the types of crops planted and crops management. Changes in production can have effects on soil erosion that may be greater than other effects of climate change. Exactly how such changes occur will be a complex function of changing precipitation and temperature regimes, atmospheric CO_2 concentrations, economics, and plant genetics, among other factors.

Southworth et al. (2002a,b) used global circulation model output (from the U.K. Hadley Centre HadCM2 model) with various crop models to evaluate potential changes in wheat, corn, and soybean production in Indiana, Illinois, Ohio, Michigan, and Wisconsin by the mid-21st century. The studies projected significant changes in planting and harvest dates, which certainly have the potential to influence erosion rates. Those results were then coupled with economic modeling (Pfeifer and Habeck 2002; Pfeifer et al. 2002) to create scenarios of producer adaptation. Taking all of this information together, O'Neal et al. (2005) conducted a study of climate change effects on projected runoff and soil erosion in the five States with changes in corn-soybean-wheat management, which included projected changes in the percentage of the three crops grown across the region, biomass production, planting dates, tillage dates, andharvest dates, as well as changes in temperature and precipitation patterns themselves. The results of the simulations projected runoff increases from 10% to 310% and soil loss increases from 33% to 274% from 2040-2059 relative to 1990-1999 for 10 of the 11 sub-regions of the study area due to reduction in projected corn biomass (and hence reduced crop residue) production and a shift in crop percentages toward soybeans, which are much more erodible crops than either corn or wheat (Wischmeier and Smith 1978). These projections are uncertain, however they indicate the large potential magnitudes of erosion rate changes that could occur with changes in production.

The primary region of concern for wind erosion on U.S. croplands stretches across the Great Plains, from Texas north to Montana, North Dakota, and western Minnesota.

Enhanced Atmospheric Carbon Dioxide

Recent findings from open-top chamber and free-air research systems show that soil organic matter (SOM) turnover appears to accelerate under elevated CO_2, and with adequate soil moisture and nutrients, plant productivity is consistently increased (Peralta and Wander 2008; Moran and Jastrow 2010; Cheng et al. 2011). This suggests that accelerated SOM turnover rates may have long-term implications for soil's productivity and C storage potential.

Adaptation

Future changes in the climatic drivers of soil erosion and farmer management adaptations to a changing climate (e.g., crop selection and dates of planting, harvest, and tillage) have the potential to greatly influence soil erosion rates, with a general trend in the United States toward higher rates of erosion. Agricultural production systems will change under a changing climate, but if production systems are implemented congruently with appropriate conservation management systems as they inevitably shift in response to climate change, the effects of most increased precipitation amounts and intensities on soil erosion can be alleviated (Delgado et al. 2011; Lal et al. 2011). The additional benefit of conservation management is the contribution to climate change mitigation by sequestering atmospheric CO_2 through increased organic matter in the soil and by reducing emissions of nitrogen trace gases such as N_2O through improved rate, timing, and method of fertilizer application (Delgado and Mosier 1996; Eagle et al. 2010; Lal et al. 2011).

Conservation tillage, crop residue management, cover crops, and management of livestock grazing intensities have the potential to reduce much or all of the acceleration of soil erosion rates that might occur under a more intense rainfall regime associated with climate change (Delgado et al. 2011). In addition, these techniques in general enhance soil quality by increasing SOM content and improving soil structure (Karlen et al. 1994a, 1994b; Lal 1997; Reicosky 1997; Weltz et al. 2003; Weltz et al. 2011), both of which improve the water-holding capacity of soils and hence could be key to adaptation for water management during drought.

A newer method in the conservation toolbox is the use of precision conservation, an approach that targets conservation practices to places on the landscape where they will be most effective. Precision conservation takes into account the temporal variability of weather events, the variability of surface flows,

the variability of slope gradient and length, and the variability of soil and chemical properties of soil across the landscape (Berry et al. 2003; Mueller et al. 2005; Schumacher et al. 2005; Pike et al. 2009; Luck et al. 2010; Tomer 2010). Precision conservation techniques may be particularly well adapted to application under the increased variability and rainfall intensities associated with climate change. Among the expected effects of climate change is greater frequency of extreme precipitation events. Since soil variability, variations in hydrology, and variability in surface terrain affect erosion rates, extreme precipitation events will accentuate variation in erosion rates across any given field, increasing the erosion rates at given locations across the field where surface flows will be spatially more concentrated.

Agricultural Water Resources and Irrigation

Changing climate conditions over the coming decades are likely to significantly affect water resources, with broad implications for the U.S. crop sector. Rising temperatures and shifting precipitation patterns will alter crop-water requirements, crop-water availability, crop productivity, and costs of water access, resulting in differential effects across the agricultural landscape. The resulting shift in crop regime competitiveness, in turn, will drive changes in cropland allocations and production systems. Regional production effects will depend on climate-induced changes to hydrologic systems and on the sensitivity of current cropping regimes to changes in water requirements and water availability.

Dryland production (i.e., farming occurring in semi-arid areas) may be particularly sensitive to shifting climatic conditions, as changes in growing season precipitation and soil water evaporation directly affect soil-moisture reserves essential for dryland crops. The effect of a warming climate on soil-moisture would vary regionally, depending on the net effect of higher evaporative losses and changes in precipitation. Increased precipitation variability may also have important implications for dryland production. An increase in field runoff due to heightened storm intensity would reduce the fraction of precipitation infiltrating into the crop root zone (SWCS, 2003).[3] Coupled with changes in precipitation will be

---

[3] This report did not examine potential increases in flood risk due to climate change, however increased crop losses and yield declines due to excessive water are significant concerns in low-lying areas that are subject to periodic flooding (DOI, 2011).

an increase in the atmospheric moisture demand due to the projected increases in temperature and higher saturated vapor pressure, which in turn will reduce the availability of water for crops (Hatfield et al. 2001). Areas prone to warmer and drier conditions may see greater and more severe drought frequency, increasing variability in annual dryland yield.

Under irrigated production, natural soil-moisture deficits may be replenished during the growing season through applied irrigation. In arid areas of the Western United States, where soil-moisture reserves are generally low and crop-water demands high, irrigation provides a significant share of crop-water requirements in most years. In more humid areas of the United States, irrigation supplements available soil-moisture reserves, particularly during periods of below-average rainfall. While irrigation reduces the risk from variable seasonal rainfall associated with dryland production, irrigators may be at greater risk from the cost and availability of purchased water supplies.

Climate change effects on the intensification of the hydrologic cycle will have consequences for agricultural production and soil conservation across many U.S. regions. Common to most regions are projected increases in precipitation amounts, along with increased intensity and frequency of extreme events. Drought frequency and severity will increase, rain-free periods will lengthen, and individual precipitation events will become more erratic and intense, leading to more runoff.

The U.S. Irrigated Sector Under a Changing Climate

Climate change has important implications for the extent and distribution of future U.S. irrigated crop production. Irrigated lands in the United State are located in many different climatic regimes and utilize a range of water resources (e.g., groundwater and surface water). Regional adjustments in irrigated acreage will depend on changes to regional water balances under a warming climate and the resulting effects on the viability and competitiveness of irrigated production. In this report, we do not consider regional shifts in the proportion of acreage irrigated and briefly discuss three important determinants of acreage response to irrigation, including: (1) agricultural water requirements; (2) water-supply availability; and (3) relative returns to irrigated and dryland production.

Agricultural Water Requirements

Climate change can alter regional water requirements for crop production through two pathways: crop-level changes in water demand (i.e., biophysical responses) and land-use changes from producer adjustments in terms of which and how many crops to grow, as well as how best to grow them (i.e., adaptation responses).

The potential interactions of a changing climate on crop-level water requirements are highly complex. As noted earlier, carbon enrichment in isolation increases crop water-use efficiency (i.e., yield per unit of evapotranspiration (ET) through both reduced transpiration and increased photosynthetic efficiency. That effect may be offset, however, by the rising temperatures associated with increased carbon concentrations, which increase plant transpiration and associated water loss. Furthermore, research suggests that the magnitude of the CO_2-related reduction in ET may also be tied to temperature, where CO_2 enrichment effect declines as temperature increases (CCSP 2008).

Potential changes in irrigation water demand will depend on how climate-induced adjustments in crop-water requirements compare with adjusted precipitation levels in that region. Where crop ET rises relative to the change in growing season precipitation, irrigation requirements for that crop will increase. However, an increase in growing-season precipitation above crop-water demand may reduce crop-level irrigation requirements, although soil-moisture levels would depend on the timing of rainfall. How changes in crop-level water demand aggregate up to regional changes in irrigation demand will depend on shifts in land use and crop allocations in response to climate change.

Water-Supply Availability

Nationally, 58% of irrigation water withdrawals are from surface water (Kenny et al. 2009). Climate change is likely to have an effect on surface-water resources, with temperature and precipitation shifts expected to alter the volume and timing of storm and snowmelt runoff to surface-water bodies (Nayak et al. 2010). Annual streamflow may increase in the Northern and Eastern United States, where annual precipitation is projected to increase, while precipitation declines for the Southern Mountain and Southern Plains regions will likely result in reduced streamflow and a shift of seasonal flow volumes to the wetter winter months in this irrigation-dominated area (DOI, 2011).

In arid areas of the Western United States, where soil-moisture reserves are generally low and crop-water demands high, irrigation provides a significant share of crop-water requirements in most years.

Snowpack is an especially important factor in the magnitude and timing of seasonal runoff and stored water reserves used for irrigated crop production. This trend is a particular concern in the West, where much of the surface-water runoff comes from mountain snowmelt. Higher temperatures will restrict the snow storage season, resulting in reduced snow accumulations and earlier spring meltoff (Knowles et al. 2006; Nayak et al. 2010). Stored water reserves are projected to decline in many river basins, especially in critical summer growing months when crop-water demands are greatest (DOI, 2011).

The effect of precipitation changes on surface-water flows may be offset or compounded by temperature-induced shifts in potential ET (PET). Higher temperatures are projected to increase both evaporative losses from land and water surfaces and transpiration losses from non-crop land cover, potentially lessening annual runoff and streamflow for a given precipitation regime.[4]

Ground water is a primary water source for irrigation in the Plains States and an important irrigation water supply for the Eastern United States, as well as areas of the Mountain and Pacific West regions, however, relatively less research attention has focused on climate effects on ground water systems. In the Southwest, one study focusing on Arizona's San Pedro Basin projected substantial decreases on ground water recharge based on multiple downscaled climate models and scenarios (Serrat-Capdevila et al. 2007). While ground water aquifers are generally less influenced in the short term by weather patterns, changing climate effects on precipitation, streamflow, and soil-water evaporation can affect ground water systems over time through effects on ground water recharge (Dettinger and Earman 2007).

In arid areas of the Western United States, regional water supply changes will affect irrigated acreage response. Agriculture may become increasingly water constrained across the central and southern portions of the Mountain and Pacific West regions (DOI 2011), where reduced mountain snowmelt will affect the stored surface-water reserves that provide much of the region's irrigation supply. Projected precipitation increases in the northern Rockies and Pacific Northwest, on the other hand, could experience

improved surface-water supplies (DOI 2011). Heavy reliance on ground water from the Ogallala Aquifer may shield the Plain States' irrigated sector from annual and seasonal water availability shifts. Ground water pumping at greater-than-natural recharge rates has caused a significant decline in water-table levels over much of the region, and the likely increase in water demands due to climate change may intensify pressures on ground water resources. Water supplies in the Southern Plains may be further constrained over the long term, while shifting precipitation patterns may increase soil moisture and surface-water availability in the Northern Plains (DOI 2011; Ojima et al. 1999). In the more humid Eastern United States, projected precipitation increases may sustain surface and ground water supplies across the Central and Northern regions. Potentially drier conditions in the Delta and Southeast regions, however, could tighten water supplies. Future irrigation expansion may depend on potential shifts in drought event frequency and severity.

Returns to Crop Production

Changing climate patterns may alter returns to irrigated and dryland production through differential adjustments in production costs and crop yields. Where precipitation is generally adequate to support dryland production in most years, a shift in relative returns may have more influence on irrigated acreage response than adjustments in regional water supplies.

In general, production costs for irrigated crop enterprises are substantially higher than dryland production costs, reflecting both the additional costs of irrigation water access and distribution and the more intensive use of inputs in irrigated production. Where climate change results in increased water-supply scarcity, the cost of irrigation is likely to increase. Regional effects on irrigation returns will vary depending on climate interactions with surface and ground water systems and the cost of applied water in irrigated production. Energy cost adjustments attributable to climate change would also have a large effect on irrigation returns, reflecting the costs of water pumping and pressurization as well as increased energy needs associated with operations (e.g., harvesting) and inputs (e.g., petroleum-based nitrogen fertilizer) in irrigated production.[5]

Where climate change results in increased water-supply scarcity, the cost of irrigation is likely to increase.

---

[4] Other factors, including precipitation, radiation, cloud cover, humidity, wind velocity, and atmospheric carbon, affect ET rates, and our understanding of how factors would interact under a changing climate is incomplete.

[5] Climate change could influence energy costs through adjustments in aggregate energy demand, changes in hydropower generation caused by altered flow regimes, and climate mitigation policies to reduce carbon emissions and expand renewable energy sources.

Changes in relative crop returns may also reflect the sensitivity of dryland and irrigated yields to climatic factors. The projected change in dryland yields relative to a change in irrigated yields is an indicator of the relative competitiveness of irrigation under alternative climate scenarios and potential directional shifts in irrigated acreage. Dryland production may continue in Northern regions, for example, where projected precipitation increases may supplement soil-moisture reserves. In Southern regions facing a potential decline in growing season soil moisture, a relative decline in dryland yields would suggest the potential for irrigation expansion. Actual irrigated acreage response, however, will depend on the availability of regional water supplies to support irrigation.

Farm-level adaptation to climate change can help mitigate potential costs to irrigated agriculture, particularly for areas of the U.S. West facing a potential contraction of irrigated acreage due to growing water scarcity (Adams and Peck 2008; Howitt et al. 2010). Cropland allocations are likely to favor higher valued or less water-intensive irrigated crops. Improved irrigation technologies can improve water conveyance and field application efficiency, enhancing productivity in the face of limited water supplies. Expanded ground water withdrawals may offset surface-water shortfalls in deficit years. Changes in water resource infrastructure and institutions may facilitate the optimal allocation of limited water supplies under a warming climate. Potential infrastructure improvements include improved water-supply forecasting, more efficient water-storage and delivery operations, expanded use of water market transfers, and water-supply enhancement through reservoir storage, aquifer storage and recovery, and wastewater reuse.

Adaptation

Adoption and implementation of soil conservation practices by producers and land managers depend on attitudes about and participation in the stewardship of soil and water resources. Today's agricultural economy often forces farmers to make decisions that may be necessary for survival of their business, but are less protective of soil and water resources. Additionally, a substantial fraction of croplands are now leased on short-term contracts, such that operators lack incentives for investments in soil conservation. Increasing technical assistance, financial incentives, education, and awareness of the effects of climate change may encourage more farmers to adopt soil-conserving behaviors that mitigate the effects of intensified climate regime on soil erosion. Examples

of soil conservation programs that could be refined to promote additional adaptive actions to climate change effects include:

- The Environmental Quality Incentives Program (EQIP), a cost-sharing assistance program aimed at promoting production and environmental quality;

- The Conservation Reserve Program (CRP), which consists of annual rental payments and cost-share assistance to establish long-term, resource-conserving ground covers;

- The Conservation Stewardship Program (CSP), which rewards producers for practices and systems that protect the environment and natural resources; and

- Conservation Innovation Grants (CIG) to stimulate the development and adoption of innovative conservation approaches and technologies.

Ecosystem Services

Agricultural systems offer a range of potential ecosystem services, including pollination, biological pest control, nutrient cycling, hydrological cycling, greenhouse gas and carbon sequestration, and biodiversity. More than simply providing services, agricultural systems also utilize the available ecosystem services and processes for their function, which increases system complexity. Hatfield (2006) showed the need to examine agriculture from the viewpoint of multi-functionality of outcomes rather than a singular focus on productivity of feed, forage, fruit, or fiber. Power (2010) states that agricultural ecosystems provide humans more than just food, forage, bioenergy, and pharmaceuticals, these ecosystems are also essential to human well-being. In addition to the landscape, the range of agroecosystem goods and services is also expressed at the watershed and airshed (similar to a watershed, an airshed encapsulates the ways in which air flows across the landscape) scales. Expressions of ecosystem health at atmosphere and water endpoints can be represented by water quality, air quality, biodiversity, and recreation.

While the benefit of agroecosystem services is clear, the biological effect and interactions with social values are not easily expressed in monetary terms (Heal 2000). Further, lacking direct studies that relate environmental goods and services to climate change scenarios, assessment of the effects of climate change is not currently possible. Fischlin et al. (2007)

Farm-level adaptation to climate change can help mitigate potential costs to irrigated agriculture, particularly for areas of the U.S. West facing a potential contraction of irrigated acreage due to growing water scarcity.

provide an overview of agroecosystem goods and services and potential linkages to climate change, concluding that the ability of ecosystems to function within the bounds of their ability to adapt will be exceeded by the combination of climate change coupled with disturbances in the ecosystem (e.g., flooding, droughts, insects, and changes in land use). The linkage of agricultural systems within the ecosystem context under the pressure of climate change will require increased emphasis on quantifying the role of agriculture as a component of the ecosystem and the feedbacks among the components.

Potential increases in soil erosion occurring with increases in rainfall intensity show that runoff and sediment movement from agricultural landscapes will likewise increase (Nearing 2001). Changes in precipitation event intensity are already occurring and are expected to continue to increase throughout the remainder of this century (Kunkel et al. 1999). Increases in surface runoff lead to potential increases in sediment transport of herbicides and phosphorus from the surface. Shipitalo and Owens (2006) showed that extreme events were responsible for a large amount of the herbicide loss from fields. Extreme events will play a large role in affecting the linkage between agricultural systems and offsite effects caused by the potential effect of increased precipitation.

Pollinators

Ecosystem services reliant on biological interactions may be particularly vulnerable to climate change if the interacting species respond differently to environmental change (Tylianakis et al. 2008; Hegland et al. 2009). Crop pollination is an important biologically mediated service, because 75% of the leading global food crops are pollinated by animals (Klein et al. 2007). The phenology of many ecological processes is modulated by temperature, making them potentially sensitive to climate change. Mutualistic interactions such as pollination may be especially vulnerable due to the potential for phenological mismatching (i.e., asynchrony in its activity period) if different taxa do not respond similarly to temperature changes (Root et al. 2003). In particular, if pollinators and flowering plants respond differently to warming temperatures, this could result in phenological mismatches with negative outcomes for both groups of organisms.

An analysis was conducted on climate-associated shifts in the phenology of wild bees, the most important pollinators worldwide, and compared to published studies of bee-pollinated plants over the

same time period (Bartomeus et al. 2011). Over the past 130 years, the adult activity period of 10 bee species from northeastern North America has advanced by a mean of 10.4 ±1.3 days. Most of this advance has taken place since 1970, paralleling global temperature increases. When compared to the shifts in plant phenology over this time period, the changes in phenological rates are not distinguishable from those of bees, suggesting that bee emergence is keeping pace with shifts in host-plant flowering, at least among the generalist species investigated in this study. However, the case could be different for bees that specialize on particular plants, and plants that specialize on particular bees; such taxa have not yet been investigated.

In addition to shifts in bee phenology, climate change may also affect the daily activity patterns of bees. Potential future effects of climate warming on crop pollination services were evaluated utilizing data from 18 watermelon farms in New Jersey and Pennsylvania between 2005 and 2010 (Rader 2012, personal communication). To assess this interaction, pollen deposition and daily activity patterns of seven dominant pollinator taxa were evaluated as a function of temperature and time of day. Future plant-pollinator interactions were then simulated based on two Intergovernmental Panel on Climate Change (IPCC) climate change scenarios (one assuming low greenhouse gas (GHG) emissions, the other assuming high emissions) at two future time periods (2050 and 2100) to determine the effect of rising temperatures on pollinator activity patterns and subsequently on crop pollination services. Under current conditions, pollinators differ in their activity patterns at varying temperatures within a day. Model predictions suggest that under future, warmer climate scenarios, five of the seven taxa should provide increased pollination services. Conversely, the honeybee, which is the dominant crop pollinator worldwide, and one native bee species, are predicted to provide less pollination under projected future warmer conditions. The differential responses among bee species to rising temperatures should help stabilize pollination services, as the decline in services by some taxa is buffered by the increase in others. It is important to note that native pollinator species provide this buffering effect and that the study system where the work was done has high levels of crop pollination (about 60%) from native bees. In other, more intensive agricultural systems where native bees are absent, the honey bee is the primary crop pollinator. The results of this study suggest that in such systems, pollination will decline as the climate warms.

Changes in precipitation event intensity are already occurring and are expected to continue to increase throughout the remainder of this century.

Adaptation

Ecosystem services represent the interaction among agricultural systems across the landscape scale and are interlinked with time. Effective adaptation strategies will have to account for these interactions, modifying the various components of the agricultural system to change their response to climate stressors so as to ensure the multifunctionality of the various endpoints. As an example, development of habitat conducive to the survival of pollinators would provide an adaptation strategy for this system that would allow it to better cope with climate change.

U.S. Agricultural Production

The effects of climate change on plants and livestock are critical to the future of efficient and profitable agricultural production. Changes in CO_2, temperature, precipitation, and evaporative demand directly influence plants and animals. Production systems will also be altered through the effects of climate change on insects, weeds, and diseases. The direct effects of climate change have similarities, such as rising temperatures causing rapid development, increased water use, and altered productivity. The end results are different, however, because each crop and livestock type has specific thresholds (see Table 5.1 for examples) in response to each of these variables. These specific responses

will determine the efficacy of adaptation practices and the potential change of plant or livestock distribution as climate change occurs. The following sections will detail responses of select production systems to climate variables.

Corn and Soybean

During 2011, 91.9 million acres of corn and 75.0 million acres of soybean were planted in the United States (www.nass.usda.gov). These two crops are often grown in rotation, with the major production region of both crops concentrated in the Midwest. Iowa and Illinois account for approximately one-third of the U.S. corn crop, and more than 80% of soybean acreage is concentrated in the upper Midwest. According to the USDA Economic Research Service, corn grain typically accounts for more than 10% of U.S. agricultural exports, and the value of soybean oilseed exports currently exceeds $20 billion.[6] Thus, understanding the implications of global environmental change on current and future corn and soybean production has profound economic implications for the United States and the world.

Temperature Effects

For both corn and soybean, effects of rising temperature depend upon current mean temperatures

---

[6] For more information, see http://www.ers.usda.gov.

Table 5.1. Cardinal base and optimum temperatures (°C) for vegetative development and reproductive development, optimum temperature for vegetative biomass, optimum temperature for maximum grain yield, and failure (ceiling) temperature at which grain yield fails to zero yield, for economically important crops. The optimum temperatures for vegetative production, reproductive (grain) yield, and failure point temperatures represent mean temperatures from studies where diurnal temperature range was up to 10°C.

Crop	Base Temp Veg	Opt Temp Veg	Base Temp Repro	Opt Temp Repro	Opt Temp Range Veg Prod	Opt Temp Range Reprod Yield	Failure Temp Reprod Yield
Corn	8[1]	34[1]	8[1]	34[1]		18-22[2]	35[3]
Sorghum	8[16]	34[16]	8[16]	31[17]	26-34[18]	25[17,19]	35[17]
Bean					23[28]	23-24[28,29]	32[28]
Cotton	14[20]	37[20]	14[20]	28-30[20]	34[21]	25-26[22]	35[23]
Peanut	10[24]						
Rice	8[12]	36[13]	8[12]	33[12]	33[14]	23-26[13,15]	35-36[13]
Soybean	7[4]	30[4]	6[5]	26[5]	25-37[6]	22-24[6]	39[7]
Wheat	0[8]	26[8]	1[8]	26[8]	20-30[9]	15[10]	34[11]

Sources: [1]Kiniry and Bonhomme (1991), Badu-Apraku et al. (1983); [2]Muchow et al. (1990); [3]Herrero and Johnson (1980); [4]Hesketh et al. (1973); [5]Boote et al. (1998); [6]Boote et al. (1997); [7]Boote et al. (2005); [8]Hodges and Ritchie (1991); [9]Kobza and Edwards (1987); [10]Chowdury and Wardlaw (1978); [11]Tashiro and Wardlaw (1990); [12]Alocilja and Ritchie (1991); [13]Baker et al. (1995); [14]Matsushima et al. (1964); [15]Horie et al. (2000); [16]Alagarswamy and Ritchie (1991); [17]Prasad et al. (2006a); [18]Maiti (1996); [19]Downs (1972); [20]K. R. Reddy et al. (1999, 2005); [21]V. R. Reddy et al. (1995); [22]K. R. Reddy et al. (2005); [23]K. R. Reddy et al. (1992a, 1992b); [24]Ong (1986); [25]Bolhuis and deGroot (1959); [26]Prasad et al. (2003); [27]Williams et al. (1975); [28]Prasad et al. (2002); [29]Laing et al. (1984).

during critical reproductive growth phases (Table 5.1 contains cardinal and optimal temperatures for growth and yield). A rise in temperature of 0.8 °C over the next 30 years in the Corn Belt is estimated to decrease corn yields by 2% to 3%, assuming no interacting effects from soil moisture deficits (Hatfield et al. 2011). This trend is largely based on observations of geographic variation in maximum corn yields, so is likely an underestimate because it does not consider the interaction of temperature and water availability, nor does it incorporate potential effects of increasing temperature on photosynthesis, respiration, or reproductive parameters (Hatfield et al. 2011).

Lobell and Field (2007) estimate an 8.3% decrease in corn yield per each 1°C increase in average growing season temperature. For soybean, the mean growing-season temperature in the upper Midwest is approximately 22.5°C, so a 0.8°C increase in temperature may increase yields (Hatfield et al. 2011), but this conclusion is not supported by recent historical analysis (e.g., Hatfield et al. 2011). For the Southern United States, growing-season temperatures are higher, such that midcentury warming of 0.8°C is estimated to decrease yields by approximately 2.4% (Hatfield et al. 2011). This estimate is greater than the projected value based on extrapolating the global historical temperature/yield relationship, which predicts approximately a 1.3% decrease in soybean yield per 1°C increase in temperature (Lobell and Field 2007). One limitation of using historical relationships to project future crop performance is that such relationships cannot account for steady genetic improvements in yield potential over time.

Enhanced Atmospheric Carbon Dioxide

The physiological basis for corn and soybean response to rising CO_2 is fundamentally different (Leakey et al. 2009a). As a C_4 plant, corn photosynthesis is saturated at current levels of CO_2, so increasing concentrations of CO_2 over this century are unlikely to stimulate photosynthetic gain, except during times of drought (Leakey 2009). Conversely, soybean is a C_3 plant, and increasing atmospheric CO_2 increases intercellular CO_2 concentrations, which leads to an increased rate of photosynthesis and a lower respiration rate, resulting in a net increase of photosynthesis and growth (Bernacchi et al. 2006).

Across three growing seasons and two contrasting nitrogen (N) treatments, corn yield was not significantly increased by growth at elevated CO_2 (Leakey et al. 2004; Leakey et al. 2006; Markelz et al. 2011). However, soybean yields were significantly increased by growth at elevated CO_2 (550

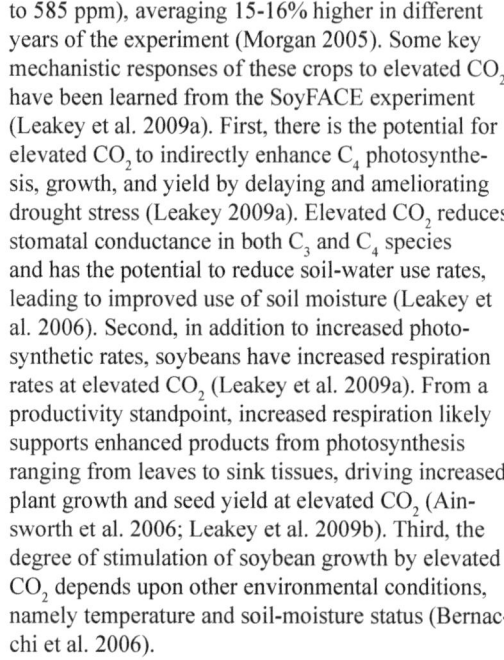

to 585 ppm), averaging 15-16% higher in different years of the experiment (Morgan 2005). Some key mechanistic responses of these crops to elevated CO_2 have been learned from the SoyFACE experiment (Leakey et al. 2009a). First, there is the potential for elevated CO_2 to indirectly enhance C_4 photosynthesis, growth, and yield by delaying and ameliorating drought stress (Leakey 2009a). Elevated CO_2 reduces stomatal conductance in both C_3 and C_4 species and has the potential to reduce soil-water use rates, leading to improved use of soil moisture (Leakey et al. 2006). Second, in addition to increased photosynthetic rates, soybeans have increased respiration rates at elevated CO_2 (Leakey et al. 2009a). From a productivity standpoint, increased respiration likely supports enhanced products from photosynthesis ranging from leaves to sink tissues, driving increased plant growth and seed yield at elevated CO_2 (Ainsworth et al. 2006; Leakey et al. 2009b). Third, the degree of stimulation of soybean growth by elevated CO_2 depends upon other environmental conditions, namely temperature and soil-moisture status (Bernacchi et al. 2006).

Adaptation

Over the past 30 years, both corn and soybean have been planted increasingly earlier in the spring (Sacks and Kucharik 2011). Across the United States, corn planting dates have advanced by 10 days and soybean by 12 days from 1981 to 2005. This earlier planting has been accompanied by a longer growing season, especially for corn. Trends in early planting coupled with the shift to longer season cultivars have together contributed to the yield increase observed over the past three decades (Bruns and Abbas 2006; Kucharik 2006; Sacks and Kucharik 2011). Simulations of corn yield potential predicted a 2-Megagram-per-hectare (Mg ha^{-1}) increase for a 7-day longer maturity (119-day versus a 112-day) hybrid (Yang et al. 2006). However, the trend in earlier planting is not necessarily related to wide-ranging springtime warming, which only occurred over a small portion of the U.S. Midwest from 1981 to 2005, but rather to the development of genotypes tolerant of suboptimal early season temperatures, planting equipment improvements, and adoption of conservation tillage (Kucharik 2006). Reduced tillage practices and advanced equipment capable of completing several tasks in one pass lessen the time and resources needed to prepare soils for spring planting (Kucharik 2006).

Recent analyses of historical yield data and growing-season temperatures indicate a negative relationship, meaning that yields of both corn and soybean are depressed during warmer years (Lobell and Field 2007; Kucharik and Serbin 2008). A trend toward

cooler U.S. growing-season temperatures predomi-
nated from 1980 to 2008, which likely contributed to
yield gains of both corn and soybean over that period
(Lobell and Asner 2003). Globally, the United States
appears to be an anomaly, as many other countries
world-wide have shown a clear warming trend during
the growing season (Lobell et al. 2011). Therefore,
although recent climate trends have had only a small
effect on corn and soybean yields in the United
States, climate trends have reduced corn yields over
the past 30 years (Lobell et al. 2011).

Another favorable trend over the past 90 years is that
crop-growing seasons have become wetter in parts
of the Midwest (Illinois and Indiana), and droughts
have become more localized over that same time
(Mishra and Cherkauer 2010). From 1980 to 2007,
soybean and corn yields were well correlated to
meteorological drought during grain filling periods
and to daily maximum air temperature (Mishra
and Cherkauer 2010). Thus, it is likely that some
of the negative effect of warming on yields has
been counter-balanced by increases in precipitation
(Kucharik and Serbin 2008). A limitation of these
historical relationships is that even within relatively
small regions, such as the State of Wisconsin, there
is significant spatial variability of climate trends
(Kucharik and Serbin 2008). Additionally, it is often
difficult to separate the effects of increasing temperature from effects of moisture stress, as the two are
intrinsically linked.

The changes in climate over the past century have
been driven by changes in the atmosphere, notably
an increase in atmospheric CO_2 and a variable concentration of tropospheric ozone (Chapter 3). Over
the past 50 years, atmospheric CO_2 has increased by
~73 ppm, which is estimated to have increased corn
yields by 9% and soybean yields by 15% in dry years
(McGrath and Lobell 2011). Other estimates of the
CO_2 fertilization effect over the past 50 years in both
wet and dry years range from 0-13% for corn and
3% to 17% for soybean (reviewed by McGrath and
Lobell 2011).

Adapting Corn and Soybean Production to Future Growing Conditions

Understanding the interactions of direct and indirect
climate change effects on corn and soybean production in the United States is a challenge. Regional
projections for future temperature and drought stress
vary, as do incidences of weeds, pests, and pathogens
(Luck et al. 2011). How these stresses combine to
affect productivity can be complex; for example,
elevated CO_2 alone or in combination with elevated
ozone significantly reduced downy mildew disease
severity by 39% to 66% in Midwestern soybean, but

the same conditions also increased brown spot severity (Eastburn et al. 2010). Hatfield et al. (2011) predicted that with adequate water, corn in the Midwest
would see a net yield response of -1.5% in 30 years
due to temperature changes alone. However, corn
production in the South would be more negatively
affected by future climate because corn is closer to
its optimal temperature in that region, and temperature increases would be expected to reduce yields
more profoundly (Hatfield et al. 2011). Midwestern
soybean was projected to show an increase in seed
yield of 9.1% with the climate projected for 30 years
from now; however, this again assumes sufficient
water availability (Hatfield et al. 2011). In the South,
soybean yields are projected to decrease as the negative effects of higher temperature will outweigh the
benefits of rising CO_2 (Hatfield et al. 2011). A major
limitation of these projections is a paucity of experimental data to validate the conclusions.

Rice

Rice is widely acknowledged as a significant source
of food for roughly 2 billion people, principally in
Asia. U.S. rice production, centered in the Mississippi
Delta region, currently occupies approximately
3 million acres, with the United States being the
world's fourth largest rice exporter (Livezey et al.
2004). At present, it is estimated that to keep pace
with projected population increases, rice production
must increase globally by approximately 1%
annually (Rosegrant et al. 1995). Although dramatic
yield increases were observed after the successful
introduction of short-statured (semi-dwarf) rice
varieties in the 1980s, recent trends indicate that U.S.
rice yields have, in fact, stabilized (Figure 5.1). The
gap between current rice production levels and future
needs represents a growing challenge for agronomists
and plant breeders.

U.S. rice production, centered in the Mississippi Delta region, currently occupies approximately 3 million acres, with the United States being the world's fourth largest rice exporter.

Fig. 5.1. Recent trends in rice productivity in the United States since 1997. Source: www.nass.usda.gov.

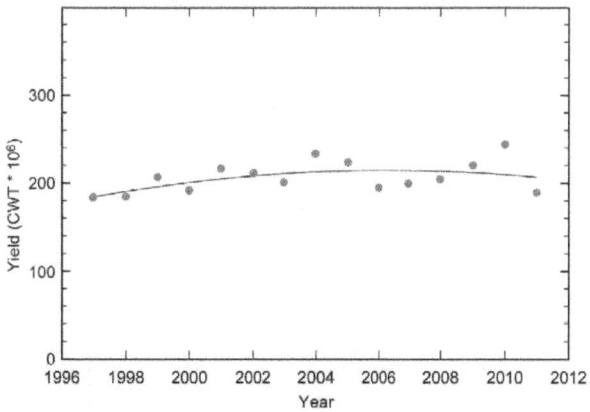

Clearly, in addressing this challenge, the climate change context will also have to be considered. Specifically, how rising atmospheric CO_2 concentrations, changing temperature, and water availability might alter future rice production. Accurate projections regarding the effect of such changes are essential in determining not only rice production but, because of the importance of rice as a basic caloric source, global food security. Overall, the challenge for rice production in the United States is twofold: increasing production, while facing a suite of direct and indirect stressors associated with climate change.

Temperature Effects

Temperatures below 20°C or above 35°C at flowering generally result in increases of floral or spikelet sterility (Satake and Hayase 1970; Satake and Yoshida 1976; Satake 1995) due to lack of anther dehiscence (i.e., failure of pollen to form normally and be released). Reproductive processes, which occur within 1 to 3 hours after anthesis (i.e., dehiscence of the anther, shedding of pollen, germination of pollen grains on stigma, and elongation of pollen tubes), are disrupted by daytime air temperatures exceeding 33°C (Satake and Yoshida 1976). Since anthesis occurs in most rice cultivars between about 9 and 11 a.m., exceeding such air temperatures may become more prevalent in the future and affect rice grain yields. Cultivars that shed their pollen earlier in the day would avoid exposure to high temperature.

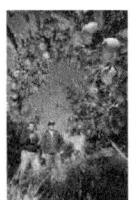

Tian et al. (2010) observed in rice that a combination of high temperatures (>35°C), coupled with high humidity and low wind speed, caused the panicle temperatures to be as much as 4°C higher than air temperature, creating a situation inducing floret sterility. Some rice-growing regions could compensate for these losses; a warmer environment could support northward expansion of growing regions for japonica varieties. With yield currently limited by cold temperatures, warmer temperatures have the potential to generate greater yields in these areas. The majority of global rice production, however, is located in tropical and semi-tropical regions that would be negatively affected by higher projected temperatures, which will increase sterility and decrease yields in these areas (Prasad et al. 2006). Emerging evidence has shown that there are differences among rice cultivars for flowering time during the day (Sheehy et al. 2005). Shah et al. (2011) find that flowering at cooler times of the day would be beneficial to rice grown in warm environments and might become a phenotypic marker for high-temperature tolerance.

Additionally, increases in nighttime temperature minimums have been shown to reduce rice yields through increased plant respiration (Mohammed and Tarpley 2009). Initial investigations in outdoor, sunlit chambers indicated that higher nighttime relative to daytime temperatures could reduce seed set and grain yield (Ziska and Manalo 1996). Long-term trends in minimum temperature suggest that such an effect may already be occurring in situ for production areas in China (Peng et al. 2004) and that increasing nighttime temperatures may be strongly associated with declining rice yields and rice quality (Welch et al. 2010). Reduced grain size and increased grain chalkiness have been associated with high nighttime temperatures; this results in lower grain-milling yields and reduced crop value (Cooper et al. 2008).

Enhanced Atmospheric Carbon Dioxide

Rice, like most crop species with the C_3 photosynthetic pathway, has been shown in a number of studies to respond to increasing levels of atmospheric CO_2 (e.g., Baker et al. 1992). However, a number of contrasting spatial and temporal changes in rice development and yield variations in response to rising CO_2 levels have also been reported (Kim et al. 1996; Moya et al. 1998; Kim et al. 2003). Of particular importance may be additional information regarding intra-specific variation among rice cultivars as well as quantification of the yield response to increasing CO_2 in conjunction with other environmental factors likely to change (e.g., temperature) (Ziska and Bunce 2007).

Changing Water Availability

Although rice is grown in a number of different geographic environments, irrigated, or paddy, rice accounts for the majority of global production, including that occurring in the United States. Without adequate access to water, rice yields decline to less than one-third of that of irrigated production. Additionally, in hot dry areas (e.g., California), irrigation is needed to maintain sufficient evaporative cooling to avoid floral sterility.

Although the total area planted in rice is roughly equivalent between irrigated and non-irrigated fields, irrigated rice accounts for 75% of the total rice production (Bouman et al. 2007). Much of the surface runoff used in irrigation is derived from snow and ice melt from mountain sources. These sources may be particularly vulnerable to warmer and drier conditions (IPCC 2007b; Immerzeel et al. 2010). Ground water supplies from aquifers are also likely to be affected in arid regions, due in part to declining water tables (overdrafts) and increasing pumping costs. In the United States, more than 80% of the rice crop is grown in the Mississippi River alluvial plain. The most intense rice production occurs in the Grand Prairie region of the Mississippi River Delta, where irrigation water is primarily derived from the alluvial aquifer (ASWCC 1997). However, the alluvial aquifer is not expected to sustain current extraction rates beyond 2015 due to ground water overdraft (Scott et al. 1998; U.S. Corps of Army Engineers 2000).

Increased pumping costs and declining water levels in the alluvial aquifer have caused some farmers to install irrigation wells in the Sparta-Memphis aquifer that underlies the alluvial aquifer. Currently, about 30 new agricultural irrigation wells per year are being drilled into this aquifer (Charlier 2002). This is of concern since the Sparta-Memphis aquifer is the source of drinking water for more than 350,000 people, and it has much less capacity to sustain heavy agricultural pumping rates (ASWCC 1997). Thus, one of the consequences of intense rice production using current, water-intensive production practices is the potential for ground water depletion and reduced agricultural sustainability over the long term. In fact, four Arkansas counties, accounting for 120,000 hectares of rice production, have been declared critical ground water areas by the Arkansas Natural Resource Commission and may be in jeopardy of losing access to water needed for irrigation (Young and Sweeney 2007).

Evaluation of the other moisture extreme, flooding, should also be considered in the context of global climate change and rice productivity. While rice is tolerant of short-term water immersions, it is equally vulnerable to extended (more than 48 hours) submergence. Although major losses in rice crops are frequently reported in tropical regions of the world due to flooding, some 25,000 hectares of planted rice were lost due to flooding of the Mississippi River during 2011.

Lastly, water quality in the context of rising sea level is critical. Thailand and Vietnam currently supply the bulk of rice exports. However, the World Bank estimates that even a 1-meter rise in sea level would increase the salinity of key river deltas sufficiently to reduce rice yields in both countries by up to 50% (World Bank 2000). Some 80,000 hectares of U.S. rice production are located along the Gulf Coast. Subsidence has been observed in these coastal marshlands for decades and has resulted in salt-water intrusion that affects rice production in this region. Storm surges associated with hurricanes Katrina and Rita in 2005 increased salinity in the soil and irrigation water, putting 10% of this area out of rice production for some years. The potential for similar effects throughout the Mississippi Delta region needs investigation.

Extreme Weather Events

Because of mechanization, large areas of rice and other cereals are grown in genetically uniform, mono-cropping systems. Such systems are capable of producing large quantities of grain – if weather is stable. However, because the number of extreme climatic events is likely to increase in the future, the lack of genetic diversity in such cropping systems makes them biologically more vulnerable to such occurrences (Roberts 2008).

Insects and Diseases

There is a dearth of assessments regarding the vulnerability of rice production to climate change and pest biology. This may be due, in part, to distinguishing between pest management and climate. For example, overreliance on pesticide applications, the subsequent selection of pesticide-resistant insects, weeds, and diseases, resulting in increased pest pressure may be of more concern than would be the case under gradual climate changes (Heong et al. 1995). Global warming will certainly affect insect fecundity, by changing synchronization of growth stages and growth requirements between pest and host; for example, a plant's leaves must be at a certain stage of growth to provide certain insects a place to lay eggs. In addition, changes in geographic distribution of rice insect pests are likely to occur (Huang and Khanna 2010). Unlike other tropical regions of the world, insect pests in U.S. rice production fields have been limited due to winter-time survival. However, over the last 30 years the Mexican stem borer, which attacks sugarcane, rice, and other crops, has become a serious pest and has advanced from the southern tip of Texas to the Louisiana border, causing yield losses of up to 50%.

For rice, it is generally known that water shortages, irregular rainfall patterns, and related water stresses can increase the intensity of some diseases, including brown spot and blast. Kobayashi et al. (2006) demonstrated that rising CO_2 may affect disease directly by lowering leaf silicon content, which may have contributed to increased susceptibility to leaf blast. Assessments of rice disease under concurrent conditions of elevated CO_2 and other climatic variables such as temperature and water are currently not available.

Weeds

Weeds impose the largest single limitation on crop yields (Oerke 2006). An overview of crop and weed competitive studies indicate that weeds could limit crop yields to a greater extent with rising levels of CO_2 (Ziska 2010). To date, there have been a limited number of studies on the influence of weeds in rice

Weeds impose the largest single limitation on crop yields.

systems in the context of CO_2 and/or climate change. Alberto et al. (1996) evaluated competitive changes between rice and a C_4 weed (*Echinochloa glabrescens*) at concurrent changes in CO_2 and temperature to demonstrate that while increasing CO_2 favored rice over the C_4 weed, the combined changes in temperature and CO_2 favored the weed species. Zhu et al. (2008) also showed that rice was favored over a C_4 grass (*Echinochloa crus-galli*) with elevated CO_2, but only if the N supply was adequate. If N was limited, elevated CO_2 led to a decline in rice relative to the grass. Data from this experiment, as with that of Alberto et al. (1996), indicate that crop-weed competition in response to CO_2 increases may be contingent on other soil-related (edaphic) and physical parameters. Research conducted to compare the response of a widely grown southern U.S. rice cultivar with red rice, a common weed in rice production fields, at recent and projected increases of atmospheric CO_2 (300, 400 and 500 ppm, respectively) demonstrated that the weedy red rice produced a more dramatic increase in seed yield and biomass with rising CO_2 compared to the commercial cultivar (Ziska et al. 2010). Overall, while additional data are needed, the information to date indicates that weed infestation and rice-weed competition may impose a greater limitation on rice production in the context of a changing climate (Ziska et al. 2010).

Adaptation

Little research is underway in the United States to address rice production vulnerabilities to climate change and/or opportunities associated with rising CO_2 levels. Nevertheless, a number of potential adaptation strategies can be employed to maintain rice productivity.

Cultivar Selection

To promote adaptation to high temperature, plant breeders have suggested phenotypic traits including heat tolerance during flowering, high harvest index, small leaves, and reduced leaf area per unit of ground area as adaptive strategies that reduce canopy temperatures. Shifting peak flowering times to cooler periods may also be beneficial (Prasad et al. 2006). Selection of traits related to extremes of water availability (drought and flooding) is underway by the International Rice Research Institute (IRRI) and elsewhere (Wassmann et al. 2009).

One additional means of adaptation to global climate change may lie in recognizing that CO_2, the principal anthropogenic gas, also provides the raw material (carbon) needed for photosynthesis and growth. Because 95% of all plants currently lack optimal levels of CO_2 for photosynthesis (i.e., those with the

C_3 photosynthetic pathway), this anthropogenically driven increase in CO_2 represents a rapid rise in an available plant resource. Differential responses to such a changing resource, in turn, could provide a basis for human selection within crop lines for improved yields. Hence, selecting for CO_2 responsiveness among rice lines may provide an opportunity by which breeders and agronomists could adapt to climate change while maintaining both food security and economic stability (Ziska and McClung 2008).

Agronomic Practices

Production site adaptation for rice can include shifting planting dates, choosing cultivars with different growth duration, changing crop rotations, and utilizing different soil treatment applications (e.g., till vs. no-till). Adjustment of planting dates could be used to avoid temperature-induced spikelet sterility, provided that such shifts do not interfere with crop rotation or double-cropping practices. Such adjustments would be aided by improved climate forecasting (Gadgil et al. 2002).

It is clear that among climate drivers in the Southern United States, reductions in water availability will require new management methods to reduce water use in rice production. Such methodologies must be economically viable and account for resource savings without significant loss in grain yield. New management techniques like intermittent irrigation appear particularly promising, with a 50% reduction in water application and no concomitant loss of production (Massey et al. 2003). Further water savings may be possible with other irrigation practices, but quantification of water use and yields are not available.

Improved Pest Management

An obvious need exists to assess the vulnerability of rice to climate and CO_2-driven changes to pest biology. Vulnerability can be defined as the measure of the potential effects of a given change, minus the adaptive capacity to respond to that change within the system being affected (Sutherst et al. 2007). Potentially, innovations such as simulation modeling can be used to assess regional variability of rice and other cereals to demographic changes of pest distribution with projected climate. Some potential adaptation strategies would include development of pest-resistant cultivars; breeding with wild, related species of rice to select for genes and/or phenotypes that may be well suited to changing climate/CO_2; greater reliance on integrated pest management; and a greater understanding of how climate is likely to change pest management.

....crop-weed competition in response to CO_2 increases may be contingent on other soil-related (edaphic) and physical parameters.

Wheat

Occupying 54.9 million acres of U.S. farmland in 2011, total wheat production amounted to 2.69 billion bushels (www.nass.usda.gov). The grain is primarily used to make flour for bread, pasta, cookies, and other foods. The United States exports more than 1.2 billion bushels of wheat annually (ERS 2012). While wheat can be grown throughout the continental United States, production is concentrated in the Great Plains and the Columbia River Basin. Wheat varieties are classified as having a winter or spring habit, depending on whether the plants require a cold period to flower (vernalization). Different end-uses require different types of grain characteristics, and market type and grain quality are important in understanding regional differences.

Temperature Effects

The foremost effect of temperature on wheat is an increase of the rate of development, thus reducing the length of the crop cycle, most notably duration of grain filling. The optimal temperatures for development are 20°C to 30°C (expressed as daily average temperatures), with grain filling having an upper maximum of 35°C (Porter and Gawith 1999). Estimates of a lower limit for development up to anthesis are from -1°C to 5 °C (Porter and Gawith 1999). Leaf photosynthesis shows a broad optimum from 15°C to 30°C (instantaneous temperature), but ceases by 45°C (Bindraban 1999), agreeing with estimates that the lethal temperature for growth is around 47°C. Photosynthesis ceases near 1°C.

During vegetative growth, winter wheat can acclimate to temperatures below -10°C, enabling survival during harsh winter conditions. Spring wheat lacks this capability, the difference relating to action of the vernalization loci that control growth habit (Dhillon et al. 2010). Frost events after jointing can sterilize the development of exposed spikelets, resulting in severe yield reductions (Marcellos 1977; Thakur et al. 2010).

Heat stress disrupts sexual reproduction, and stress appears to disrupt multiple aspects of development, including pollen and ovule formation and early embryo development (Zinn et al. 2010). Wheat grown at 20°C showed reduced grain set when transferred to 30°C for 1 day (Saini and Aspinall 1982). Ferris et al. (1998) showed that increasing the number of hours of exposure to temperatures above 31°C resulted in reduction of grain numbers and lower grain biomass at harvest. Data from a temperature gradient tunnel (Wheeler et al. 1996) suggested that grain-set ceased when the air temperature exceeded 40°C for at least 30 minutes during a 5-day period ending at anthesis each day.

Enhanced Atmospheric Carbon Dioxide

Wheat has the C_3 photosynthetic mechanism, and photosynthesis responds strongly to short-term exposure to elevated CO_2. An increase of CO_2 from 360 ppm to 720 ppm typically increases photosynthetic rates of well-lit leaves by 30% to 40%. However, with longer term exposure to elevated CO_2, responses are less pronounced as the plant acclimates. Mechanisms for the lessened response involve multiple adaptations of the photosynthetic pathway (Osborne et al. 1998).

Water Deficits

Water transpired as CO_2 is taken up for photosynthesis, additional water being lost through evaporation from the soil surface or deep drainage. Biomass production and grain yield typically increase linearly with water consumed by a crop (transpiration plus evaporation), but the quantitative relationship varies with climate and effects of other factors such as pests, weeds, and tillage practices. In the west-central Plains, Stone and Schlegel (2006) estimated that wheat yield (grain per unit of water transpired) increased 138 kilograms per hectare for every centimeter (i.e., water efficiency is 138 kg ha^{-1}cm^{-1}) of water used with no-till and 86 kg ha^{-1}cm^{-1} with conventional tillage. In the southern Plains, Musick et al. (1994) found that the response was 122 kg ha^{-1}cm^{-1}, combining data from dryland and irrigated systems, and in the Pacific Northwest, the response was 154 kg ha^{-1}cm^{-1} (Schillinger et al. 2008). Much of U.S. wheat production occurs in areas where water deficits limit yields in most seasons, and droughts can cause crop failures over large regions.

Excess water

Waterlogging can reduce wheat yields 20% to 50% (Collaku and Harrison 2002), and prolonged flooding will kill a wheat crop. Flooding also can limit the area planted. Untimely rains delay plantings and harvests, and rains prior to harvest can cause pre-harvest sprouting, which lowers yield but more importantly, reduces grain quality (Nielsen et al. 1982).

Ozone

Wheat has shown yield reductions under elevated atmospheric ozone that are considered intermediate among major crops (Heagle 1989). In studies with open-top chambers, yield responses have varied greatly, making it difficult to predict potential effects (Heagle 1989; Bender et al. 1999; Feng et al. 2008). Biswas et al. (2008) reported that wheat cultivars with higher stomatal conductance were more sensitive to ozone, which is consistent with the known mode of action of ozone, but suggests that apparent tolerance will decrease with yield potential.

Pests, Diseases, and Weeds
Potential effects of pests, diseases, and weeds include both direct effects on yield and wheat quality plus effects on production costs through the need for control measures (Coakley et al. 1999). Predicting climate change effects on organisms that interact with the wheat crop is essentially an order of magnitude more complex than direct effects on wheat crops per se. Few studies have examined interactions of climate change factors with biotic constraints affecting wheat.

Pests
Warmer temperatures typically increase rates of insect population growth, and warming is expected to extend the growing season in most U.S. agricultural regions, allowing pest populations to breed over a longer period each season. For many insect pests, cold winter weather severely reduces populations. Warmer winter temperatures could increase survival, leading to more rapid reestablishment each spring, further increasing severity of pest effects on agriculture (Bale et al. 2002). While warming also could aid beneficial species that feed on wheat pests, Hance et al. (2007) argued that warming may increase pest outbreaks because of disruption of co-evolved temporal or geographical synchronization.

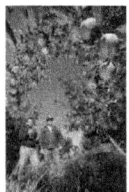

Hessian fly, the most important pest of wheat in the United States, is controlled in part by planting winter wheat when low temperatures reduce the activity of egg-laying flies (Harris et al. 2003). With warming, the onset of the "fly-free" period would be later, requiring farmers to plant later, which reduces yield potential. Numerous other insects affect wheat (Hatchett et al. 1987), both through direct feeding and as vectors for viruses (e.g., the barley yellow dwarf virus (BYDV)).

Evidence for effects of CO_2 on pest damage on wheat is scarce. A review by Sun et al. (2011), which included examples from wheat, noted that elevated CO_2 can reduce the nutritional value of the plant's sap to sucking insects, inducing greater feeding by aphids, but can increase the production of secondary plant-defense compounds that protect plants against insect damage.

Diseases
Wheat diseases differ in their temperature optima, and warming may alter the relative importance of major diseases. Among wheat rusts, stem rust reportedly prefers warmer temperatures than stripe or leaf rust (Garrett et al. 2006), but there is evidence for races (i.e., strains) of stripe rust that tolerate warmer conditions (Luck et al. 2011). Similar to pests, warming may increase overwintering of wheat diseases. Nonetheless, there is great uncertainty concerning the factors that determine whether or not a given wheat disease becomes established. For example, there are concerns that warming may allow Karnal bunt, a disease occurring in the head of wheat and barley, to become established in Europe (Peterson 2009).

Pests and Management Effects
As with pests and diseases, predicting crop-weed interactions under climate change has high uncertainty. The large effect of elevated CO_2 on wheat growth and water use implies an elevated CO_2 interaction with most environmental factors. Elevated CO_2 will not offset temperature effects on development but, by reducing photorespiration, effects of high temperatures on photosynthesis are partially mitigated. Elevated CO_2 can partially reduce water deficits through CO_2-induced reduction of stomatal conductance (Kimball et al. 2002).

The effect of CO_2 on stomatal conductance also explains why elevated CO_2 can reduce the effect of ozone on wheat yields (Feng et al. 2008). Wheat has shown yield reductions under elevated atmospheric ozone that are considered intermediate among major crops (Heagle 1989). In studies with open-top chambers, yield responses have varied greatly, making it difficult to predict potential effects (Bender et al. 1999; Feng et al. 2008). Relative yield responses to CO_2 under low N have been similar to responses with adequate N (Kimball et al. 2002), but clearly if potential grain yields increase under elevated CO_2 plus warming, N inputs through fertilization will have to increase to match the increased crop N requirement.

Elevated temperatures exacerbate water deficits by increasing ET. However, in regions where higher temperatures allow earlier resumption of growth in winter wheat, warming may enable crops to make better use of winter moisture and avoid larger, end-of-season atmospheric water demand. Earlier planting of spring wheat may result in similar benefits.

Adaptation

Crop genetics

Time of anthesis in wheat is affected by genetic controls on day-length sensitivity, vernalization requirement, and intrinsic earliness. The inheritance from generation to generation of these plant parameters is reasonably well understood, so cultivars with modified flowering times could be bred to match changed planting dates and temperature regimes. However, the genetic controls of the duration of grain fill and of heat tolerance in growth processes are much less well understood. Genetic variation for response to CO_2 is known (Manderscheid and Weigel 1997; Ziska et al. 2004), but breeding for increased responsiveness is very difficult given the lack of screening environments to select responsive varieties. Increases in responsiveness to CO_2 likely would reduce water-saving benefits associated with elevated CO_2.

In the United States, dating to the Nation's colonial period, farmers have continually adapted wheat and other crops to new or changing environments and circumstances (Olmstead and Rhode 2011). Recent examples of major changes in production practices include adoption of semi-dwarf wheats and no-till technologies. Thus, Olmstead and Rhode (2011) suggest that North American wheat farmers, who have long shown a remarkable ability to adapt, will be well positioned to manage in the face of climatic challenges.

Crop management

If provided credible guidance, including assessment of risk, especially from frost damage, producers can readily change planting dates and cultivars. Fertilizer regimes also will need to be adjusted according to changes in yield expectations. However, changes in pest, disease, or weed pressure would likely have to be managed on a case-by-case basis.

Crop distribution

An extreme adaptation to changing climate for a specific location is to change crops. Ortiz et al. (2008) suggested that by 2050 the spring wheat belt in North America might shift more than 10 degrees latitude northward, into western Canada. Although not explicitly discussed, presumably winter wheat would move north into former spring wheat regions and portions of the southern-most winter wheat lands would become unsuitable for wheat. Hubbard and Flores-Mendoza (1995) predicted that warming would substantially increase land used for growing wheat. One intriguing option is that the Southern United States might become more suitable for winter-sown spring wheat.

Cotton

Cotton is the principal fiber crop grown in the United States with more than 14.7 million planted acres during the 2011 growing season (NASS 2011). Upland cotton (Gossypium hirsutum L.) is grown throughout the entire U.S. cotton production belt and makes up the bulk of U.S. cotton production (14.4 million acres planted). Pima cotton (Gossypium barbadense L.) production constitutes the remaining portion of U.S. cotton (300,000 acres planted), with production primarily confined to the Western States of California and Arizona. In 2011, Upland cotton production contributed approximately $6.6 billion to the U.S. farm economy, while Pima production

> An extreme adaptation to changing climate for a specific location is to change crops.

Estimated Effects: Empirical Evidence

Free-Air CO_2 Enrichment (FACE) studies are thought to provide the most reliable estimates of wheat response to CO_2. Yield increases for 200 ppm above ambient CO_2 average +12% (Kimball 2011), with response as high as 25% under water deficits (Kimball et al. 2002). Field-based estimates of the impacts of warming on wheat come primarily from analyses of historical yield data and from field experiments where planting dates or other manipulations were used to alter the temperature regime. Estimates of yield decreases have ranged 4% to 6% per degree warming. In recent work combining infrared heating and altered planting dates (Ottman et al. 2012), grain yield of spring wheat declined about 6% per degree warming during grain filling. Lobell et al. (2011) analyzed historic wheat yield data and found a 5% decrease in yield per 1°C increase. These substantial effects largely result from the impact of warming on crop duration and thus exclude potentially offsetting benefits from management adaptations such as changing in planting dates or cultivars. They also ignore the beneficial effect of elevated CO_2 on biomass production and yield.

Estimated Effects: Process-based Modeling

More than 35 peer-reviewed studies have simulated potential impacts of climate change on wheat in the United States. This technical document largely relies on recent studies (2003 or newer) that consider adaptation through changes in planting dates and cultivar phenology. Much of this research assesses methodological issues in combination with potential impacts and generally highlighted uncertainties caused by differences among GCMs and downscaling approaches.

For the Pacific Northwest, results from Stöckle et al. (2010) suggest that winter wheat yields are likely to increase about 20% by 2040 and by 30% by 2080. For spring wheats, in contrast, yields would increase only 7% by 2040 and 3% by 2080. For both growth habits, responses varied among sites, demonstrating the importance of local temperature and precipitation regimes.

In the southern Great Plains, Zhang et al. (2011) reported no change in winter wheat yields for the period 2010-2039. Simulations of winter wheat in the Southeast using GCM and regional climate model scenarios corresponding roughly to 2090 (Tsvetsinskaya et al. 2003), showed 25% and 21% yield reductions, respectively. The largest reductions were for Florida (58% and 50%). The study assumed a fixed cultivar over the region; the results from other regions suggest leads to overestimation of negative impacts of climate change.

Overall, the trend seems to be one of beneficial long-term effects of concurrent climate change and elevated CO_2 on winter wheat yield at higher U.S. latitudes, with declining benefits turning into negative effects towards lower latitudes.

contributed approximately $670 million. Thus, cotton production is a major contributor to the U.S. farm economy and any effect that climate change has on cotton production will be felt throughout the U.S. economy.

Upland and Pima cotton are both indeterminate perennial crops, that is, they bear produce over the season, but they are cultured as annuals. As such, their ability to flower over an extended period of time during the growing season might buffer reproductive response to climate change. However, their growth, development, and performance will be affected by the changing environmental landscape.

Temperature Effects

Cotton in its native state grows as a perennial shrub in a semi-desert habitat and requires warm temperatures. However, despite originating in hot climates, cotton does not necessarily grow and yield best at excessively high temperatures, and a negative correlation has been reported between yield and high temperature during flowering and early boll development (Oosterhuis 1999). Although cotton is sensitive to high temperature at all stages of development, it is particularly sensitive during

reproductive development, and environmental stress during floral development represents a major limitation to crop development and productivity (Snider et al. 2010; Oosterhuis and Snider 2011). Furthermore, the effects on growth from elevated temperatures during the night may be of more importance than during the day. High temperatures can have direct inhibitory effects on both growth and yield and can create high evaporative demand leading to more intense water stress (Hall 2001).

No clear consensus exists about the optimum temperature for cotton, as plant response varies with plant developmental stage, plant organ, and the environment in which the cultivar was developed (Burke and Wanjura 2009). The ideal temperature range for cotton shown from growth chamber studies in Mississippi is from 20°C to 30°C (Reddy et al. 1991), and the optimal thermal kinetic window for enzyme activity in which metabolic activity is most efficient for Upland cotton was from 23°C to 32°C (Burke et al. 1988). Cotton growth and reproductive development are severely inhibited at temperatures in excess of this optimal day/night temperature regime. These higher temperatures commonly occur in the U.S. Cotton Belt during flowering and boll development and, depending on the timing and severity of the stress, can represent a serious limitation to yield. With projected temperature increases from climate change, this is likely to become even more important. However, cotton is successfully grown at temperatures in excess of 40°C in India and Pakistan, indicating some tolerance to high temperature in cotton germplasm (living tissue from which new plants can be grown, e.g., a seed).

High temperature plays a vital role in germination and emergence, and in subsequent stand development, fruiting patterns, and final yield. As maximum temperatures increase, cotton developmental events occur much more rapidly (Reddy, K.R. et al. 1995). The optimum temperature for stem and leaf growth of cotton is about 30°C (Hodges et al. 1993), and once temperatures are above 35°C, leaf area declines (Reddy et al. 1992a; Bibi 2010). Decreases in shoot biomass of Upland and Pima cotton occur with temperatures exceeding 30°C (Reddy et al. 1991). The number of vegetative and fruiting branches produced per plant is strongly influenced by temperature, with an increase in vegetative branches and a decrease in fruiting branches with high temperatures (Hodges et al. 1993). Roots generally have a lower optimum temperature range for growth than shoots, with optimum temperatures reported to be 30°C (Arndt 1945; Pearson et al. 1970). McMichael and Burke (1994) showed that root growth was enhanced when the root temperatures were within or below cotton's thermal kinetic window (i.e., optimal temperature range).

Reproductive growth is generally much more sensitive to high temperatures than vegetative growth (Singh et al. 2007). The flowering period of cotton is reported to be the most sensitive phase to elevated temperatures (Reddy et al. 1996; Oosterhuis 2002). This is because a number of reproductive processes must occur in a highly sequential fashion during pollination to fertilization for successful fertilization and seed production to occur (Reddy et al. 1996). Successful pollination, pollen germination, pollen tube growth, and subsequent fertilization of the ovule are prerequisites for seed formation in cotton; seeds with their associated fibers are the basic components of yield.

Depending upon the duration, timing, and severity of the heat stress, fertilization could be limited by poor pollination, decreased pollen germination, or limited pollen tube growth. Sensitivity of reproductive organs to heat stress was attributed to the sensitivity of pollen grains to high temperature extremes. A positive correlation exists between anther sterility and the maximum temperatures at 15 and 16 days prior to anthesis (Meyer 1966). However, Barrow (1983) reported that pollen viability and germinability were unaffected by pretreating pollen with temperatures as high as 40°C, whereas penetration of the stigma, style, and ovules was negatively affected at 33°C and above. Snider et al. (2011) confirmed that pollen tube growth rate was more sensitive to high temperature than any of the processes occurring during anthesis. The optimal temperature range for cotton pollen germination is between 28°C and 37°C, whereas the optimal temperature for pollen tube growth is from 28°C to 32°C for a range of Upland cultivars (Burke et al. 1988; Kakani et al. 2005). Poor fertilization efficiency under high temperature accounts for the decline in seed number observed for cotton exposed to high temperature conditions in both the field (Pettigrew 2008) and the growth chamber (Snider et al. 2009; Bibi 2010).

The sequence of reproductive development is also hastened as temperatures increase (i.e., the time to the appearance of first square (fruiting bud), first flower, and first mature open boll decreased as the average temperature for each event increased) (Reddy et al. 1996). In addition, the time required for development of flowers up the main stem and the vertical flowering interval increase with increasing temperature (Hodges et al. 1993). The total number of fruiting sites produced increased approximately 50% as the temperature increased from 30°C to 40°C, whereas at temperatures above 35°C abscission of bolls increased sharply with near zero retention of bolls at 40°C (Hodges et al. 1993). Boll retention decreases significantly under high temperature (Reddy et al. 1991; Reddy et al. 1992b; Zhao et al. 2005) and is reported to be the most heat sensitive component of cotton yield, with enhanced abortion of squares and young bolls at temperatures above 30°C for both Pima and Upland cotton (Reddy et al. 1991).

Final cotton yield has also been shown to be strongly influenced by temperature (Wanjura et al. 1969), with a negative correlation between cotton lint yield and high temperature reported for the Mississippi Delta (Oosterhuis 2002). High, above average temperatures during the day can decrease photosynthesis and carbohydrate production (Bibi et al. 2008), and high night temperatures will increase respiration and further decrease available carbohydrates (Gipson and Joham 1968; Loka and Oosterhuis 2010), resulting in decreased seed set, reduced boll size and decreased number of seeds per boll, and reductions in number of fibers per seed (Arevalo 2008).

Boll number and boll size, cotton's basic yield components, are negatively affected by high temperature. Boll retention has been shown to decrease significantly under high temperature (Reddy et al. 1991; Reddy et al. 1992a; Zhao et al. 2005), with temperatures in excess of a 30°C/20°C day/night temperature regime resulting in significantly lower boll retention due to enhanced abortion of squares and young bolls (Reddy et al. 1991). Zhao et al. (2005) found that cotton plants exposed to a 36°C/28°C day/night growth temperature regime retained approximately 70% fewer bolls than plants grown under a 30°C/22°C day/night temperature regime. In addition to negatively affecting boll retention, temperatures in excess of the optimum also resulted in decreased boll size (Reddy et al. 1999; Pettigrew 2008).

The number of seeds per boll is an important basic component of cotton yield (Groves 2009) and accounts for more than 80% of total yield variability. High-temperature stress is a major factor negatively affecting seed development. Pettigrew (2008) reported that slight elevations in daily max-min temperatures of approximately 1°C under field conditions were not sufficient to cause a decline in seed weight, but were sufficient to cause a significant decline in seed number per boll, which was the primary cause of reduced yield under high temperature conditions. This confirmed observations of Lewis (2000), who showed that a lower average number of seeds per boll (23.6 seeds/boll) developed in a hot year (mean maximum daily temperature of 36.6°C for July) compared to 28 seeds/boll in a cool year (mean maximum daily temperature of 32.2°C for July). He concluded that about 99% of the variation in seed numbers per area in his 3-year study was explained by changes in the mean maximum July temperatures when flowering occurred. Although Pettigrew (2008) observed declines in boll size and lint percent, boll size was more negatively affected than was lint percent; therefore, the author concluded that decreased seed number caused a decline in boll size and lint yield. Furthermore, Pettigrew (2008) speculated that heat stress may have decreased seed number by compromising ovule fertilization, which was subsequently confirmed by Snider et al. (2009).

Higher temperatures adversely influence the growth, development, and yield of cotton, and with the increased concern about climate change, this has focused attention on the need for enhanced thermotolerance in commercial cultivars. A number of researchers have documented genotypic thermotolerance in cotton (Taha et al. 1981; Brown and Zeiher 1998; Cottee et al. 2007; Snider et al. 2010). However, although genotypic variation exists in the cotton germplasm pool, this has generally not been exploited in Upland cotton breeding programs. Oosterhuis et al. (2009) reported that breeding trials do not indicate substantial genotypic differences in Upland cotton grown in the U.S. Cotton Belt that may be exploited by plant breeders for improved thermotolerance. However, substantial thermotolerance exists in foreign cultivars from warmer climates (Snider et al. 2010; Snider et al. 2011b), as well as in wild type cotton strains (Bibi 2010).

Pima cotton appears to be more tolerant to higher temperatures than Upland Delta-type cotton (Hodges et al. 1993), and breeders have improved yields in Pima cotton by increasing high-temperature tolerance (Kittock et al. 1988). Although little progress has been made in improving high-temperature tolerance in U.S. commercial Upland cotton cultivars, there appears to be substantial thermotolerance in wild type Upland genetic material collected from areas where cotton grows under conditions of extreme heat such as southern Mexico (Bibi 2010). It has been speculated that modern cotton cultivars are more sensitive to environmental stress conditions compared to obsolete (older than 30 years) cultivars. Brown and Oosterhuis (2010) showed that modern Upland cultivars had improved physiological responses (leaf photosynthesis, chlorophyll fluorescence, and membrane integrity) under ideal temperature environments (30°C), whereas obsolete cultivars were less sensitive in high temperature (38°C) conditions. A similar conclusion was reached by Fitzsimons and Oosterhuis (2011) in an analysis of long-term temperatures and yields in the eastern Arkansas. Snider et al. (2010) showed that genotypic differences in reproductive thermotolerance of Upland cotton is closely associated with the thermal stability of the subtending leaf, and the energetic status (carbohydrates and Adenosine triphosphate (ATP)) and pre-stress antioxidant enzyme activity of the pistil are strong determinants of reproductive thermotolerance in cotton.

The consequences of increased temperatures during the growing season are that cotton seed and fiber yields are likely to be reduced. However, the cotton crop does have some innate thermosensitivity through acclimation to higher temperatures, within the limits of the thresholds of temperature effects on physiological processes. Also, even though the cotton crop is particularly sensitive to high temperature during flowering, due to its perennial nature and indeterminate growth habit, compensation can occur for short periods of heat stress. For example, variation in temperatures during the cropping season allows some flowers during the flowering period to escape exposure to damaging temperatures such

that some bolls are produced; also, boll retention following periods of heat stress will be increased. Higher temperatures together with associated changes in precipitation patterns are likely to change the geographical areas suited to cotton production.

Moisture Stress

Many of the climate change projection scenario outcomes show altered precipitation distribution patterns. These alterations in distribution can result in an area receiving more or less precipitation than has occurred historically. These disruptions also mean an area may now experience precipitation extremes with cotton encountering either flooding or drought stress during one of the critical stages of growth.

Considerable literature exists on research conducted over the years on the effects of drought stress on cotton production; most of this literature has been summarized in a recently published book chapter (Loka et al. 2011). Moisture deficit stress reduces overall plant stature, resulting in plants with less leaf area production (Turner et al. 1986; Ball et al. 1994; Gerik et al. 1996; Pettigrew 2004a). Reduced photosynthetic activity and increased leaf senescence are also caused by moisture deficit stress in cotton (Constable and Rawson 1980; Perry and Krieg, 1981; Marani et al. 1985; Faver et al. 1996; Pettigrew 2004a). This reduced photosynthetic activity, coupled with reduced photosynthesizing leaf area, decreases the total amount of available assimilates for drought-stressed cotton plants to utilize for further growth, vegetative or reproductive.

Drought stress can ultimately lead to lint yield reductions, with the production of fewer bolls per unit of ground area being the principal yield component affected (Stockton et al. 1961; Bruce and Shipp 1962; Grimes et al. 1969; Gerik et al. 1996; Pettigrew 2004b). Fewer bolls are produced primarily because of reduced flower production, but also because of increased boll abortions when the stress is extreme and occurs during reproductive growth (Grimes and Yamada 1982; McMichael and Hesketh 1982; Turner et al. 1986; Gerik et al. 1996; Pettigrew 2004b). Flowering and reproductive growth occur over a more extended period of the growing season compared to the determinate crop species that flower during a brief period. Depending upon the severity and duration of the stress, cotton can tolerate some moisture stress better than determinate crops by somewhat compensating for any reproductive loss during the stress period with production in the remaining unstressed reproductive period.

Fiber quality can also be compromised through exposure to moisture deficit stress, and fiber length can be reduced when moisture deficit stress is severe and occurs shortly after flowering (Eaton and Ergle 1952; Bennett 1967; Marani and Amirav 1971; Pettigrew 2004b). Fiber micronaire, an estimate of fiber fineness, can be inconsistently affected by drought. Depending upon when the stress occurs and its duration, micronaire can either be decreased (Eaton and Ergle 1952; Marani and Amirav 1971; Ramey, 1986; Pettigrew 2004b) or increased (Bradow 2000; McWilliams 2003). These micronaire variations are thought to be tied to how the drought stress affects the relationship between the photo-assimilate supply (i.e., nutritional energy source) and the boll load (nutritional energy sink) (Pettigrew 1995).

A second consequence of altered precipitation distribution patterns is the potential for intermittent flooding events to occur during the growing season. Unfortunately, not as much research has been conducted on flooding or waterlogged conditions as has occurred for drought stress conditions. Yield losses are often associated with flooding events due to a reduced number of bolls being produced (Bange et al. 2004). Overall, plant dry matter was reduced by flooding because of a reduction in the efficiency of solar radiation use rather than a reduction in solar radiation interception. Reinforcing the lower radiation use efficiency observed with flooding, Conaty et al. (2008) reported reduced leaf photosynthetic rates when a flooding event occurred. Flooding that occurred during the early squaring period was more detrimental to yield production than when the flooding occurred during the peak green boll period (Bange et al. 2004).

Enhanced Atmospheric Carbon Dioxide

Almost all the future climate change forecasts call for and are based upon greater concentrations of atmospheric CO_2. Because cotton is a C_3 plant, it exhibits a positive photosynthetic and growth response as the ambient CO_2 concentration is elevated. In controlled environment chambers, K.R. Reddy et al. (1995, 1997) and V.R. Reddy et al. (1995) were able to demonstrate increased leaf photosynthesis, with decreased stomatal conductance, greater total biomass production, and increased boll yield when CO_2 levels were increased from 350 to 700 ppm. This increased photosynthesis coupled with the reduced stomatal conductance under elevated CO_2 conditions also led to increased water-use efficiency. An increase in CO_2 levels, in and of itself, was not observed to have profound effects on the quality of the lint produced (Reddy et al. 1999).

Many of the phenomena observed in these controlled environment chamber studies were confirmed in an Arizona field situation under a FACE environment, with CO_2 concentration elevated to 550 ppm. Photosynthesis (Hileman et al. 1994; Idso et al. 1994) and water-use efficiency (Mauney et al. 1994) increased in elevated CO_2 conditions. Water-use efficiency increased due to increased above-ground biomass production, rather than reduced water use (Mauney et al. 1994). The CO_2 enrichment increased cotton height, leaf area, above-ground biomass, and reproductive output, but not the total root biomass (Derner et al. 2003). Cotton grown in these FACE environments accumulated more total plant nutrients, but had lower tissue nutrient concentrations than cotton grown in ambient CO_2 levels (Prior et al. 1998). The increased total nutrient uptake but decreased nutrient concentration occurred because while above-ground biomass production increased, thereby increasing total uptake, the increased amount of nutrients was diluted across a greater biomass. The researchers also observed that under elevated CO_2 conditions, nutrient uptake from the soil increased, as did nutrient-use efficiency when compared to cotton grown under ambient CO_2; yields increased 43% in these FACE environments (Mauney et al. 1994).

Genetics may offer additional tools for producers to deal with elements of projected climate change.

Adaptation

The front line of defense for cotton producers, if projected climate change scenarios play out, will be an evaluation and appropriate alteration of production practices. Water management capabilities could play a major role in enabling producers to maintain economically viable operations as the climate changes. Obviously, irrigation will be enormously important in areas where alterations in rainfall distribution lead to episodes of moisture-deficit stress occurring. Irrigation is also important in helping the plant mitigate the detrimental effects from excessively high temperature. The plant's ability to lower tissue temperature through transpirational cooling is dependent upon an adequate moisture supply. However, water use for cotton irrigation will have to compete with industrial and urban municipal use for the dwindling ground and surface water supply in many areas. Advances in subsurface drip-irrigation, low energy precision application (LEPA) irrigation, and furrow-dikes can aid cotton producers in making more efficient use of this limited resource (Bordovsky 1992; Sorensen et al. 2011). In areas where flooding could become more problematic as the climate changes, land-forming procedures may be needed on some fields to promote rapid water runoff.

One of the consequences of warmer temperatures on cotton production is an extension of the growing season length. This opportunity provides producers with more flexibility with planting and production decisions. Research has demonstrated improved yield potential when cotton in the Mississippi Delta is planted earlier than during the period traditionally considered as the optimum planting window (Pettigrew 2002). This approach allows much of the cotton crop to be produced prior to the onset of many late-season stresses (high temperature, moisture deficit, and heavy insect infestations). Alternatively, the longer growing season also allows for expansion of the area in which cotton can be successfully double-cropped (two crops per year) behind other crops such as wheat (Wiatrak et al. 2005; Wiatrak et al. 2006).

Another consequence of a warming climate is that cotton can be grown further north than its traditional planting region. For instance, cotton acreage in Kansas has increased from 1,500 acres in 1990 to 38,000 acres in 2009, with the acreage peaking in 2006 at approximately 115,000 acres (USDA 1990-2010). However, most of this Kansas acreage increase occurred due to economic considerations rather than a dramatic shift in climatic conditions. Nevertheless, a warmer climate could allow cotton to encroach further north into regions traditionally used for corn, soybean, or wheat production.

Genetics may offer additional tools for producers to deal with elements of projected climate change. Many cotton genetics programs are trying to develop germplasm with tolerance to various direct (abiotic) stresses (Allen and Aleman 2011). Efforts have been put forth using molecular markers to identify and characterize quantitative trait loci (QTL) associated with abiotic stress tolerance in cotton (Paterson et al. 2003; Saranga et al. 2004). For the most part, these efforts have focused on mining traits and genetic variability within the cotton germplasm pool and other closely related Gossypium species.

Alternatively, genes associated with targeted biochemical pathways involved in conveying a stress tolerance that come from a completely different source could be introduced into the cotton genome through transgenic technologies. Although the use of transgenic technology can provide a more focused approach to genetic manipulations, it also comes with its own set of problems, such as how the inserted foreign DNA could affect native physiological processes. Nevertheless, many private and public breeding programs are devoting resources to select for drought and temperature stress tolerance. However, these traits are highly complex, which dictate that progress will be slow to occur. Most of the initial screening and selecting of lines has occurred in controlled environments, such as greenhouse or growth chambers. Field testing and confirmation of these stress tolerance traits has not proceeded as fast. As of now, no cotton varieties with consistently demonstrable abiotic stress tolerance are available for cotton producers to utilize in production systems. It may be many years before any such varieties with useful stress tolerance are available on the market.

Annual Specialty Crops

Specialty crops are defined in law as "fruits and vegetables, tree nuts, dried fruits, and horticulture and nursery crops, including floriculture." Annual specialty crops include many vegetable and fruit plants, each with their own environmental preferences. The primary annual specialty crops, for which production data are annually collected by the USDA's National Agricultural Statistics Service (USDA-NASS) are listed in Table 5.2. Of these 29 crops, most are considered popularly to be vegetables, and the remainder – strawberries and three kinds of melon – are considered fruits. The USDA recommends that the human diet include half vegetables and fruits (http://www.choosemyplate.gov/print-materials-ordering/dietary-guidelines.html), thus potential effects of climate change on this group of crops are of high interest.

The primary States that produce annual specialty crops are, in order of total production, California, Florida, Arizona, Georgia, and New York (USDA-

Table 5.2. Principal annual specialty crops for fresh market, 2010 production acreages, values, and primary production States obtained from the USDA-NASS, Vegetables 2010 Summary, January 2011. 2009 values for potatoes, sweet potatoes, and dry beans were obtained from the USDA-National Agricultural Statistics Service, Agricultural Statistics, Chapter IV, Statistics of Vegetables and Melons, United States, Government Printing Office, Washington: 2010). Only includes estimates for the selected crops in the NASS annual program. These crops are not estimated for all states that might produce them. *Includes processing total for dual usage crops.

	Production 1,000 Cwt	Acres harvested	Value total dollars	Value dollars/acre	No. states listed	Main production states by acres planted 1	2	3	4	5
Artichokes*	900	7,200	$46,350,000	$6,438	1	CA				
Asparagus*	799	28,000	$90,777,000	$3,242	3	CA	MI	WA		
Beans, dry edible	25,360	1,463,000	$793,722,000	$543	19	ND	MI	MN	NE	ID
Beans, Snap	5,062	88,500	$303,679,000	$3,431	11	FL	GA	TN	CA	NY
Broccoli *	18,219	121,700	$648,886,000	$5,332	2	CA	AZ			
Cabbage	22,797	66,400	$378,404,000	$5,699	14	CA	NY	FL	TX	GA
Cantaloupes	18,838	74,730	$314,379,000	$4,207	9	CA	AZ	GA	TX	IN
Carrots	22,777	68,000	$597,362,000	$8,785	3	CA	MI	TX		
Cauliflower*	6,281	36,360	$247,456,000	$6,806	3	CA	AZ	NY		
Celery*	20,285	28,500	$398,854,000	$13,995	2	CA	MI			
Corn, Sweet	29,149	247,200	$750,467,000	$3,036	26	FL	CA	GA	NY	OH
Cucumbers	8,482	43,900	$193,643,000	$4,411	11	FL	GA	NC	MI	CA
Garlic*	3,737	22,750	$265,510,000	$11,671	3	CA	NV	OR		
Honeydews	3,204	14,700	$49,608,000	$3,375	3	CA	AZ	TX		
Lettuce (total)	87,189	267,300	$2,249,998,000	$8,418	4	CA	AZ			
Head	50,750	139,000	$1,205,575,000	$8,673	2	CA	AZ			
Leaf	11,180	48,000	$429,432,000	$8,947	3	CA	AZ			
Romaine	25,259	80,300	$614,991,000	$7,659	4	CA	AZ			
Onions*	73,213	149,670	$1,383,595,000	$9,244	12	CA	WA	OR	GA	NY
Peppers, Bell*	15,739	52,700	$637,113,000	$12,089	7	CA	FL	GA	NC	NJ
Peppers, Chile*	4,502	22,500	$135,364,000	$6,016	4	NM	TX	CA	AZ	
Potatoes	431,425	1,045,000	$3,452,276,000	$3,304	30	ID	WA	ND	WI	CO
Pumpkins*	10,624	48,500	$116,539,000	$2,403	6	IL	MI	OH	NY	PA
Spinach	6,133	38,900	$256,924,000	$6,605	4	CA	AZ	NJ	TX	
Squash*	6,542	43,500	$203,592,000	$4,680	12	FL	MI	CA	NY	GA
Tomatoes	28,916	104,500	$1,390,754,000	$13,309	14	CA	FL	TN	OH	VA
Strawberries*	28,501	56,990	$2,245,319,000	$39,398	10	CA	FL	OR	NY	NC
Sweet potatoes	19,469	96,900	$410,361,000	$4,235	9	NC	CA	LA	MS	FL

NASS, 2011), though for any particular crop the number and identity of States considered to contribute significantly to the annual harvest varies (Table 5.2). As many as 30 States contribute to the annual potato harvest, while only California contributes significantly to the artichoke supply. California is the leading producer of 19 of the 29 listed annual specialty crops and is among the top 5 production States for all specialty crops except for potatoes, dry beans, and pumpkins. Projected climatic changes, including changed precipitation regimes and increased temperature for agriculturally important parts of California, are of great importance to future production of annual specialty crops.

Temperature Effects
Temperature is a major environmental change expected to affect production of annual specialty crops. Warm-weather crops such as tomato have different temperature responses than cool-weather crops such as potato, lettuce, and onion (McKeown et al. 2004; Else and Atkinson 2010). In addition, the various crops are sensitive to specific forms of stress, such as periods of hot days, overall growing season climate, minimum and maximum daily temperatures, and timing of stress in relationship to developmental stages (Ghosh 2000; Pressman et al. 2002; McKeown et al. 2005; Sønsteby and Heide 2008; Dufault et al. 2009).

For mild heat stress (a 1°C to 4°C increase above optimal growth temperature), a common result is moderately reduced yield (Sato 2006; Timlin et al. 2006; Wagstaffe and Battey 2006; Tesfaendrias et al. 2010). Plants were most sensitive to heat stress 7 to 15 days before anthesis, consistent with a critical time in pollen development. More intense heat stress (generally greater than 4°C increase over optimum) leads to severe yield loss up to and including complete failure of marketable produce (Ghosh 2000; Sato et al. 2000; Kadir et al. 2006; Gote and Padghan 2009; Tesfaendrias et al. 2010). Ample evidence exists that temperature effects on yield loss varies among crops. For example, tomatoes under heat stress struggle to produce viable pollen, though their leaves remain active. The dysfunctional or non-viable pollen does not properly pollinate flowers, causing a failure in fruit set (Sato et al. 2000). If the same stressed plants are cooled to normal temperatures for 10 days before flower pollination, and then returned to high heat, they are able develop fruit. Alternately, the reason some heat tolerant tomatoes perform better than others appears to be, in part, related to a superior ability to create successful pollen even in adverse conditions (Peet et al. 2003; Sato 2006). At least one report identifies a similar

role for pollen development as a facet of strawberry heat tolerance (Ledesma and Sugiyama 2005).

Water Deficits
For many annual specialty crops, ample water is essential to achieve high yields while maintaining a quality acceptable to consumers, and drought is highly detrimental to yield and quality. Depending on the cultivated variety, strawberries produced with less than optimal amounts of water have reduced leaf area, root development, and reduced berry size and yield (Bordonaba and Terry 2010; Klamkowski and Treder 2008). Yields of potato in drought conditions are reduced, especially when temperatures and wind speeds are high (Wolf 2002). The amount of water needed to produce a crop varies according to how the crop is managed and environmental factors such as temperature, light, and wind. The ranges in the amount of water needed to raise a crop based on management and environment are very large (tomato, 2.58-11.88 kg·m^{-3}; potato, 1.92-5.25 kg·m^{-3}; melon, 2.46-8.49 kg·m^{-3}; watermelon, 2.70-14.33 kg·m^{-3}; and cantaloupe, 4.18-8.65 kg·m^{-3}) and call attention to the need for continued research on water management in crop production (Rashidi and Gholami 2008). Even apparently minor differences in furrow orientation resulted in yield reduction of onion, a crop not known as terribly drought sensitive except at the seedling establishment stage; changes in row direction can change plant evapotranspiration and potentially lead to greater accumulation of harmful salts in the soil (Villafañe and Hernández 2000).

To compensate for the uncertainties of precipitation, many annual specialty crops are grown with irrigation. For these crops, drought is a less pressing issue as long as there is an ample water reservoir for agriculture and other users and the cost to irrigate is affordable. Under circumstances where water use is restricted or costly, as is the case in California and Arizona (two major annual specialty crop production States), production of some crops may become less profitable if current climate and trends continue. This may shift cultivation to the other States with more available water.

Excess Water
Other specialty crop States, such as Florida, Georgia, and New York, and much of the East Coast, have been receiving increased precipitation. Since precipitation has been and is expected to occur in more extreme events, the primary benefit will be in the form of a reservoir of irrigation water.

Severe flooding reduces yield by killing plants, while less severe flooding changes the plant in ways

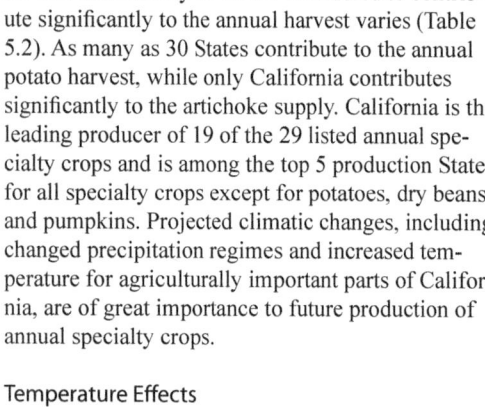

Under circumstances where water use is restricted or costly, as is the case in California and Arizona (two major annual specialty crop production States), production of some crops may become less profitable if current climate and trends continue. This may shift cultivation to the other States with more available water.

that cause it to be weaker, potentially diminishing its yield. As a consequence of moderate flooding of strawberries, for example, fruit yield, total leaf area, and weight decreased, while dead leaf area increased, with one cultivated variety (cultivar) more sensitive to flooding than others (Casierra-Posada 2007). Tomato tolerance to flooding is also cultivar dependent (Ezin et al. 2010). Even storms that produce only minor flooding or no flooding can damage marketable yield. For example, tomatoes are famous for cracking and splitting after a storm. Strawberry flavor is so strongly affected by water availability that strawberries grown in excess water have lower sugar content and taste "watery," while carefully restricting water can be used to increase sugar content and make them taste sweeter (Bordonaba and Terry 2010).

Extreme Events

Extreme precipitation events will be damaging to crops due to a combination of heavy rain that can physically injure plant parts, inject excessive water in the root zone, result in physical damage if high winds accompany rainstorms, and increase pressure from some fungal and bacterial diseases.

Many plants fall over in high winds associated with storms. Their stems can break, and plants such as tomatoes can lose all their fruit without stem support. Wind can also reduce yield of short plants without a main stem (e.g., strawberries) by causing physical damage to the plant (Peri and Bloomberg 2002), and can even reduce yield of root crops such as onion (Greenland 2000).

Solar Radiation Effects

Another environmental condition associated with increased precipitation is reduced light from overcast conditions. In the extreme, reduced light will reduce yields, and optimal light levels depend on crop, cultivar, and growing conditions. Higher light levels seem to be important for allowing maximum uptake of excess CO_2 by tomatoes (Tartachnyk and Blanke 2007). On cloudy days, tomato leaves were unable to make use of the additional photosynthetic building blocks supplied by elevated CO_2 levels, but the amount of CO_2 in a greenhouse atmosphere dropped sharply on sunny days because it was being incorporated into plant growth. In greenhouses in southern Ontario, high concentrations of CO_2 coupled with high light intensity provided measurable gains in yield compared to low light intensity. Given sufficient light and water, plants quickly benefited from increased atmospheric CO_2 (Hao 2008). Finally, a computer model simulating growth conditions of high CO_2 levels and moderate light intensity

predicted approximately a 17% increase in tomato yield. These results were corroborated by 2 years of field trials (Heuvelink et al. 2008). Therefore, light intensity is an additional critical factor in predicting plant response to increased atmospheric CO_2 concentrations.

Moderate light reduction improves both yield and quality of several annual specialty crops. Repeat-fruiting strawberries grown in the Netherlands had higher yields without shading (Wagstaffe and Battey 2004), but once-fruiting strawberries grown in Nova Scotia had higher yields under moderately reduced light (Li et al. 2010), indicating that there is a range of light levels optimum for yield. That optimum varies by strawberry type (e.g., some plants need short-day exposure to flower, while others are neutral to day length, i.e., day neutral) and even cultivar. Highest marketable tomato yields were produced under 50% shading (Gent 2007). In potato, the amount of light was more important than temperature or photoperiod in explaining differences between spring and autumn seasonal yields (Bisognin et al. 2008). The amount of light received is positively correlated with lettuce plant growth (Grazia et al. 2001), but too much light, or rather, too much light of the wrong wavelength, can cause problems with quality (Wissemeier and Zuhlke 2002; Frantz et al. 2004). Research has shown that maximum yield and quality can be achieved in protected cultivation (Oliveira et al. 2006). Cracked skin was the tomato defect most alleviated by shade (Gent 2007). Too much light can also reduce strawberry fruit quality; for example, high light and temperature levels lead to the development of strawberry fruit bronzing (damaged fruit that is bronze in color and may be desiccated or cracked on the surface) in Commander, a particularly susceptible cultivar (Larson et al. 2005). Tomato plants grown in 25%-27% reduced light tolerate and can even achieve higher yields in higher temperatures (Pino et al. 2002; Uzun 2007).

Interactions Across Climate Change Stressors

Because plants continuously integrate myriad environmental signals, tolerance to increased temperature is often dependent on the status of other environmental factors like humidity and light. It has been worthwhile to identify specific cultivar responses to interaction effects in various crops (Amadi 2009; Santos et al. 2009). Examples from tomato highlight plant responses to these interactions. Under high light conditions, tomato fruit yield reached maximum levels at 22°C, and under reduced light conditions, yield continued to increase to 25°C, so that reducing light increased tolerance to higher temperatures (Uzun 2007). A study by Peet et al.

(2003) demonstrated that plants were more sensitive to the combination of high heat and humidity than to either condition alone; reducing humidity increased tolerance to higher temperatures, and some cultivars performed better than others under both high humidity and heat. Onions grown at increased temperature hastened leaf expansion regardless of atmospheric CO_2 concentrations. However, increased CO_2 at high temperature led to the selective increase of carbohydrates in bulbs but not leaves (Wheeler et al. 2004). These interactions will become important when considering adaptation of crop production via relocation of agricultural activity.

Adaptation

For an individual crop, there are often cultivars with higher tolerances for stressful temperatures, water availability, light, and other environmental factors, just as there are cultivars that are resistant or susceptible to certain diseases. Studies of specialty crops have identified promising sources of heat-tolerant genetic material (Camejo et al. 2005; Harbut et al. 2010). This is important because borrowing superior stress tolerance mechanisms from overall inferior plants is a crucial way to improve the varieties that are grown every day in commercial production. At least one promising source of heat tolerance was identified when assessing cultivars of strawberry (Ledesma and Sugiyama 2005; Ledesma et al. 2008), tomato (Sato et al. 2000; Sato et al. 2004), lettuce (Santos et al. 2009), onion (Tesfaendrias et al. 2010), and potato (Amadi 2009). Cultivar differences in tolerance to temperature extremes appear to be greater than for any other environmental stressor. This highlights plant breeding as an important tool for adapting agriculture to future climate change.

Perennial Specialty Crops

Perennial specialty crop production is sensitive to temperature, water availability, solar radiation, air pollution, and CO_2. Furthermore, as in other C_3 plants, photosynthesis can be limited by CO_2 availability when light and other factors are not limiting (Farquhar et al. 1980). Increased atmospheric CO_2 generally increases growth rate and yield, resulting in a higher accumulation of biomass, fruit production, and quality in fruit trees (Idso and Kimball 1997; Centritto et al. 1999a; Kimball et al. 2007). However, growth enhancements in response to increasing CO_2 could diminish in the long-term due to acclimation, especially when combined with other limiting factors such as heat stress and nutrient deficiencies (Pan et al. 1998; Druta 2001; Vu et al. 2002; Adam et al. 2004).

The value of perennial specialty crops is derived from not only the tonnage but also the quality of the harvested product, for example the size of a peach, the red blush on an apple, or the bouquet of a red wine produced from a particular vineyard. In contrast to annual agronomic crop production, perennial crop production is not easily moved as the climatic nature of a region declines due to many socio-economic factors including long re-establishment periods, nearness to processing plants, availability of labor, and accessible markets. Climate change complicates the problem of food production from perennial crops.

Temperature Effects

In California, the optimum growing temperature for wine and table grapes, oranges, walnuts, and avocados is equivalent to the average temperature from 1980-2003, indicating that the current cultivars are well adapted to the contemporary California temperature regime (Lobell et al. 2006). Perennial cropping systems are commonly in place as long as 30 years, and this poses a challenge with a changing climate since the selection of a productive cultivar at planting may not be the most adapted sometime in the future. The development of new cultivars in perennial specialty crops commonly requires 15 to 30 or more years, greatly limiting the opportunity to easily shift cultivars.

In addition to the rise in global temperature, it is expected that some extreme events will increase in frequency and severity as a result of the shift in mean conditions and/or a change in climate variability (Easterling et al. 2002; McCarthy et al. 2001). These extreme events and climatic variation will also pose additional challenges to perennial cropping systems. Socio-economic factors and inability to rapidly identify adapted cultivars do not necessarily make the perennial specialty cropping systems more vulnerable to climate change, but they do call attention to the needs of the industry for new cultural and genetic tools and research to adapt in a timely and economic manner.

The value of a fruit crop is determined and limited at many points before and during the growing season because the value is based not only on biomass, but on size, color, chemical composition, firmness, and other measurable criteria. Using apple as an example, in the year prior to harvest, floral initiation occurs in June-July and high temperatures reduce the number and vigor of the potential floral buds. During the dormant winter months, extreme cold can kill buds and warming periods can de-acclimate buds, making them susceptible to later winter damage.

In the spring, frost periods can kill flowers. As the fruit are growing in the spring, high temperature can reduce cell division resulting in small fruit. During the summer months, high temperature may cause sunburn damage reducing pack-out at harvest, accelerate maturity, reduce fruit firmness and color development, and/or decrease the suitability of fruit for short- or long-term storage.

Of the many perennial specialty crops produced in the United States, apple, blueberry, cherry, citrus, grape, peach, pear, raspberry, and red maple were selected as representative perennial nursery and ornamental crops. Critical temperature and photosynthetically active radiation (PAR) thresholds for key phenological stages were identified in the scientific literature (Tables 5.3, 5.4, 5.5, and 5.6). Each crop has a range of cultivars, and so there is a range of critical thresholds for the various phenological stages. Conservative thresholds were selected from

the literature for use by crop/climate modelers and policy makers in assessing future climate change effects. The response of these crops to a proposed doubling of atmospheric CO_2 is evaluated from the scientific literature in Tables 5.7, 5.8, 5.9.

Modeling of past and future climate changes in the United States has demonstrated that warming in the historical record and future warming will affect perennial specialty cropping systems. Historically, apple mid-bloom dates in the Northeastern United States have advanced 0.20 days/year (Wolfe et al. 2005), with a temperature rise of 0.25°C/decade (Hayhoe et al. 2007). According to Stöckle et al. (2010), apple bloom will occur approximately 3 days earlier by 2020 in eastern Washington. From 1948 to 2002 in the main grape growing regions of California, Oregon, and Washington, growing seasons have warmed by 0.9°C (Jones et al. 2005). In future climate scenarios, grape bloom time in the central

Table 5.3. Critical temperature thresholds for the production of apple, blueberry and cherry at various phenological stages.

Phenological stage	Apple	Blueberry	Cherry
Winter hardiness/chill accumulation (chill units)	400-2900; 5 or 7°C base. (Swartz and Powell, 1981; Hauagge, 2010)	600-1200; 4°C base (Arora et al., 1997)	900-1500; 7°C base (Seif and Gruppe, 1985)
Freeze susceptibility of flowers	-3 to -4°C (Powell and David, 2011)	-2°C (Powell and David, 2011; Snyder and Melo-Abreu, 2005)	-2°C (Snyder and Melo-Abreu, 2005)
Pre-bloom flower development/floral initiation	>5°C prebloom detrimental to fruit set; >17°C reduces floral initiation; Elevated fall temperatures delay bloom up to 3 days. (Warrington et al., 1999; Tromp, 1976; Jackson et al., 1983; Tromp and Borsboom, 1994; Wilkie et al., 2008)	>28°C reduces initiation in highbush. (Darnell and Williamson, 1997)	30-35°C during initiation results in doubles. (Beppu et al., 2001;
Effects on pollination	High temperatures increase pollen tube growth but decrease stigma and ovule viability and converse with low temperature. In general, *O. cornuta* was active from 10 to 12°C and 200 w/m², and *A. mellifera* from 12 to 14°C and 300 w/m². (Vicens and Bosch. 2000; Way, 1995; Sanzol and Herrero. 2001)	Honey bee activity increased linearly from 18-28°C. (Danka and Beaman, 2007)	>5°C reduces ovule viability. An increase in temperature reduced pollen germination, but accelerated pollen tube growth. (Beppu et al., 1997; Hedhly et al., 2004; Postweiler et al., 1985)
Fruit set/fruit drop	Temperatures >13°C increase fruit drop. (Grauslund, 1978)	Nd	Increasing daily mean 3°C above norm decreased fruit set. Optimal temperature is ~15°C during flower development. (Beppu et al., 1997; Hedhly et al., 2004)
Chemical thinning	Temperatures >25°C can result in excessive fruit thinning. Temperatures > 27°C overthin. Temperatures < 18°C are ineffective. Thinning increased linearly from 8-24°C with an ideal range of 21-24°C. (Wertheim, 2000; Yuan, 2007; Stover and Greene, 2005; Forshey, 1976; Buban, 2000)	not a standard cultural practice	Thinning increases linearly from 16-20°C. (Olien and Bukovac, 1978)
Maturity/harvesting	Maximum volatile production occurs at 22°C during ripening. Increasing air temp 40-80 days after bloom linearly increases fruit size and soluble solids but decreases firmness. Temperatures >20°C reduces anthocyanin production. Fruit surface temperatures > 45°C induce sunburn. (Barber and Shape, 1971; Felicetti and Schrader, 2008; Warrington et al., 1999; Dixon and Hewett, 2000; Lin-Wang et al., 2011; McArtney et al., 2011)	Night temperatures > 21°C decrease fruit size. Day temperatures >29°C decrease size. (Darnell and Williamson, 1997)	Bruising decreases linearly with increasing temperature above 0°C. (Crisosto et al., 1993)

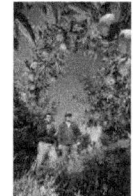

Table 5.4. Critical temperature thresholds for the production of citrus, grape and maple at various phenological stages.

Phenological stage	Citrus	Grape	Maple
Winter hardiness/chill accumulation (chill units)	not generally thought to need chilling	90-1400 ; 4°C budbreak; 7°C base leaf appearance. (Moncur et al., 1989; Reginato et al., 2010)	1000-1200; 7°C base (Wilson et al., 2002)
Freeze susceptibility of flowers	-2 to -3°C (Snyder and Melo-Abreu, 2005)	-1°C (Snyder and Melo-Abreu, 2005)	nd
Pre-bloom flower development/floral initiation	5 weeks of 10°C day or night required for initiation; Hardening -4/8°C (min/max); pre-bloom 0/14°C; flowering 10/27°C with daily mean >20°C to begin. (Bustan et al., 1996; Iglesias et al., 2007; Cole and McLeod, 1985; Moss, 1976)	High temperature pulse of 20-30°C previous year during stage 5-7 is required for initiation. (Srinivasan and Mullins, 1980; Caprio and Quamme, 2002	nd
Effects on pollination	High temperatures < pollination period. Low temperatures > time for pollination. (Iglesias et al., 2007)	12/9°C and 15/10°C (day/ night) reduced pollen growth and ovule viability. (COOMBE and MAY 1995; Srinivasan and Mullins, 1980)	nd
Fruit set/fruit drop	Temperatures >30°C increased fruit drop; optimum range 22/27°C (min/max). Iglesias et al., 2007; Cole and McLeod, 1985; Bustan et al., 1996)	Temperatures ≥26°C were associated with good production, probably because warm temperatures are required for flower bud initiation and development. (Caprio and Quamme, 2002).	nd
Chemical thinning	Temperatures >30°C removes excessive fruit. (Guardiola Garcia-Luis, 2000)	not a standard cultural practice	not a standard cultural practice
Maturity/harvesting	Brix and acid decline with increasing effective heat units with the optimum range of 13 to 27°C; > 33°C reduces size; high temperatures can lead to re-greening. (Bustan et al., 1996; Iglesias et al., 2007; Hutton and Landsberg, 2000;	Temperatures>36°C reduce production. 14.0–16.0°C best range for Pinot Noir. 16.5–19.5°C best temperature range for Cabernet Sauvignon best. Temperatures >35°C decreases anthocyanins in Cabernet Sauvignon. 15°C is optimal for color and anthocyanin development. Acidity can be halved with 10°C increase in temperature and variation in maturity increases with temperature. (Jones et al. 2005; Diffenbaugh et al., 2011; Jones, 2005; Lobell et al., 2006; Jones and Goodrich, 2008; Poudel et al., 2009; Mori et al., 2007; Woolf and Ferguson, 2000; Spayd et al., 2002; Caprio and Quamme, 2002)	

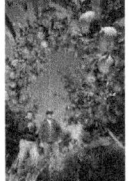

Table 5.5. Critical temperature thresholds for the production of peach, pear and raspberry at various phenological stages.

Crop	Phenological stage Pre-bloom flower development/floral initiation	Chemical thinning	Fruit
Apple	<30% full sun reduces floral initiation. (Wilkie et al., 2008)	3 days of cloudy weather greatly increase thinning at optimal temperature. (Stover and Greene, 2005)	> 1200 μmol/m²/s with fruit surface temperature > 45°C. (Chen et al., 2008)
Blueberry	nd	not a standard cultural practice	
Cherry	>20% full sun needed. (Flore and Layne, 1999)	not a standard cultural practice	
Citrus	750-1000 μmol/m²/s are required in the canopy for floral initiation. (Germana et al., 2003)	not a standard cultural practice	
Grape	10 hr of full sun/day in florescence development period. (Srinivasan and Mullins, 1981)	not a standard cultural practice	
Maple		not a standard cultural practice	
Peach		nd	< 23% full sun reduced color and soluble solids content. (Marini et al., 1999)
Pear	>30% full sun needed. (Wertheim, 2000)	nd	
Raspberry	nd	not a standard cultural practice	

valley of California declines 0.08 to 0.169 days/year (Gutierrez et al. 2006). Results of citrus production simulations without CO_2-induced response (Rosen-zweig et al. 1996; Tubiello et al. 2002) indicate that production may shift slightly northward in the Southern States due to reduced frost frequency.

Enhanced Atmospheric Carbon Dioxide

Experimental studies on perennial specialty crops have reported a sustained stimulation of photosynthesis and growth under elevated CO_2 similar to the findings from forest tree species (Curtis and Wang 1998) grown in open-top chambers (Norby et al. 1999) or FACE systems (Ainsworth and Long 2005). For example, leaf area-based net CO_2 assimilation at saturating light and growth was enhanced by an average of 44% in select fruit crops (Tables 5.7, 5.8, 5.9). Some of these crops have exhibited detectable reductions in photosynthetic rates (e.g., apple, citrus), while others show mixed (e.g., cherry) or little acclimation (e.g., grape, peach). Stomatal conductance to water vapor in general was reduced in these crops grown at elevated CO_2 by an average of 23%, which is similar to reported tree response in forest ecosystems (Medlyn et al. 2001). This increased assimilation under elevated CO_2 resulted in a considerable increase in leaf water-use efficiency (58%). A similar response was reported at the crop-level water-use efficiency in several crops (i.e., cherry, citrus, and peach) (Tables 5.7, 5.8, 5.9). However, despite a considerable increase in water-use efficiency at both leaf and crop levels, the actual amount of crop water use remained similar. This is likely because of an increase in tree leaf area in response to elevated CO_2.

On average, above-ground biomass increased by 60% in elevated CO_2 across the crops reviewed here. On the other hand, root-to-shoot ratio remained similar in apple (Chen et al. 2002) and citrus (Kimball et al. 2007), and slightly increased in cherry (Druta 2001). A rapid increase in tree leaf area during the early season accelerates early growth and biomass accumulation, especially in open canopies (referred to as "compound interest effect" by some) (Norby et al. 1999; Körner 2006). However, this accelerated growth response, such as shown in apple (Chen et al. 2001) and cherry (Centritto et al. 1999a), is likely to be less pronounced in a dense closed canopy in which the leaf area index (LAI) is more or less stable, so that competition for light and other resources are high (Norby et al. 1999). This is particularly true for natural systems where below-ground resources such as nutrients, soil moisture, and space are major limiting factors.

It has been suggested that long-term, natural responses to increasing CO_2 are likely to be less drastic than what has been reported in short-term experiments where plant-soil and/or plant-atmosphere connection have been decoupled (Körner 2006). However, many orchard and other perennial specialty cropping systems are highly managed with ample fertilization, irrigation, spacing, canopy management, thinning and pruning, and other cultural practices to realize high yield and produce quality. With relatively larger sinks for carbohydrates (e.g., fruit load and wood formation) than annual field crops, it is conceivable that initial stimulation of high CO_2 is sustained and in some cases amplified in perennial

Table 5.6. Solar radiation thresholds of perennial specialty crops at various phenological stages.

Crop	Phenological stage		
	Pre-bloom flower development/floral initiation	Chemical thinning	Fruit
Apple	<30% full sun reduces floral initiation. (Wilkie et al., 2008)	3 days of cloudy weather greatly increase thinning at optimal temperature. (Stover and Greene, 2005)	> 1200 μmol/m²/s with fruit surface temperature > 45°C. (Chen et al., 2008)
Blueberry	nd	not a standard cultural practice	
Cherry	>20% full sun needed. (Flore and Layne, 1999)	not a standard cultural practice	
Citrus	750-1000 μmol/m²/s are required in the canopy for floral initiation. (Germana et al., 2003)	not a standard cultural practice	
Grape	10 hr of full sun/day in florescence development period. (Srinivasan and Mullins, 1981)	not a standard cultural practice	
Maple		not a standard cultural practice	
Peach		nd	< 23% full sun reduced color and soluble solids content. (Marini et al., 1999)
Pear	>30% full sun needed. (Wertheim, 2000)	nd	
Raspberry	nd	not a standard cultural practice	

specialty crops. One such case study is a long-term CO_2 enrichment experiment on citrus that ran for 17 years in Maricopa, Arizona, in which the enhancement in biomass accumulation under elevated CO_2 was sustained at 70% after a peak stimulation occurring in only 2 to 4 years (Kimball et al. 2007). A less dramatic but still consistent and considerable CO_2 stimulation has been also observed in citrus grown using open-top chambers in humid Florida (Allen and Vu 2009).

While multiple studies examined biomass and allocation response to elevated CO_2, few studies report fruit yield response (Idso and Kimball 1997; Bindi et al. 2001; Ito et al. 2002). Even fewer studies have addressed the effects of elevated CO_2 on produce and product quality with the exception of wine grapes (Bindi et al. 2001; Goncalves et al. 2009; de Orduna 2010). Produce and product quality measures are likely to reflect different biochemical and physiological pathways of interactions between CO_2, nutrients

Table 5.7. Physiological response of apple, blueberry, and cherry to a doubling of atmospheric carbon dioxide.

Physiological parameter	Apple	Blueberry	Cherry
Leaf A_{max}	+39% (Ro et al., 2001)		-19% (Atkinson et al., 1997)
Canopy photosynthesis	+100% (Pan et al., 1998)		
Photosynthetic acclimation	Down-regulation (Chen et al., 2001; Chen et al., 2002a; Druta, 2001; Pan et al., 1998)		No change (Centritto, 2005); Down-regulation (Druta, 2001), (Centritto et al., 1999c) (Atkinson et al., 1997) (Wilkins et al., 1994)
Stomatal conductance			Down (Centritto et al., 1999c) -52.5% (Atkinson et al., 1997)
Stomatal density			No change (Centritto et al., 1999c)
Leaf transpiration	-27~33% (Chen et al., 2001)		Down (Centritto et al., 1999b) Down -49% (Atkinson et al., 1997)
Crop water use	+13-16% (Chen et al., 2002c)		No change (-5.3%) (Centritto et al., 1999b)
Leaf water use efficiency (WUE)			Up 66% Atkinson et al., 1997)
Crop WUE			Up 47% (Centritto et al., 1999b)
Leaf area	+64% (55.3~73%) (Chen et al., 2002c)		+25% (Centritto et al., 1999b) +27% after 10 mon (Atkinson et al., 1997)
Leaf temperature			
Biomass	+81% (Chen et al., 2002a)		+20% (Druta, 2001) +40% (Centritto et al., 1999a)
Yield			
Leaf chemistry (Non structural carbon etc.)	+40% Starch, sucrose (Chen et al., 2002b) Sucrose, starch increased (Pan et al., 1998) Leaf sucrose increased, sorbitol decreased, Phloridzin decreased, (Kelm et al., 2005)		No change except a reduction of [fructose] in leaf and root (Centritto, 2005) Starch increased (Centritto et al., 1999c)
Root/shoot	No change (Chen et al., 2002a)		increase (Druta, 2001) up (Centritto et al., 1999b)
Leaf [N]			Down. Reduction depends on DOY (Centritto et al., 1999c)
Interaction with N			No response in low N, positive response in adequate N (Wilkins et al., 1994)
Interaction with salinity			
Fruit quality	NA		
Interactions with mycorrhizae			
Interactions with pest damage			
Development	Accelerated (Chen et al., 2001)		Accelerated (Centritto et al., 1999a)

Table 5.8. Physiological response of citrus, grape and maple to a doubling of atmospheric carbon dioxide.

Crop	Citrus	Grape	Maple
Leaf A_{max}	+45% (Adam et al., 2004) +39% (Vu et al., 2002)	+34% (Mouthinho-Pereira et al., 2009)	+59% at 200ppm + (Deluca and Thomas, 2000) +69% (Groninger et al., 1996) +68% (Kubiske and Pregitzer, 1996)
Canopy photo-synthesis	+93% average of two temps (Brakke and Allen, 1995) (Baker and Allen Jr, 1993)		
Photosynthetic acclimation (+, 0, -)	Down-regulation, -25% (Adam et al., 2004) Down-regulation with Rubisco down (Vu et al., 2002) Down-regulation (Keutgen and Chen, 2001)	No change (Mouthinho-Pereira et al., 2009)	
Stomatal conductance	+33% (Adam et al., 2004) -28% (Vu et al., 2002)	-15% (Mouthinho-Pereira et al., 2009)	-28.5% (McElrone et al., 2005) (Bunce, 1992)
Stomatal density	No change (Estiarte et al., 1994)	-18.6% (Mouthinho-Pereira et al., 2009)	No change (McElrone et al., 2005)
Leaf transpiration	+26% Arizona FACE (Adam et al., 2004) -31% (Vu et al., 2002)	-12% (ns) (Mouthinho-Pereira et al., 2009)	
Crop water use	-30% avg of two temps (Baker and Allen Jr, 1993)		
Leaf water use efficiency (WUE)	Up 76% (Vu et al., 2002) +14.7% (Adam et al., 2004)	Up 69.4% (Mouthinho-Pereira et al., 2009)	
Crop WUE	Up 80% (Leavitt et al., 2003)		
Leaf area	+12% (Kimball et al., 2007)		
Leaf temperature			
Biomass	+27% total biomass (2 yrs, chamber) (Allen and Vu, 2009) +70% total biomass (17 yrs, FACE) (Kimball et al., 2007) +78%, wood biomass (Adam et al., 2004)	+45~50% (Bindi et al., 2001)	+51~92% (Norby et al., 2000) No response after 1 yr (Edwards and Norby, 1999) +33.8% greenhouse study (Groninger et al., 1996)
Yield	Fruit number (more than doubled after 6 years) and volume increased (Idso and Kimball, 1997)	+42.5% (Mouthinho-Pereira et al., 2009) +45~45% (Bindi et al., 2001)	
Leaf chemistry (Non structural carbon etc.)	+166% (Vu et al., 2002)	+14% [sugar] (Bindi et al., 2001)	C/N ratio up (+20%) Phenolics (-15%) Tannins (-14%) (McElrone et al., 2005) Soluble C up (+38.9%) (Bauer et al., 2001) TNC/N up by 31% (Williams et al., 2000)
Root/shoot	No change (Kimball et al., 2007)		
Leaf [N]	-11% (Adam et al., 2004) -18.5% (Keutgen and Chen, 2001) -10% in initial years (Penuelas et al., 1997) No change after 17 years (Kimball et al., 2007)	-9.3% (Mouthinho-Pereira et al., 2009)	-20% (McElrone et al., 2005) Reduction (Bauer et al., 2001) -14.5% (Williams et al., 2000) -25% (Norby et al., 2000)
Interaction with N	No change (Kimball et al., 2007)		
Interaction with salinity	(Garcia-Sanchez and Syversten, 2006)		
Fruit quality		No change in wine quality (De Orduna, 2010) No difference in wine quality (Bindi et al., 2001) Berry and wine quality unaffected (Goncalves et al., 2009)	
Interactions with mycorrhizae	CO_2 effect became more beneficial with AM (Syversten and Graham, 1999)		
Interactions with pest damage			Reduced (-19%) foliar disease (McElrone et al., 2005) Gypsy moth growth decline by 39% (Williams et al., 2000)
Development			

(N in particular), temperature, and pest damage. Several studies have examined leaf chemistry of fruit trees grown in elevated CO_2 (Centritto et al. 1999b; Centritto 2002; Adam et al. 2004; Moutinho-Pereira et al. 2009). In these studies, leaves grown under elevated CO_2 had about 15% lower nitrogen concentration on average. Similarly, significant increases in leaf sucrose, starch, and overall carbon-to-nitrogen ratio have also been found in several studies (Pan et al. 1998; Chen et al. 2002; Vu, Joseph C. V. et al. 2002; McElrone et al. 2005).

In summary, perennial specialty crops exhibit physiological and growth response that are similar to trees in forest and other unmanaged ecosystems. The CO_2 fertilization effect may be amplified and sustained longer for perennial specialty crops if (1) other resources (e.g., nutrients and water availability) are amply supplied, and (2) proper management options (e.g., spacing, pruning, thinning) are practiced

to facilitate the positive CO_2 effects by balancing source-sink relations for carbohydrates. This will likely require maintaining intensive cropping systems. In addition, the positive CO_2 effect may be negated by the detrimental effects of extreme temperatures on phenology, carbon sinks, and reproductive physiology.

Ozone Effects

As reviewed by Fuhrer (2009), when elevated ozone is combined with elevated CO_2, yield loss is typically considerably less than with ozone alone. The protective effect of CO_2 is primarily due to reduced stomatal conductance reducing ozone flux into the leaf, and this mechanism is associated with elevated CO_2. Consequently, ozone can also diminish the stimulating effect on yield of elevated CO_2, and the CO_2 protection from ozone effects also becomes less effective with increasing temperature.

Table 5.9. Physiological response of peach, pear and raspberry to a doubling of atmospheric carbon dioxide.

Physiological parameter	Peach	Pear	Raspberry
Leaf A_{max}	+58.5% well water, OTC (Centritto et al., 2002)	+88% estimated from Fig. 1 (Ito et al., 2002)	
Canopy photosynthesis			
Photosynthetic acclimation (+, 0, -)	No down regulation (Centritto, 2002)		
Stomatal conductance	No change (Centritto et al., 2002)	Reduced (Ito et al., 2002)	
Stomatal density			
Leaf transpiration	No change (Centritto et al., 2002)		
Crop water use			
Leaf water use efficiency (WUE)	Up 51% (Centritto et al., 2002)		
Crop WUE	Up 57% (Centritto et al., 2002)	Increased (Ito et al., 2002)	
Leaf area	No change, +3.7% (Centritto et al., 2002)	No change (Ito et al., 2002)	
Leaf temperature			
Biomass	+33% (Centritto et al., 2002)	Stem biomass increased (Ito et al., 2002)	+115% (Martin and Johnson, 2011)
Yield		Fruit size increased (Ito et al., 2002)	
Leaf chemistry (Non structural carbon etc.)			
Root/shoot			
Leaf [N]	-16.5% (Centritto et al., 2002)		
Interaction with N			
Interaction with salinity			
Fruit quality			
Interactions with mycorrhizae			
Interactions with pest damage		Variable interactions with aphids depending on genetic susceptibility (Martin and Johnson, 2011)	
Development			

Solar Radiation Effects

Perennial specialty cropping systems require high light intensity and light quality for both biomass production and fruit quality (Jackson 1980; Dokoozlian and Kliewer 1996). Pruning and training systems optimize light interception and distribution within the canopy to increase fruit quality. Excessive light can result in solar damage/sunburn, while insufficient light can reduce fruit bud formation, color development, soluble solids development, and fruit size (Table 5.6).

Disease

In eastern Washington State, a cherry powdery mildew is predicted to increase under the Community Climate System Model, version 3 (CCSM3; 2020 only) and the Coupled Global Climate Model, version 3 (CGCM3) climate projections (Stöckle et al. 2010). There will be small increases or no change in the risk from grapevine powdery mildew for all climate projections. Overall, warmer climate, but with small changes in precipitation during the growing season, tends to maintain and eventually reduce the incidence of these diseases, unless an increase in precipitation occurs early in the growing season (Stöckle et al. 2010).

In the Northeastern United States, the projected increase in short- to medium-term drought (Hayhoe et al. 2007) will tend to decrease the duration of leaf wetness and reduce some forms of pathogen attack on leaves. However, an increase in humidity and frequency of heavy rainfall events projected for the Northeast (Frumhoff et al. 2006) will tend to favor some leaf and root pathogens (Coakley et al. 1999), and the projected increased rainfall frequency (Frumhoff et al. 2006) may reduce the efficacy of contact fungicides, requiring more frequent applications. In forest ecosystems, maple is expected to have reduced (-19%) foliar disease (McElrone et al. 2005) with doubled CO_2.

Insects

A warming trend is likely to lead to increased pesticide use in the Northeast due to earlier arrival of migratory insects, higher winter-time survival of insects that currently are only marginally adapted to the region, and more generations of insects within a single season (Wolfe et al. 2008). In addition, some classes of pesticides (pyrethroids and spinosad), key to protecting perennial specialty cropping systems, have been shown to be less effective in controlling insects at higher temperatures (Musser and Shelton 2005).

In addition to increasing numbers and viability of insects, climate change may jeopardize biological control successes. For example, in California, DeBach and Sundby (1963) introduced a series of parasitoid species to control California red scale on citrus. These releases resulted in a sequence of climatically adapted parasitoids displacing each other in some areas. This displacement occurred until each species established itself in the subset of Californian environments most favorable for its development. Similarly, extensive biological control efforts are underway to control the vine mealybug (VMB), a major pest of grape production in California (Gutierrez 2005). High VMB densities occur in northern regions and in coastal regions of southern California, while VMB is less abundant in dryer warmer regions. The distribution and abundance of VMB's natural enemies is patchy across the different grape growing regions. While sucess to date has been elusive, if biological control of VMB is established, climate change could adversely affect it by changing the climatic conditions of the area. As is the case with other ecosystems, a forest ecosystem FACE study shows leaf chemistry changes under elevated CO_2 that have led to a decline in the growth rate of Gypsy moth larvae by 39%. Temperature, however, did not affect the growth or consumption rate by larvae in red maple (Williams et al. 2000). This result illustrates the complex linkage between direct and indirect effects of climate change on crops.

Effects of Changing Water Constraints

Increased drought frequency in the Northeast, together with warmer growing season temperatures will result in greater crop water requirements (Wolfe et al. 2008). Perennial specialty crops have reduced yield and quality in association with water deficits, and reduced profits as a result. While many producers of perennial specialty crops in the Northeast have some irrigation equipment, most have not invested in enough equipment to optimize irrigation scheduling and fully meet evapotranspiration requirements of all of their acreage (Wilks and Wolfe 1998).

Elsner et al. (2010) simulated the hydrology of Washington State and the Yakima River Basin, projecting April 1 snow water equivalents (SWE) to decrease by 28% to 30% across the State by the 2020s, 38% to 46% by the 2040s, and 56% to 70% by the 2080s. In the Yakima Basin, April 1 SWE will decrease by 35% to 37% by the 2020s, 47% to 57% by the 2040s, and 68% to 82% by the 2080s. The peak weekly SWE historically occurs near mid-March. Projections of weekly SWE for the 2020s indicate that SWE will

be reduced by an average of 39% to 41%. The peak week is projected to shift to early or mid-March. By the 2040s, SWE will be reduced by 50% to 58% with a peak projected to occur near early March, and by 67% to 80% by the 2080s with a peak projected to occur near mid-February.

Similarly in California, Miller et al. (2003) simulated the hydrology for the Sacramento, American, and Merced Basins. SWE decreases for most basins, and the peak discharge is earlier for all basins by 2080 to 2099. There is an early season increase in liquid water from 2010 to 2099, with earlier snowmelt seasons with a slower snow melt rate. Reductions in growing season irrigation water will greatly limit perennial specialty crop production in the arid and semi-arid production regions unless sufficient water is stored in reservoirs and made available for irrigation. Late season crops will feel this effect most because of the increased water-use later in the growing season due to higher temperatures. The effect on crop wateruse efficiency of elevated CO_2 on

select crops is outlined in Tables 5.7, 5.8, and 5.9. In general, water-use efficiency (i.e., biomass or yield per water use) in perennial specialty crops is likely to increase because of reduced stomatal conductance and growth stimulation under high CO_2. However, overall water use in many crops is likely to remain similar or even increase as a result of corresponding increase in leaf area.

In another regional analysis for the U.S. West Coast, Lobell et al. (2006) examined the effects of climate change on yields of perennial crops in California. The research combined output from numerous climate models with statistical crops models for almonds, walnuts, avocados, wine grapes, and table grapes. The results show a range of predicted temperature increases across climate models of approximately 1°C to 3°C for 2050, 2°C to 6°C for 2100, and a range of changes in precipitation from -40% to +40% for both 2050 and 2100. Wine grapes showed the smallest yield declines compared to other crops, but showed substantial spatial shifts in

Case Study of Grapes in the United States

White et al. (2006) demonstrated that U.S. premium wine grape production area could decline up to 81% by the late 21st century. They found that increases in heat accumulation will likely shift wine production to warmer climate varieties and/or lower quality wines and that while frost constraints will be reduced, increases in the frequency of extreme hot days (>35°C) in the growing season are projected to completely eliminate wine grape production in many areas of the United States. Grape and wine production will likely be restricted to a narrow West Coast region and the Northwest and Northeast – areas where excess moisture is already problematic. Jones (2007) examined suitability for viticulture in the western United States, and contrived five regions broad suitability for viticulture across cool to hot climates, as well as the varieties that grow best in those regions. The cooler region (I) occurs at higher in elevation, and more coastal, and more northerly regions (e.g., the Willamette Valley), while the warmest region (V) areas are mostly confined to the Central Valley and further south in California (e.g., the San Joaquin Valley). Based on the historical record, 34% of the western United States falls into regions I-V, with 59% being too cold and 7% too hot. Region I encompasses 34.2%, Region II 20.8%, Region III 11.1%, Region IV 8.7%, and Region V 25.2%. According to Jones (2007), using projections for average growing season temperatures from the Community Climate System Model (CCSM) of 1.0°C to 3.0°C for 2049 results in a range of increases in growing degree-days of 15% to 30%. For a 15% increase in growing degree days by 2049, the area of the western United States in Regions I-V increase from 34% to 39%, and at the higher range of a 25% increase in growing degree days, increases by 9% to 43%. Overall the changes show a reduction in the areas that are too cold from 59% to 41%, while the areas that are too hot increase from 7% to 16% in the greater warming scenario. Within the individual regions there are shifts to predominately more land in Region I (34.2% to 40.6%), smaller changes to Region II (20.8% to 23.4%), Region III (11.1% to 14.2%), and Region IV (8.7% to 10.1%), and a reduction of Region V area from 25.2% to 11.6%, which shifts the regions toward the coast, especially in California, and upwards in elevation (most notably in the Sierra Nevada Mountains).

suitability to more coastal and northern counties. For oranges, walnuts, and avocados, not only are areas with the potential for high yields dramatically reduced, but the areas with appropriate climate tend to be in dry or mountainous regions with limited opportunities for agriculture. Less than 5 percent of simulations for almonds, table grapes, walnuts, and avocados indicated a zero or positive response to climate change by midcentury. Two main factors contribute to this result: (1) all of these crops are either at or above their optimum temperatures in current climate, and all climate models project at least some climate warming; and (2) all of these crops are irrigated, so precipitation projections have a relatively minor effect. The authors also note that historical increases in yield have low attribution to climate trends and were due more to changes in cultural and genetic technology.

At a higher emissions scenario, within just the next few decades (2010–2039), a 5-to-10-day increase in the number of July heat stress days is projected for the southern half of the U.S. Northeast (i.e., much of Pennsylvania, New Jersey, Delaware, Connecticut, and southern New York). Under a lower emissions scenario, the climate change effect does not become substantial until midcentury (2040-2069). By the end of the century (2070–2099), with higher emissions, most days in July are projected to exceed the 32°C heat stress threshold for most of the Northeast. Even assuming relatively lower emissions, much of the Northeast is projected to have 10 to 15 more days of heat stress in July by the end of the century, except for some northern areas (e.g., northern Maine and Vermont), where the increase is in the range of 5 to 15 days. The projected increase in summer heat stress will be particularly detrimental to many cool temperature-adapted crops (e.g., apple) that currently dominate the Northeast agricultural economy. For many high value horticultural crops, very short term (hours or a few days), moderate heat stress at critical growth stages can reduce fruit quality by reducing visual or flavor quality even when total tonnage is not reduced (Peet and Wolfe. 2000).

An increase in winter temperatures will affect the Northeast perennial specialty cropping systems. Mid-winter warming can lead to early bud-burst or bloom of some perennial plants, resulting in frost damage when cold winter temperatures return. Yields will be negatively affected if the chilling requirement is not completely satisfied because flower emergence and viability will be low. All perennial specialty crops have a winter chilling requirement ranging from 200 to 2,000 cumulative hours. Wolfe et al. (2008) analyzed the future chill requirements of the Northeast and found that a 400-hour chilling

Case Study of Apple Production in the Northeastern United States

According to Wolfe et al. (2008), an extended frost-free period as projected for the Northeast (Frumhoff et al. 2006; Hayhoe et al. 2007) will tend to benefit perennial specialty cropping systems requiring a relatively long growing season such as apples, peaches, and grape varieties. However, projections for an increase in summer heat stress and drought can reduce yield and crop quality. In contrast, Wolfe et al. (2008) found that apple yields for western New York (1971–1982) were lower in years when winters were warmer than average (based on accumulated degree days >5°C from January 1 to budbreak). This was likely related to more variable fruit set following warmer winters. Wolfe et al. (2008) compared projections of summer heat stress frequency (increase in number of days with maximum temperature exceeding 32°C) for the increase in number of heat-stress days in the month of July at early-, mid-, and late-21st century.

requirement will continue to be met for most of the Northeast during this century regardless of emissions scenario. However, crops with prolonged cold requirements (1,000 or more hours) could be negatively affected, particularly in southern sections of the Northeast, and at the higher emissions scenario, where less than 50% of years satisfy the chill requirement by mid-21st century. The effect on crops will vary with species and variety since each species has a range of cultivars with widely varying chill requirements (Tables 5.3, 5.4, and 5.5).

There is a historical trend for increased frequency of high-precipitation events (>5 cm in 48 h) (Wake 2005) in the Northeast, and this trend is expected to continue with a further increase in the number of high-precipitation events of 8% by midcentury and 12% to 13% by the end of the century (Frumhoff et al. 2006). More spring rainfall concentrated into high-precipitation events, combined with stable to modest reductions in summer and autumn rainfall and increased temperatures, leads to a projection for more short- (1- to 3-month) and medium-term (3- to 6-month) droughts for the region, particularly in the northern and eastern parts of the region (Frumhoff et al. 2006; Hayhoe et al. 2007). Drought

frequency is projected to be much greater at the higher Intergovernmental Panel on Climate Change (IPCC) emissions scenario (A1F1), as compared to lower (B1) emissions scenario, according to Wolfe et al. (2008). By the end of the century and with higher emissions, short-term droughts are projected to occur as frequently as once per year for much of the Northeast, and occasional long-term droughts (>6 month) are projected for western, upstate New York, where perennial specialty crops are a major industry (Wolfe et al. 2008).

Adaptation

Development of adapted cultivars is the long-term solution of perennial specialty cropping systems in a changing climate. There is wide variety of adapted cultivars that can be evaluated for new regions. Typical breeding programs require 10 to 30 years to confirm and improve a cultivar. Recent technology demonstrates how this breeding hurdle can be overcome using molecular approaches (Kean 2010; Srinivasan et al. 2010) to reduce perennial crop generation time to months instead of years. Since perennial specialty crops have a chill requirement (the minimum period of cold weather after which a fruit-bearing tree will blossom), it is necessary to induce and end dormancy at times in the growing season that minimize killing frosts both in spring and fall. This requires that the plant react to day length instead of temperature patterns. Research on possible adaptation focused on day length includes work done by Wisniewski et al. (2010), who transformed apple from temperature-induced dormancy to photoperiod-induced dormancy using a technology that may be adaptable for other perennial specialty crops. In addition to macroscale research, molecular biology

is identifying genes associated with climate change (Hancock et al. 2011) that will benefit perennial specialty crops in the future. While projections of future climate indicate average warmer temperatures will affect crops, in today's environment, increased temperatures already reduce plant productivity. To deal with current temperature issues, technology such as application of reflective particle films (Glenn 2009) has been developed and commercialized that reduces canopy and fruit temperature, increasing yield and quality in the face of increasing growing-season temperatures (Figure 5.2).

In addition to these adaptations, perennial specialty crop growers have a wide assortment of management tools that will help them adjust to climate change. These include crop load adjustment, canopy pruning, irrigation, increased use of mechanization, and automation technology. As examples, overhead irrigation effectively reduces canopy temperature and is effective in frost mitigation although it is water-use inefficient, and the rotatable cross-arm trellis system of bramble production provides a cultural means to protect sugar cane from winter damage, frost damage, and sunburn damage by manipulating cane crop orientation (Takeda and Phillips 2011).

Grazing Lands and Domestic Livestock

The livestock industry makes a significant contribution to most rural economies. It accounts for 40% of the world's agriculture Gross Domestic Product (GDP) and in developing countries can account for as much as 80% of GDP (World Bank 2007a, b). In the United States, the livestock industry has more than 1 million operations, with annual

Case Study of Apple Production in the Pacific Northwest

According to Stöckle et al. 2009, climate change is predicted to slightly decrease the production of apples by 1%, 3%, and 4% for the 2020, 2040, and 2080 scenarios with no elevated CO_2 effect. Under a warmer climate, crop development will proceed at a faster rate, reducing the opportunity for biomass gain. However, when the effect of elevated CO_2 and warming is modeled, yields are projected to increase by 6%, 9%, and 16% for 2020, 2040, and 2080 scenarios compared to current levels, assuming the availability of varieties able to use the extended season or other adaptive technologies. Although average temperatures are projected to increase for all climate scenarios, the frequency of frost events may limit cropping due to earlier flowering. Under the projected climate change, flowering will occur about three days earlier in the 2020 scenario, which will slightly increase the frequency of frost events, increasing yield loss from frost damage or increase the need and expense for frost protection. Limited chill accumulation is not projected to limit apple production in eastern Washington. Water supply was assumed sufficient for irrigated crops, but other studies suggest that it may decrease in many locations due to climate change.

Case Study of Citrus in the United States

Tubiello et al. (2002) simulated U.S. citrus production. Overall, yields increased 20-50%, while irrigation water use decreased. Crop loss due to freezing was 65% lower on average in 2030 and 80% lower in 2090, at all sites. In the primary citrus production areas, Miami, FL, experienced the smallest increases, 6% to 15% and the other major production sites in Arizona, Texas, and California, increases were 20-30% in 2030 and 50% to 70% in 2090. All sites experienced a decrease in crop loss from freezing. Potential for northward expansion of U.S. citrus production was small because results indicated that in 2030 and 2090 northern sites of current marginal production would continue to have lower fruit yield, higher risk of crop loss due to freezing, and lower water availability than the southern sites.

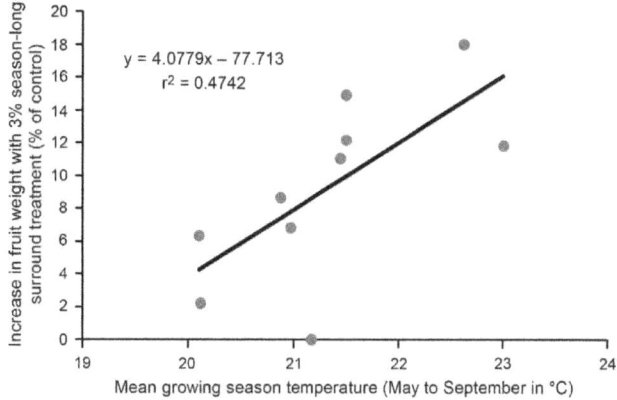

Fig. 5.2. Relationship between the percentage increase in mean fruit weight of particle film treated fruit of 'Empire' apple compared to the control treatment and the mean growing season temperature from 1998-2007 at Kearneysville, West Virginia.

$y = 4.0779x - 77.713$
$r^2 = 0.4742$

sales totaling $153.6 billion according to the USDA National Agricultural Statistics Service. Livestock sales comprise 51.7% of all agricultural commodity sales. Total number of beef cattle, dairy, swine, and poultry in the United States, in millions of animals, are 96, 9, 68, and 9,560, respectively.

Rötter and Van de Geijn (1999) suggest that shifts in climatic conditions could affect animal agriculture in four primary ways, through change in (1) feed-grain production, availability, and price, (2) pastures and forage crop production and quality, (3) animal health, growth, and reproduction, and (4) disease and pest distributions. The ensuing discussion focuses on the implications for livestock production systems and potential adaptive responses to climate change (such as the utilization of different species and genotypes of animals and forages), changes in facilities utilized for care and management of livestock, and a redistribution of livestock in a region (Gaughan et al. 1999; Gaughan et al. 2009).

Effect of Climate Change on Animal Productivity
Livestock production occurs under a variety of management scenarios and environmental conditions. Livestock production systems that provide partial

or total shelter to mitigate thermal environmental challenges can reduce the risk and vulnerability associated with adverse environmental events. In general, livestock such as poultry and swine are largely managed in housed systems where airflow can be controlled and housing temperature modified to minimize or buffer against adverse environmental conditions. In recent years, these industries have moved to utilizing more semi-controlled environmental systems to ameliorate production problems associated with changing and/or extreme environmental conditions. While shifts in these industries were made largely independent of climate change concerns, they can be adapted for use in an expected warmer future. However, despite modern heat-abatement strategies, summer-induced poor performance still costs the American swine industry more than $300 million annually (St-Pierre et al. 2003).

Greater concerns with regard to climate change are for animals managed in unsheltered and/or unbuffered environments. The majority of American domestic livestock managed in more extensive outdoor facilities are ruminants (goats, sheep, beef cattle, and dairy cattle). Within limits, these animals can adapt to and cope with most gradual thermal challenges. However, the rate at which environmental conditions are projected to change, the extent to which animals are exposed to extreme conditions, and the inability of animals to adequately adapt to sudden and/or dramatic environmental changes, are always a concern. Lack of prior conditioning to rapidly changing or adverse weather events most often results in catastrophic deaths in domestic livestock and losses of productivity in

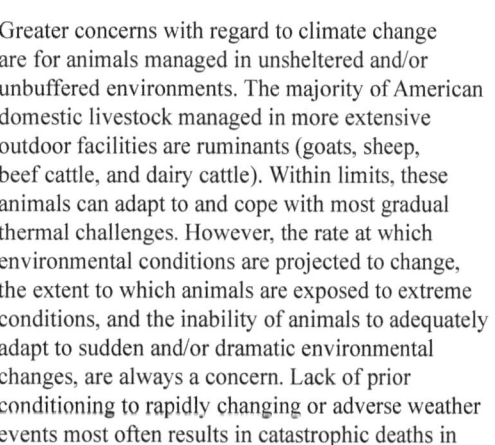

surviving animals (Mader 2003). Animal phenotypic and genetic variation, management factors (facilities, stocking rates, and nutrition), physiological status (stage of pregnancy, stage of lactation, growth rate), age, and previous exposure to environmental conditions may exacerbate the effect of adverse environmental conditions.

The optimal environmental conditions for livestock production are comprised of a range of temperatures and other environmental conditions for which the animal does not need to significantly alter behavior or physiological functions to maintain a relatively constant core body temperature. Ambient environmental conditions directly affect mechanisms and rates of heat gain or loss by all animals (NRC 1981). In many species, 5°C to 7°C deviations from core body temperature can cause significant reductions in productive performance and may lead to death (Gaughan et al. 2009).

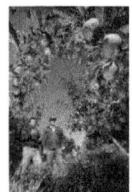

As environmental conditions result in core body temperature approaching and/or moving outside normal diurnal boundaries, the animal must begin to conserve or dissipate heat to maintain homeostasis (Davis et al. 2003; Mader and Kreikemeier 2006). This is accomplished through shifts in short-term and long-term thermoregulatory processes (Gaughan et al. 2002a, b; Mader et al. 2007). The onset of a thermal challenge often results in declines in physical activity and an associated decline in eating and grazing (for ruminants and other herbivores) activity. In addition, environmental stress may affect hormonal changes that in turn result in shifts in cardiac output, redistribution of blood flow to extremities, altered metabolic rates, and slowed digesta passage rate.

The risk potential associated with livestock production systems due to changing climatic conditions can be characterized by levels of vulnerability as influenced by animal performance and environmental parameters (Hahn et al. 2005). When performance level and environmental influences combine to create a low level of vulnerability, little risk exists. However, as performance levels decrease, the vulnerability of the animal increases. When coupled with an adverse environment, the animal is at greater risk.

Inherent genetic characteristics or management scenarios that limit the animal's ability to adapt to or cope with environmental change also puts the animal at risk. At very low performance levels, any environment other than near-optimal increases animal vulnerability. For example, the modern high-producing dairy cow begins to experience heat stress

at a thermal heat index (THI) of 68, this is at least four THI units lower than was the case for cows 40 years ago when environmental stresses were lower (Zimbleman et al. 2009).

The potential effects of climate change on overall performance of domestic animals can be determined using defined relationships between climatic conditions and dry matter intake (DMI), climatological data, and General Circulation Model (GCM) output. Because ingestion of feed is directly related to heat production, any change in DMI and/or energy density or nutrient profile of the diet will change the amount of heat produced by the animal (Mader 2003; Mader and Davis 2004). Environmental conditions influence heat transfer by the animal; however, animals exposed to the same environmental conditions will not exhibit the same reduction in DMI. Body weight, body condition, and level of production also affect DMI; having a better understanding of what contributes to the variation in heat-induced DMI decrease is of obvious interest. In addition to reduced feed intake, heat stress also directly affects post-absorptive metabolism (Rhoads et al. 2009), which results in a reprioritization of nutrient utilization. This altered metabolic hierarchy and reduced nutrient intake primarily explains why animals produce less during the warm months.

In the Central United States, a modeling exercise based on the Canadian Global Coupled Model (CGC) projections for 2040 and associated with changes in DMI (Frank et al. 2001; USDA 2008) indicate that days to slaughter-weight for swine increased by an average of 3.7 from the baseline of 61.2 days. Potential losses under this scenario averaged 6% and would cost swine producers in the region $12.4 million annually. Losses associated with the Hadley scenario (United Kingdom Meteorological Office/ Hadley Center for Climate Prediction and Research) are less severe. Increased time to slaughter-weight averaged 1.5 days or 2.5%, costing producers $5 million annually.

For confined beef cattle reared in the Central United States, time-to-slaughter-weight associated with the CGC 2040 scenario increased by 4.8 days (above the 127-day baseline value) or 3.8%, costing producers $43.9 million, annually. Climate changes projected by the Hadley 2040 model resulted in a loss of 2.8 days of production or 2.2%. For dairy, the projected CGC 2040 climate scenario would result in a 2.2% (105.7 kg/cow) reduction in milk output and cost producers $28 million annually. Production losses associated with the Hadley scenarios would average 2.9% and cost producers $37 million annually. Across the entire United States, percent increase in days to market for swine and beef and the percent decrease in dairy milk production for the 2040 scenario averaged 1.2%, 2.0%, and 2.2%, respectively, using the CGC model, and 0.9%, 0.7%, and 2.1%, respectively, using the Hadley model. For the 2090 scenario, respective changes averaged 13.1%, 6.9%, and 6.0%, using the CGC model, and 4.3%, 3.4%, and 3.9%, using the Hadley model. For these scenarios it should be noted that production losses for the dairy sector were generally not as great as those found for beef and swine in the U.S. South and Southeast.

Projected animal production responses based on a doubling (2040) and tripling (2090) of atmospheric greenhouse gas CO_2 levels for the period June 1 to October 31 were obtained for the Central United States (Mader et al. 2009). For swine, a slight northwest (Montana) to southeast (Louisiana) gradient was evident. The west side (Montana to New Mexico) of the Central Plains showed few production losses with some benefits under the doubling CO_2 scenario, however, losses up to 22.4% were found under the tripling scenario. On the east side (Minnesota to Louisiana), few to no losses in productivity were found in the North, but losses between 40% and 70% were found in the South under the tripling scenario. For beef, small changes were found in the Western Plains with increasing temperatures, although a northwest to southeast gradient was also evident. Production losses never exceeded 20% for any location or under any scenario. For dairy, no positive benefits in milk production were found due to climate effects. Projected production declines ranged from 1% to 7.2%, depending on location. However, ranges in predicted differences were less than those simulated for beef and swine. These simulations suggest that regional differences in animal production due to climate change will be apparent. For small changes in climatic conditions, animals will likely be able to adapt, while larger changes in climate conditions will likely dictate that management strategies be implemented. Exploration of the effects of climate change on livestock should allow producers to adjust management strategies to reduce potential effect and economic losses due to environmental changes.

International studies may shed some light on what U.S. agricultural producers might expect from future climate changes. Seo and McCarl (2011) project that under the hotter and drier conditions anticipated for Australia, sheep would increase in number by 122%, beef cattle by 211%, dairy cattle by 29%, and pigs by 71%. On the other hand, sheep would increase by only 22% if summers become wetter. Livestock revenue is anticipated to increase by 47% by 2060.

In the above analyses, a hotter and drier climate is beneficial for livestock because it is projected to alter the landscape from croplands to pasture suitable for livestock. At the same time, however, these changes could lead to reduced feed available from grain production, lower stocking rates on pasture, and reduced forage quality. In addition, a number of pastoral ecosystems in Australia are already marginal for livestock production, some of which would potentially become even less usable for grazing under projected changes in climate.

In the United States, negative effects of hotter weather in summer likely will outweigh benefits of warmer winters (Adams et al. 1999). Thus, only a portion (estimated to be about 50%) of the declines in domestic livestock production during hotter summers can be offset by milder winter conditions. Climate change likely will affect high-producing animals more than low producers. However, positive winter effects will not offset summer declines in conception rates, particularly in cattle that breed primarily in spring and summer. Hahn (1995) reported that conception rates in dairy cows were reduced 4.6% for each unit change in the THI. Conception rates of *Bos taurus* cattle declined by more than 2 percent for each unit increase in THI, and by 1.5% to 3.8% for each degree Celsius increase in minimum temperature (Amundson et al. 2006). Animal productivity, body condition, geographical location, and seasonal breeding patterns also influence conception rates (Sprott et al. 2001).

Effect of Climate Change on Grasslands

The United States has nearly 480 million acres of range and pastureland. Approximately one-third of U.S. lands, or 777 million acres, are grazing lands. These include 614 million acres of grassland pasture and rangeland, 36 million acres of cropland pastures, and 127 million acres of forested rangelands (Nickerson et al. 2007). Grazing land acreage has steadily declined from 1,061 million acres in 1945 when the USDA Economic Research Service began its major land use surveys. Loss of grazing lands occurs for a variety of reasons. Cropland pastures convert to croplands when commodity prices are high. Recreation, wildlife, and environmental applications have claimed many of these lands. Favorable weather can cause shifts toward forestry, especially in the South. Urbanization has resulted in substantial losses of grazing lands throughout the country.

Grazing land changes differ notably by geography. For instance, non-forested grazing lands grew by 28 million acres in the Southern Plains, and by 1 million acres in the Southeast between 1949 and

In the United States, negative effects of hotter weather in summer likely will outweigh benefits of warmer winters.

2007, while decreases in large tracts of Federal lands for wilderness resulted in large reductions in grazing lands in the West. In general, climate change would add to the challenges and uncertainty posed by a growing population in the United States that is putting additional pressures and expectations on the goods and services expected from grazing lands (Morgan et al. 2008; Havstad et al. 2009).

Temperature Effects

Temperature exerts two basic, controlling effects on grazing-land ecology by regulating (1) rates of chemical reactions, and (2) exchanges of energy between the agroecosystem and the atmosphere, with water supply often modulating the influence of these temperature-driven effects. This is most pronounced for rangelands, where water is the primary ecological driver (Noy-Meir 1973; Sala et al. 1988). Thus, an understanding of the combined effects of rising temperature and changing precipitation patterns is necessary to forecast effects on grazing lands.

Results from recent warming and observational experiments support earlier work (Dukes et al. 2005; Klein et al. 2007) indicating that compensating effects of temperature result in earlier spring green-up (Cleland et al. 2006; Sherry et al. 2007; Hovenden et al. 2008), increased nitrogen mineralization (Luo et al. 2009), and higher early-season growth rates with more severe mid- and late-season desiccation (Cleland et al. 2006).

While aboveground net primary production (ANPP) is relatively stable in grassland species (Xia et al. 2009; Fay et al. 2011; Morgan et al. 2011; Pendall et al. 2011), warming can reduce ANPP by increasing desiccation, as it did in a cool temperate climate for grassland assemblages (De Boeck et al. 2008), or increase ANPP, as it did in an Oklahoma tallgrass prairie, where deep soils contained sufficient soil moisture to support a 21% ANPP increase (Luo et al. 2009). Because grazing lands are highly dependent on inherent environmental conditions, warming effects on these lands likely will differ regionally. In the Southwestern United States where water already exerts a major limitation on ANPP, rising temperature in combination with altered precipitation is expected to increase droughts (Seager and Vecchi 2010), with negative effects on grazing land productivity. In the northern Great Plains, where low temperatures can sometimes restrict growing season length, warmer temperatures alone or in combination with increased annual precipitation amounts should increase forage production (Morgan et al. 2008). In the Southeast, warmer temperatures are expanding the northern range of species once limited to the Gulf Coast Region (Gates et al. 2004) and may increase

the length of growing season of C_4 grasses while limiting the productive period and economic benefit of over-seeded C_3 grasses and legumes.

Precipitation Effects

Our capacity to predict precipitation patterns is limited, but it is clear that changes in precipitation could dramatically affect grazing lands. Annual precipitation amount is the key driver affecting ANPP in native grasslands (Sala et al. 1988), although seasonal distribution of precipitation can be as important as total precipitation. The anticipated change in precipitation into fewer but larger events may increase both the frequency of drought and the probability of flooding (Knapp et al. 2008). In general, grazing land response to precipitation depends on complex interactions among quantity, frequency, and size of precipitation events (Fay et al. 2008). Local or regional differences in evaporative demand, plant communities, and soil type regulate effects of more variable precipitation patterns on soil water dynamics, plant utilization, and species responses (Noy-Meir 1973; Bates et al. 2006; Knapp et al. 2008; Craine et al. 2010; Debinski et al. 2010; Whitford and Steinberger 2011). Even though ANPP and biodiversity can respond strongly to these altered dynamics (Bates et al. 2006; Robertson et al. 2010; Derner et al. 2011), results differ substantially among grazing land ecosystems. For instance, larger but less frequent precipitation events tend to decrease ANPP and other critical ecosystem functions in tallgrass prairie, but enhance ANPP in shortgrass steppe (Knapp et al. 2008; Heisler-White et al. 2009; Fay et al. 2011).

Based on results from a unique multi-factor climate change experiment, Fay et al. (2011) propose the following conceptual model of grassland responses to warming and altered precipitation:

- Inter-annual climate variation, mainly related to growing-season rainfall, drives inter-annual variation in average soil moisture and rates of key ecosystem processes.

- Increased growing-season rainfall variability reduces rates of most ecosystem processes, resulting in lower ecosystem rainfall-use efficiency.

- Warming stimulates rates of ecosystem processes active during cooler parts of the growing season, but increased rainfall variability and warming during the middle, warmer, and water-limited portions of the growing season likely will reduce rates of ecosystem processes.

Enhanced Atmospheric Carbon Dioxide

In addition to its effects as a greenhouse gas, CO_2 directly influences plants. Plant response to increased CO_2 is driven by two basic mechanisms: a direct stimulation of photosynthesis and an indirect stimulation of plant water-use efficiency resulting from partial stomatal closure (Morgan et al. 2004). The direct photosynthetic response is much stronger in C_3 than C_4 plants; photosynthesis is nearly saturated at present atmospheric CO_2 concentrations in C_4 plants, but unsaturated in C_3 plants (Polley 1997; Anderson et al. 2001; Reich et al. 2001; Poorter and Navas 2003). However, stomatal responses to CO_2 are similar in some C_3 and C_4 plants (Wand et al. 1999).

Recent research supports the notion that grasslands are particularly responsive to changes in CO_2 due to their sensitivity to water. For tallgrass prairie species growing in three soil types of central Texas, CO_2 acted as a surrogate for water by contributing to shifts in species abundances that mimic those observed along a precipitation gradient (Polley et al. 2011). In addition, CO_2 enrichment promoted water savings in a northern mixed-grass prairie in Wyoming by completely off-setting desiccating effects of moderately warmer temperature (Morgan et al. 2011). The water-savings effect of CO_2 enrichment appears to be robust, having been detected in native vegetation around natural springs that have long been exposed to elevated CO_2 (Onoda et al. 2009).

Interactive Effects and Plant Community Responses

Our ability to predict responses to global changes is limited by our incomplete understanding of how ecosystem effects of climate change factors interact (Shaw et al. 2002; Morgan et al. 2011). Leuzinger et al. (2011) demonstrate that the magnitude of ecosystem responses to climate change treatments usually declines as the number of factors considered increases. Mechanisms responsible for partially compensating effects of climate change treatments are not completely understood, but appear to differ among ecosystems and treatment combinations. Warming and CO_2 enrichment have offsetting effects on soil water availability (Morgan et al. 2011). Limitations in soil nitrogen (Newman et al. 2006; Reich et al. 2006) or phosphorous (Gentile et al. 2011) may constrain plant responses to CO_2, while plant species shifts in response to nitrogen additions can favor plant species that respond limitedly to CO_2 (Langley and Megonigal 2010).

Climate change effects often are interpreted in the context of a relatively stable plant community and unchanging disturbance regimes. It is becoming clear, however, that we can reliably predict climate change effects on productivity and other ecosystem processes only by accounting for interactions between environmental changes and other regulators of ecosystems, including soil resource supply, major functional groups of organisms, and disturbance regimes (Polley et al. 2011). Changes in these 'interactive regulators' can feed back to dampen or amplify ecosystem responses to climate change factors. Most feedbacks will be negative and dampen global change effects on ecosystems (Luo et al. 2004). Indeed, ecosystem responses to global change treatments generally decline over longer time periods and larger spatial scales (Leuzinger et al. 2011) because negative feedbacks from interactive regulators promote homeostasis in ecosystem processes. Changes in plant species and functional group composition can amplify ecosystem responses to global changes and thus contribute to beneficial (Zavaleta et al. 2003; Niu et al. 2010) or negative effects of global changes (Morgan et al. 2007; Suttle et al. 2007).

Vegetation changes of greatest concern on extensively managed grazing lands are those that are essentially irreversible within the constraints of traditional management, and that fundamentally alter ecosystem structure and functioning. Such shifts between "alternate stable states" of vegetation usually occur when changes in soil properties, disturbance regimes, or animal populations remove limitations on increasing plants or create limitations on current dominants.

Vegetation changes tend to occur gradually, as when woody plants replace grasses as a result of prolonged grazing, but also can occur abruptly, as when a threshold of soil loss or water content is crossed that prevents continued dominance by current species occupants of a site (Friedel 1991; Polley et al. 2011). Developmental changes, like flowering date, can exhibit threshold responses to precipitation that can have long-term and possibly transforming effects on plant community composition (Craine et al. 2010; Fay et al. 2011). Climate extremes can have significant effects on biogeochemical functions like water relations and nutrient cycling, although long-term alterations in fundamental ecosystem attributes like net primary productivity (NPP) or functional group composition may involve more long-term changes (Arnone et al. 2011; Jentsch et al. 2011). Thus, global changes will more often influence the susceptibility of vegetation to disturbances and other factors like fire that directly influence the state of vegetation (Bond 2008). For more intensively managed mesic pastures (lands with well-balanced moisture supply), vegetation changes will involve

Recent research supports the notion that grasslands are particularly responsive to changes in CO2 due to their sensitivity to water.

considerations of which forage species and/or combinations will perform better in a changing environment (Sanderson et al. 2009).

Effects on Forage Nutritive Value

Grazing lands are managed to produce forage or fodder for livestock. The nutritional quality, as well as the quantity, of the forage/fodder resource is of interest. Nutritive value, in turn, depends on chemical and physical characteristics of each of the plant species that contribute to the resource and species composition. Global changes likely will cause modest changes in the forage quality of individual plant species. Most studies indicate rising CO_2 and temperature reduce nutritive value of plants (Henderson and Robinson 1982; Akin et al. 1987; Newman et al. 2005; Morgan et al. 2008; Craine et al. 2010; Gentile et al. 2011), although complex interactions of global change factors on soil, available nutrients, and plant responses suggest that both increases and decreases in nutritive value are possible (Craine et al. 2010; Dijkstra et al. 2010). On the other hand, global changes could cause substantial shifts, either beneficial or negative, in forage nutritive value by contributing to vegetation change. For instance, CO_2 enrichment increased the nutritional value of grass biomass by shifting the relative abundance of tallgrass prairie species (Polley et al. 2011). By contrast, global changes that facilitate a shift in vegetation from forage to non-forage species, such as from grasses to weeds or woody plants (Morgan et al. 2007; Morgan et al. 2008), will substantially reduce forage nutritive value.

Effects of Climate Change on Animal Health

Climate change may indirectly affect animal production by altering the frequency, intensity, or distribution of animal diseases and parasites. Climate affects microbial density and distribution, the distribution of vector-borne diseases, host resistance to infections, food and water shortages, or food-borne diseases (Baylis and Githeko 2006; Gaughan et al. 2009; Thornton 2010). Earlier springs and warmer winters may allow for greater proliferation and survivability of pathogens and parasites. For example, bluetongue was recently reported in Europe for the first time in 20 years (Baylis and Githeko 2006).

Regional warming and changes in rainfall distribution may lead to changes in the spatial or temporal distributions of diseases sensitive to moisture, such as anthrax, blackleg, hemorrhagic septicemia, and vector-borne diseases (Baylis and Githeko 2006). Climate change also may influence the abundance and/or distribution of the competitors, predators, and parasites of vectors themselves (Thornton 2010). Hotter weather may increase the incidence of ketosis, mastitis, and lameness in dairy cows and enhance growth of mycotoxin-producing fungi, particularly if moisture conditions are favorable (Gaughan et al. 2009). However, there is no consistent evidence that heat stress negatively affects overall immune function in cattle, chickens, or pigs.

Adaptation

Adaptation is defined as an adjustment in natural or human systems in response to actual or expected global changes or their effects (IPCC 2007a). Adaptation to global changes will necessitate adjustments at the enterprise to regional scales and likely will include changes in management, livestock species or breeds, pest management strategies, or even enterprise structure (Morgan 2005; Morgan et al. 2008).

Animal Adaptation

In an effort to optimize animal production, producers likely must select breeds and breed types that are genetically adapted to changed climate conditions. Climate change and associated variation in weather patterns may also require that livestock be managed in or near facilities in which the microclimate can be modified (Mader et al. 1997; Mader et al. 1999; Gaughan et al. 2002b; Mader et al. 2007). Environmental management for all domestic livestock, but especially for ruminants, needs to consider (1) general short- and long-term changes in environmental conditions, (2) changes in nighttime conditions that do not allow for adequate cooling, and (3) increases in the occurrence of extreme events (e.g., hotter daily maximum temperature and more/longer heat waves).

Rötter and van de Geijn (1999) suggest that effects of heat stress may be relatively minor for the more intensive livestock production systems where some control can be exercised over the exposure of animals to climate. In general, domestic livestock are remarkable in their ability to mobilize coping mechanisms when challenged by environmental stressors. Breeding and selection criteria for domestic livestock need to be considered in the context of climate change, especially for those systems in which livestock are routinely exposed to the environment. Adapting to climate change is certain to entail costs, such as application of environmental modification techniques, use of more suitably adapted animals, or even shifting animal populations.

Climate change and associated variation in weather patterns may also require that livestock be managed in or near facilities in which the microclimate can be modified.

Depending on the domestic species of livestock, adaptive responses may include hair coat gain or loss through growth and shedding processes, respectively. As a survival mechanism, voluntary dry matter intake increases (after a 1-to-2-day decline) under cold stress, and decreases almost immediately under heat stress (NRC 1987, 1996). Depending on the intensity and duration of the environmental stress, DMI can average as much as 30% above normal to as much as 50% below normal. Under extremely hot conditions, animals may completely lose appetite, while under extreme cold conditions animals may find comfort in maintaining a huddled position with other animals or remain lying. Due to the discomfort levels associated with standing and accessing feed, DMI and related performance is further compromised under cold stress. However, many adaptive and behavioral adjustments made by the animal, when exposed to moderate to extreme environmental conditions, often result in lowered animal productivity and are generally unfavorable to economic interests of humans. However, these changes are often essential for survival of the animals (Stott 1981; Gaughan et al. 2009).

Beede and Collier (1986) suggest three management options for reducing the effect of thermal stress in cattle, which have application for all livestock and poultry. The options include (1) physical modification of the environment, (2) genetic development of breeds with greater heat tolerance, and (3) improved nutritional

management during periods of high heat load. As is the case in most livestock systems today, housing and microclimate modification considerations (sunshades, or evaporative cooling by direct wetting or in conjunction with mechanical ventilation), improvements in nutritional management and disease control, and use of new technologies will need to be assessed as change dictates (Gaughan and Mader 2007; Mader et al. 2008; Gaughan et al. 2009). Included in that assessment will have to be cost of implementation of altered or new processes, which will be particularly pertinent in less developed and less intensive production systems. An additional consideration is that modifying management and/ or genetics for one environmental extreme may have adverse effects if the livestock are exposed to the opposite environmental extreme. In addition, appropriate environmental stress thresholds are needed that are flexible and can reflect stress levels based on environmental conditions, management levels, and physiological status.

Mader et al. (2010, 2011) developed a Comprehensive Climate Index (CCI) and comparable thresholds framework that incorporate multiple environmental variables into a continuous index that adjusts temperature for the combined effects of relative humidity, wind speed, and solar radiation (Table 5.10). CCI's purpose is to provide a relative indicator of the environmental conditions surrounding an animal and quantify how solar radiation, wind speed, and relative humidity interact

Table 5.10. Comprehensive Climate Index thermal stress thresholds.[1]
Source: Mader et al. 2011.

Environment	Hot conditions[2]	Cold conditions	
		Animal susceptibility level	
		High[3]	Low[4]
No stress	< 25	> 5	> 0
Mild	25 to 30	0 to 5	0 to -10
Moderate	> 30 to 35	< 0 to -5	< -10 to -20
Severe	> 35 to 40	< -5 to -10	< -20 to -30
Extreme	> 40 to 45	< -10 to -15	< -30 to -40
Extreme danger	> 45	< -15	< -40

Source: Mader et al. 2011

[1]Threshold levels indicate intensity of climatic stress experienced by the animal.

[2]Modified from indices developed by Mader et al. (Mader et al. 2006), Gaughan et al. (Gaughan et al. 2008), and the Livestock Weather Safety index (LCI 1970) with severe thresholds capable of causing death of animals and extreme thresholds having a high probability of causing death of high risk animals.

[3]Generally, young and/or non-acclimated animals cared for under sheltered (housed) or modified environmental conditions.

[4]Generally, unsheltered animals which have had adequate time to acclimate to outdoor environments through acquisition of additional external and/or tissue insulation and are receiving nutrient supplies compatible to the level of environmental exposure.

Table 5.11. Apparent temperature estimates as derived from primary environmental characteristics and the Comprehensive Climate Index equations.[1] Source: Mader et al., 2010

Tempera-ture, °C	Wind speed of 1 m/s									Wind speed of 9 m/s								
	SR of 100 W/m²			SR of 500 W/m²			SR of 900 W/m²			SR of 100 W/m²			SR of 500 W/m²			SR of 900 W/m²		
	RH of 20%	RH of 50%	RH of 80%	RH of 20%	RH of 50%	RH of 80%	RH of 20%	RH of 50%	RH of 80%	RH of 20%	RH of 50%	RH of 80%	RH of 20%	RH of 50%	RH of 80%	RH of 20%	RH of 50%	RH of 80%
-30	-32.8	-33.9	-35.0	-29.0	-30.0	-31.2	-25.3	-26.4	-27.5	-41.8	-42.9	-44.1	-38.0	-39.1	-40.2	-34.4	-35.4	-36.6
-25	-27.4	-28.6	-30.0	-23.8	-25.0	-26.3	-20.3	-21.5	-22.9	-36.4	-37.7	-39.0	-32.8	-34.0	-35.4	-29.3	-30.6	-31.9
-20	-22.0	-23.3	-24.7	-18.5	-19.9	-21.3	-15.2	-16.5	-17.9	-31.0	-32.3	-33.8	-27.6	-28.9	-30.3	-24.2	-25.5	-27.0
-15	-16.6	-17.9	-19.3	-13.3	-14.6	-16.0	-10.0	-11.4	-12.8	-25.6	-26.9	-28.4	-22.3	-23.6	-25.1	-19.1	-20.4	-21.8
-10	-11.2	-12.4	-13.8	-8.0	-9.2	-10.6	-4.8	-6.1	-7.4	-20.2	-21.5	-22.8	-17.1	-18.3	-19.6	-13.9	-15.1	-16.5
-5	-5.8	-6.8	-8.0	-2.7	-3.7	-4.9	0.4	-0.7	-1.8	-14.8	-15.9	-17.1	-11.7	-12.8	-14.0	-8.7	-9.7	-10.9
0	-0.4	-1.2	-2.1	2.6	1.9	1.0	5.7	4.9	4.0	-9.4	-10.2	-11.1	-6.4	-7.2	-8.1	-3.4	-4.2	-5.0
5	5.0	4.6	4.1	8.0	7.6	7.1	11.0	10.6	10.1	-4.1	-4.5	-5.0	-1.0	-1.5	-2.0	2.0	1.5	1.0
10	10.4	10.4	10.4	13.4	13.4	13.4	16.4	16.4	16.4	1.3	1.3	1.3	4.4	4.4	4.4	7.4	7.4	7.4
15	15.8	16.3	16.9	18.8	19.4	20.0	21.8	22.4	23.0	6.7	7.2	7.8	9.8	10.3	10.9	12.8	13.3	13.9
20	21.1	22.3	23.6	24.3	25.4	26.7	27.3	28.5	29.8	12.1	13.3	14.6	15.2	16.4	17.7	18.3	19.4	20.7
25	26.5	28.4	30.5	29.7	31.6	33.8	32.8	34.7	36.8	17.5	19.3	21.5	20.7	22.6	24.7	23.8	25.6	27.8
30	31.9	34.6	37.7	35.2	37.9	41.0	38.4	41.1	44.2	22.8	25.5	28.6	26.2	28.9	32.0	29.3	32.0	35.1
35	37.3	40.9	45.0	40.8	44.4	48.5	44.0	47.6	51.8	28.2	31.8	36.0	31.7	35.3	39.5	34.9	38.5	42.7
40	42.6	47.2	52.6	46.3	50.9	56.3	49.7	54.3	59.6	33.6	38.2	43.5	37.3	41.9	47.2	40.6	45.2	50.6
45	48.0	53.7	60.3	51.9	57.6	64.3	55.4	61.1	67.7	38.9	44.6	51.3	42.9	48.6	55.2	46.3	52.0	58.7

Mader et al. (2010); [1]SR = solar radiation; RH = relative humidity

with ambient temperature (Ta) to produce an "apparent temperature" and identify thresholds that assist with assessing levels of stress (Table 5.11). A multi-factor index is superior to a single factor index for determining environmental effects on animal well-being. For strategic decisionmaking, the goal should be to have an index that is broadly applicable across life stages and species to maximize the utility of probability information (Hahn et al. 2003). Aside from assessing environmental effects on animal health, comfort, welfare, maintenance, and productivity, the CCI could be adapted to calculate projected effects of climate change year-round.

Other useful indices that have merit for assessing environmental stress in animals (FASS 2010), include the recently revised wind-chill index (Tew et al. 2002) and modifications to the temperature-humidity index (Eigenberg et al. 2005; Mader et al. 2006). In addition, Hahn and Mader (1997), Hahn et al. (1999), and Gaughan et al. (2008) have developed classification schemes to assess the magnitude (intensity x duration) of extreme heat events that place animals at risk.

A final management issue related to climate change is water availability and utilization. Water has been recognized as one of the most important necessities for life. It plays a key role in virtually all biochemical reactions in the body and is considered to be one of the quickest and most efficient methods to reduce body temperature during warmer periods. During heat waves, normal heat exchange is impeded in livestock, which affects the thermal equilibrium of the animal and its performance.

Per unit of feed intake, water intake is generally two to three times greater under hot conditions than under cold conditions (Kreikemeier and Mader 2004; Arias et al. 2011). The interaction among climatic factors, type of diet, animal breed, animal weight, production status, and physiological strategies adopted by each animal all influence an individual animal's water intake. In addition, drinking behavior is complex and influenced by a number of social and physical factors, including degree of competition for water space, group social order, water availability and accessibility, and water quality.

Grassland Adaptation
Three aspects of plant community production determine the economic viability of livestock enterprises on lands that are managed primarily to produce forage for grazing animals. These factors are the seasonal distribution and quantity of forage, the inter-annual reliability of forage production (inverse of variability), and forage nutritive value, as affected by nutritional and physical properties of individual plant species and plant species composition. Warming, CO_2 enrichment, and altered precipitation regimes can affect each aspect of community production. Modest shifts in the seasonal distribution of forage production and quality and increases in inter-annual variability of production likely can

be accommodated by adjusting stocking rates and varying the season of grazing (Morgan 2005; Morgan et al. 2008; Torell et al. 2010). Adaptation also could include practices that lessen soil erosion, maintain vegetative cover, and promote plant regeneration after vegetation is removed or lost.

Innovative changes in management, such as a shift in livestock species, may be required to deal with changes in species abundance, forage quantity, and nutritive value. In intensively managed pastures, more reliance on species that may be better adapted to future warmer, CO_2-enriched conditions, like legumes or C_4 grasses (Nolan et al. 2001; Gates et al. 2004; Morgan 2005; Hopkins and Del Prado 2007; Morgan et al. 2011), may be advantageous.

Domestic livestock are remarkable in their adaptive ability when challenged by moderate levels of environmental stressors. Adaptive responses to climate change could involve a shift to livestock types with greater tolerance of relatively high temperatures, which better utilize existing vegetation and are more resistant to livestock pests (Morgan 2005). Livestock managers will need to be proactive and consider resource availability (e.g., feed, water, health care, fiscal, animals, land base, human) when adopting climate change mitigation strategies. According to Gaughan et al. (2009), the most important element of proactive environmental management is to reduce risk through preparation. Included in the preparation process is appropriate education and training, development of strategic plans for adjusting to changing conditions, recognition of animal needs and potential stress levels, adopting strategies to minimize and/or mitigate the stress, and selection of animals and management strategies that are compatible with the production enterprise.

Monitoring of pasture and rangeland conditions will become even more important as managers deal with novel climatic conditions (Morgan et al. 2008). Certainly a shift in current enterprise structure will occur. For example, change from grassland to woodland vegetation may require diversification of land uses, perhaps including a shift from livestock production only to ecotourism, hunting, wind energy, or carbon sequestration (Morgan 2005; de Steiguer 2008; Morgan et al. 2008). However, caution should be exercised that overcompensation to changing climatic conditions does not occur. An approach is needed that will allow appropriate changes to occur in a timely manner while avoiding undo disturbance of the socio-economic structure of the livestock and grassland production systems. A greater

understanding of the animal and grassland responses to environmental challenges is essential to successful implementation of strategies to ameliorate negative effects of climate change.

Conclusions

The direct and indirect effects of changing climate create threats and opportunities for U.S. agriculture. The direct effects of changing temperature and precipitation patterns are widely acknowledged and investigated. Producers and researchers have traditionally faced challenges of temperature and moisture changes with success. However the short-term high variability of weather events currently being experienced are outside of the realm of experience for the agricultural community. Given a continued trend of this variability, a shift of management focus from mostly average conditions to that of focus on managing average plus extreme conditions may well be advised. The addition of "event duration" or "maximum tolerable change per day," especially for sensitive growth stages, are potential additions to threshold tables defining the temperature and moisture limits for specific crops. Dealing with the weather manifestations of climate change will be integral to decisionmaking for future producers, more so than for that of past generations.

The complex nature of the agroecosystem means that effects of climate change on system components will vary broadly across geographies and temporal scales. Assessing the full effect of climate change on U.S. agricultural products will require integrated studies that incorporate the nuances of ecosystem function such as soil make-up, changes in timing of runoff, and effects of changing temperature patterns and CO_2 concentrations, together with factors related to production economics, management strategy approaches and implementation, and adaptation practices. Such studies will also feed creation of models that may more accurately project future changes and assess effects of land-use or water-resource changes that may affect crops, and assist with developing strategies that can provide insights on increasing efficient use of available resources. In addition, adaptation management practices would benefit from further research on adaptive cultivars and crop genetics so as to mitigate projected declines in future yields by taking better advantage of climate-driven shifts in ecosystem characteristics through breeding for physiological pathways that increase resilience to climate stressors. Lastly, managing for changing climate will benefit from further research

Domestic livestock are remarkable in their adaptive ability when challenged by moderate levels of environmental stressors. Adaptive responses to climate change could involve a shift to livestock types with greater tolerance of relatively high temperatures, which better utilize existing vegetation and are more resistant to livestock pests.

into technologies that improve management of
agricultural products through further automation
of processes and tools, sensor development, and
enhancement of information technologies. Advancing
these research needs will assist those working in
the realm of U.S. agriculture by providing both
pragmatic solutions while potentially reducing costs
related effects of climate change on agricultural
production.

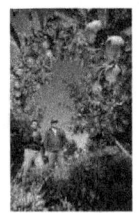

Chapter 6

Climate Change Effects on the Economics of U.S. Agriculture

The economic impacts of climate change occur at multiple scales and with a complex array of feedback loops. While the biophysical effects of climate change play out locally through the direct and indirect (abiotic and biotic) stress factors described earlier, the economic implications of those effects are shaped by an array of local, national, and global institutions, from commodity markets to systems of research, development, education, communication, and transportation. These institutions define the opportunities and constraints within which stakeholders can adjust their behavior to minimize losses and take advantage of new opportunities for gain associated with changing climate conditions. Potential adaptive behavior can occur at any level within a highly diverse agricultural system, including consumption, production, education, and research. The aggregate effects of climate change will therefore ultimately depend on a web of diffuse adaptive responses to local climate stressors, from farmers adjusting planting patterns in response to altered crop yields to seed producers investing in drought-tolerant varieties to nations changing trade restrictions in response to food security concerns.

The complexity of possible adaptive response pathways makes it extremely difficult to characterize all of the potential steps and feedback loops leading from local climate effects on yield (or on increased costs necessary to maintain yields) to regional or global effects on economic indicators such as prices, production, trade volume, consumer expenditures, or producer income and financial viability. U.S. and global agricultural markets are highly interconnected, and trade will result in a redistribution of agricultural products from regions of relative surplus to regions of relative scarcity (Adams et al. 1998). The economic implications of climate change for the United States will therefore be sensitive to yield effects and adaptation opportunities and constraints both within the United States and worldwide. An even broader set of social and political variables is required to

Economic versus Biophysical Impacts on Agricultural Productivity

Biophysical impacts on productivity are localized phenomena that are largely driven by local variations in weather impacts and mediated by local soil and water conditions. Economic impacts, on the other hand, are embedded within a complex and regionally diffuse web of production, price, consumption, and trade responses to those local productivity impacts. U.S. and global agricultural markets are highly interconnected, so economic impacts within the United States are sensitive to biophysical impacts, behavioral responses among consumers and producers, and adaptation opportunities and constraints both within the United States and worldwide. Managing the impacts of climate change on U.S. agriculture is an interdisciplinary challenge that may be most effectively addressed using systems research strategies to integrate and develop disciplinary knowledge.

Example: Climate change can impact the livestock sector along a number of pathways: directly through impacts on productivity and performance and indirectly through price and availability of feed grains, competition for pasture land, and changing patterns and prevalence of pests and diseases. These pathways parallel those of crop production impact, and the two sectors are strongly linked through feed grain markets and competition for land.

explore the implications of climate change for critical social issues such as food security and the incidence of hunger.

A comprehensive analysis of climate effects requires bringing together state-of-the-art knowledge from multiple disciplines and areas of expertise (Beach et al. 2010; Tubiello et al. 2007a; Hertel et al. 2010). Developing economic-impact estimates for climate change requires input from disciplines as diverse as climate, crop, and soil science, as well as the tools and data to represent a wide variety of potential adaptive and economic behaviors. While research is advancing on disciplinary elements of the system, transdisciplinary efforts have struggled with inconsistent data, poor communication between disciplines, and the resource challenges of developing new data sets and analysis tools to incorporate information from different disciplines. Efforts such as the Agricultural Model Intercomparison and Improvement Project (see AgMIP sidebar pg. 108) have been initiated to tackle such challenges while developing and validating scenarios, research tools, and analysis results characterizing changes to the risk of hunger and world food security due to climate change.

The biophysical effects of climate change on yields and production costs are regionally variable and have the potential to significantly alter patterns of agricultural productivity in the provision of food, feed, fiber, and fuel products worldwide. Because the agricultural economy is a complex, self-adjusting set of relationships, ultimately climate change effects will depend on how production and consumption systems adjust, or adapt, in response to those biophysical effects. This chapter reviews efforts to quantify the economic impacts of climate change to date and explores issues related to the scope and scale of those analyses. Capacity for economic impact quantification is evolving, however assessment results remain highly sensitive to elements of research scope such as exclusion versus inclusion of international effects, selective treatment of climate stressors when assessing yield and production cost impacts, selective representation of adaptation opportunities in response to yield and production cost impacts, and limited consideration of potential constraints to adaptation, including natural and financial resource constraints.

Economic Impacts and Agricultural Adaptation

Agricultural production is chronically vulnerable to stress factors like dry spells, weed competition, and insect damage. Local farm production patterns and practices have evolved in response to weather conditions and stress factors that have historically prevailed for that region. As growing conditions and stress factors change, so too will farm production decisions. Adaptation behaviors such as changing crops and crop varieties, adjusting planting and harvest dates, and modifying input use and tillage practices can lessen yield losses from climate change in some regions and potentially increase yields in others where climate change creates expanded opportunities for production (Adams et al. 1998; Malcolm et al. 2012). Several regional and national studies have predicted that U.S. cropland agriculture will be fairly resilient to climate change in the short term, with expansion of irrigated acreage, regional shifts of crop acreage, and other adjustments to inputs and outputs partially compensating for yield effects caused by changing climate patterns (Adams et al. 1990; Mendelsohn et al. 1994).

Capacity for adaptation is therefore a critical determinant of the net economic effects of climate change and of the regional distribution of those effects (Antle and Capalbo 2010; Malcolm et al. 2012). Adaptive behavior can significantly mitigate the potential effects of climate change on food production, farm income, and food security by moving agricultural production out of regions with newly reduced comparative advantage in specific production sectors and into areas with improved relative productivity (Rosenzweig and Parry 1994; Darwin et al. 1995; Adams et al. 1998; Mendelsohn and Dinar 1999; Malcolm et al. 2012; Beach et al. 2010). Darwin et al. (1995) estimated that farmers adjusting inputs and outputs on existing farmland could offset from 79 to 88% of the 19-30% reductions in world cereals production (wheat plus other grains) that they attribute to the direct crop growth and development effects of climate change. In that analysis, market adjustments further increased the percentage of yield decline offset to 97%, and expansion of cropland actually resulted in an increase in world cereal production relative to the "no climate change" scenario (Darwin et al. 1995). Reilly et al. (2007) estimate that with adaptation, the production effects of climate change are reduced to one-fifth to one-sixth of the initial yield effects. While such analyses highlight that the economic impacts of climate change will be sensitive to all such opportunities for, and constraints to, technological and behavioral adaptation, they have their strengths

and weaknesses with respect to treatment of those factors when translating climate effects into economic impacts.

Estimating Economic Impacts of Climate Change

An assessment of the economic impacts of climate change on agriculture begins with a set of assumptions or projections about future climate conditions, generally including some combination of information on patterns and magnitude of temperature and precipitation change (Tol 2009). Local climate conditions must be then translated into local yield and production cost impacts based on a subset of stressors and simultaneously into a set of economic indicators through representation of a portion of the potential production, price, consumption, technology development, and trade responses to those productivity effects.

Economic impacts of climate change can occur at many levels and to different stakeholders. Farmers (producers) are affected by initial yield and production cost effects, which they respond to through adaptive strategies and, subsequently, by the price effects that emerge from the market adjustments responding to widespread influences on productivity and adaptive behaviors. Consumers are affected by market price changes and also have adaptation options including changing consumption patterns to substitute relatively low-priced products for products that have become higher priced due to the effects of climate (Adams et al. 1998).

Efforts to quantify economic impacts are sensitive to research elements that define input assumptions and scale and scope of analysis, including:

- Climate and Yield Projections (and associated time horizon): Biophysical and economic impact assessment results are highly sensitive to the choice of climate model and projection used and to the spatial resolution of those climate scenarios (Malcolm, 2012; Beach et al. 2010; Adams et al. 1990; Adams et al. 1995; Adams et al. 2003). Climate analyses that project farther into the future generally show greater effect on yields and economic indicators, though there is also greater uncertainty about future emissions trajectories, projected changes of climate variables, and available adaptive technologies. Treatment of CO_2 fertilization effects (i.e., whether, and how, potential yield-enhancing effects of increased atmospheric CO_2 are included in the study) is also an important determinant of results (Adams et al. 1990; Adams

et al. 1995; Adams et al. 1998; Antle et al. 2004; Sands and Edmonds 2005; Cline 2007).

- Scope of the Assessment: Potential adaptive responses to climate change occur at many scales and across multiple sectors, interacting across land, commodity, and agricultural input markets. Researchers must decide how to condense the complex detail associated with the true scope of climate effects on agriculture and the adaptation response into a simplified version of reality that captures particularly significant dynamics. Those decisions include, for instance, the types of available adaptation options and whether the assessment includes consumer response and effect as well as that of producers, livestock, forest production, cropland agriculture, and international and domestic interests (Adams et al. 1998; Sands and Edmonds 2005; Hertel et al. 2010b). The exploration and identification of relevant dynamics, and compilation of the data necessary to represent these dynamics in impact analyses is an ongoing process.

- Socioeconomic and Technology Projections and Treatment of Adaptation Constraints: Climate change effects and opportunities for adaptation will unfold within a future economic, policy, and technology environment that is inherently uncertain. Variables relevant to agriculture's response to climate change range from broader social variables about economic and income growth to sector-specific assumptions including future crop and livestock productivity, farm policy, farm size, input and output prices, and availability of technical resources to facilitate adaptation (Claessens et al. 2012). Potential constraints to adaptation such as regional land and water availability, as well as constraints related to farm finances and viability, have received relatively little research attention yet have been shown to significantly affect the results emerging from both integrated assessments and statistical analyses of climate change effects (Adams et al. 1995; Darwin et al. 1995; Howden et al. 2007; Schlenker et al. 2007).

- Estimation methodology Used and Model Specification: Methods used for climate change assessment include expert opinion, hedonic and production function approaches, and integrated assessment modeling (Schlenker et al. 2005, Antle and Capalbo, 2010).

Results from a limited set of economic impact studies are presented in Table 6.1. These results, and the source of variability among them, are discussed in more detail in the sections below.

Table 6.1. Results from a limited set of studies exploring the domestic economic impacts of climate change.

Study	Climate Models Used	Economic Estimation Method Used	Climate Change Condition or Year	Economic Impacts: Producers	Economic Impacts: Consumers	Total Economic Impact	Impacts on Prices or Price Index	Climate Elements Changed	Including CO₂ Impact on Crop Yields	Include International Production Impacts
Adams et al, 1990	NASA/GISS	Simulation (ASM)	Doubled effective CO₂ (630 ppm)	+1.59 billion 1982$	+9.30 billion 1982$	+10.89 billion 1982$	crops - (.83) livestock - (.84)	temp, precip, incident solar rad	yes	no
Adams et al, 1990	GFDL	Simulation (ASM)	Doubled effective CO₂ (600 ppm)	+3.55 billion 1982$	-13.89 billion 1982$	-10.33 billion 1982$	crops + (1.34) livestock + (1.08)	temp, precip, incident solar rad	yes	no
Adams et al, 1995	NASA/GISS	Simulation (ASM)	Doubled effective CO₂ (555 ppm) (~2060)	+10.79 billion 1990$	-22.12 billion 1990$	-11.33 billion 1990$ (-1.01%)	+ (1.12)	temp, precip	no effect	no
Adams et al, 1995	NASA/GISS	Simulation (ASM)	Doubled effective CO₂ (555 ppm) (~2060)	+12.74 billion 1990$	-2.54 billion 1990$	+10.20 billion 1990$ (+.91%)	+ (1.01)		yes, 555 ppm	no
Adams et al, 1995	NASA/GISS	Simulation (ASM)	Doubled effective CO₂ (555 ppm) (~2060)	+12.56 billion 1990$	-1.74 billion 1990$	+10.82 billion 1990$ (+.96%)	+ (1.01)		yes, 555 ppm	yes, through trade adjustment
Adams et al, 1995	GFDL	Simulation (ASM)	Doubled effective CO₂ (555 ppm) (~2060)	+16.84 billion 1990$	-35.93 billion 1990$	-19.09 billion 1990$ (-1.70%)	+ (1.21)	temp, precip	no effect	no
Adams et al, 1995	GFDL	Simulation (ASM)	Doubled effective CO₂ (555 ppm) (~2060)	+7.22 billion 1990$	-2.65 billion 1990$	+4.57 billion 1990$ (+.41%)	+ (1.01)		yes, 555 ppm	no
Adams et al, 1995	GFDL	Simulation (ASM)	Doubled effective CO₂ (555 ppm) (~2060)	+6.61 billion 1990$	-2.24 billion 1990$	+4.37 billion 1990$ (+.39%)	+ (1.01)		yes, 555 ppm	yes, through trade adjustment
Adams et al, 1995	UKMO	Simulation (ASM)	Doubled effective CO₂ (555 ppm) (~2060)	+114.97 billion 1990$	-181.98 billion 1990$	-67.01 billion 1990$ (-5.96%)	+ (2.09)	temp, precip	no effect	no
Adams et al, 1995	UKMO	Simulation (ASM)	Doubled effective CO₂ (555 ppm) (~2060)	+41.52 billion 1990$	-59.11 billion 1990$	-17.58 billion 1990$ (-1.57%)	+(1.33)		yes, 555 ppm	no
Adams et al, 1995	UKMO	Simulation (ASM)	Doubled effective CO₂ (555 ppm) (~2060)	+44.44 billion 1990$	-35.41 billion 1990$	+9.03 billion 1990$ (+.80%)	+ (1.35)		yes, 555 ppm	yes, through trade adjustment
Adams et al, 2003	CSIRO (coarse resolution)	Simulation (ASM)	540 ppm	-3.31 billion 2000$ (no adapt.) -3.87 billion 2000$ (with adapt.)	+6.36 billion 2000$ (no adapt.) +9.66 billion 2000$ (with adapt.)	+3.05 billion 2000$ (no adapt.) +5.69 billion 2000$ (with adapt.)		temp, precip, incident solar radiation	yes	no
Adams et al, 2003	RegCM (finer resolution)	Simulation (ASM)	540 ppm	-3.41 billion 2000$ (no adapt.) -4.67 billion 2000$ (with adapt.)	+3.73 billion 2000$ (no adapt.) +8.27 billion 2000$ (with adapt.)	+.32 billion 2000$ (no adapt.) +3.61 billion 2000$ (with adapt.)		temp, precip, incident solar radiation	yes	no
Alig et al, 2002	Hadley Center Model	Simulation (FASOM)	avg climate conditions 2070-2100; economic projection 100 years	-7.1% (Forestry) -15.9% (Agriculture)	+1.3% (Forestry) +2.0% (Agriculture)	+.7% (both sectors)		temp, precip	not clear	no
Alig et al, 2002	Canadian Climate Model	Simulation (FASOM)	avg climate conditions 2070-2100; economic projection 100 years	-5.5% (Forestry) -7.6% (Agriculture)	+1.0% (Forestry) +1.0% (Agriculture)	+.4% (both sectors)		temp, precip	not clear	no
Reilly et al, 2003	Canadian Climate Model	Simulation (ASM)	2030/2090	-.1 to -5 billion 2000$ (range includes both GCMs)	+$2.5 to +$13 billion 2000$ (range includes both GCMs)	+$0.8 billion 2000$ (2030) +$3.2 billion 2000$ (2090)	Prices generally drop	temp, precip	yes	Trade results presented in Reilly et al (2001)
Reilly et al, 2003	Hadley Center Model	Simulation (ASM)	2030/2090	-.1 to -5 billion 2000$ (range includes both GCMs)	+$2.5 to +$13 billion 2000$ (range includes both GCMs)	+$7.8 billion 2000$ (2030) +$12.2 billion 2000$ (2090)	Prices generally drop	temp, precip	yes	Trade results presented in Reilly et al (2001)
Sands et al, 2005	UIUC (Univ of Illinois)	Simulation (ASM)		-4.2%	-2.6%	-6.8% (drop in primary agricultural output)	prices generally increase	temp, precip	no	yes
Sands et al, 2005	UIUC (Univ of Illinois)	Simulation (ASM)		+6.8%	+2.2%	+9.0%	prices generally drop	temp, precip	yes	yes

Sensitivity of Economic Impact Estimates to Climate and Yield Projections

Projections of the economic impacts of climate change are highly sensitive to assumptions made about the production cost and yield effects associated with changing climate conditions, which, in turn, vary widely with climate change projections used, time horizon of analysis, and assumptions made about the effects of uncertain processes such as CO_2 fertilization.

Several studies of climate change effects within the United States have suggested that moderate levels of climate change will increase crop yields on average, resulting in net positive estimates of welfare change in the United States (Reilly et al. 2003; McCarl 2008; Sands and Edmonds 2005). Reilly et al. (2003) and McCarl (2008) estimated an increase in U.S. consumer welfare in response to climate change because productivity increases resulted in price drops and reductions in consumer cost. However, producer welfare in the United States declined because the drop in prices offset producer benefits accruing from yield increases. Yield increases were regionally variable, however, with yields and producer returns in the South more negatively impacted than in the North. Sands and Edmonds (2005) found that the observed price decline did not always fully erode the bump in producer returns arising from increased yields, and that both producer and consumer welfare in the United States increases in two out of three future climate projections.

Projections suggesting that climate changes in temperate regions will increase yields in agriculturally important regions such as the Corn Belt are consistent with the Intergovernmental Panel on Climate Change (IPCC 2007b) assessment that "moderate climate change will likely increase yields of North American rain-fed agriculture" and its more general projection that crop productivity will increase slightly at mid to high latitudes for local mean temperature increases of up to 1 to 3°C, but are inconsistent with the results of other studies concluding that recent patterns of climate change have already had adverse effects on corn and soybean production in agriculturally important regions (Lobell and Asner 2003; Kucharik and Serbin 2008; Ainsworth and Ort 2010). In fact, the net effects of climate change on average U.S. yields will vary by crop and be sensitive to both the effect of the climate projection selected and to regional shifts in crop acreage and irrigation practices that arise through market adjustments responding to effects on yield

(Sands and Edmonds 2005; Izaurralde et al. 2011; Malcolm et al. 2012). In their analysis of the effects of climate change on crop insurance, Beach et al. (2010) projected increasing average national yields for crops such as barley, hay, oats, rye, and hard red winter wheat; decreasing average national yields for cotton, grapefruit, oranges, potatoes, soft white wheat, and Durum wheat; and mixed yield-effect results for corn, rice, silage, sorghum, soybeans, sugarcane, and hard red spring wheat, depending on the climate scenario used.

Yield effects are a critical determinant of economic impact estimates, but yield projections under climate change projections are highly uncertain. Estimates of effects may misrepresent likely yield because most analyses have not included a comprehensive treatment of the stress factors arising from climate change that can affect yields (see Chapter 4 of this report). Studies often focus on the effects of a subset of direct stress factors, usually changes in average temperature and precipitation, while excluding the potential direct effects of other changing climate conditions, such as ozone exposure and solar radiation. Most studies also fail to consider the additional effects of indirect stress factors, such as changes in pest, weed, and disease pressure, arising from community-scale, agro-ecological adjustments to changing climate (Gornall et al. 2011). Management strategies to deal with changing biotic stress can significantly affect crop and livestock production costs; a failure to consider such costs may overstate farms' financial viability in the face of changing climate conditions. Although few such studies have estimated the effects of indirect stressors on crop productivity and management costs, available research has shown that these have a significant effect on the economic estimates of climate change within crop agricultural sectors (Malcolm et al. 2012).

Economic impact results are also highly sensitive to whether and how yield-enhancing effects of atmospheric on crop yields are considered in the analysis (Sands and Edmonds 2005; Reilly et al. 2007; Cline 2007;). Nevertheless, only a limited number of studies have assessed the yield implications of CO_2 fertilization across crops under actual growing conditions; its effects, particularly in the presence of interacting changes in other factors such as temperature and soil moisture, are therefore highly uncertain (Adams et al. 1995; Long et al. 2005; Tubiello et al. 2007a; Gornall et al. 2010). Sands and Edmonds (2005) found that when uncertain CO_2 fertilization effects were excluded from the calculation of crop-yield effects, crop yields in the United States declined under the three climate scenarios they explored, as did indicators of both U.S. consumer and producer welfare. Such uncertainty about effects on crops translates into substantial uncertainty about economic impacts as well.

Because climate is projected to continue changing throughout the 21st Century, yield and economic impact assessment results are sensitive to the time horizon used in the analysis and to the rate of change assumed by the climate projection(s) relied upon for the analysis. Global analyses of the effects of climate change on Gross Domestic Product (GDP) that are not limited to the agricultural sector often find near-term economic benefits associated with modest changes in climate that are followed by losses further in the future as temperatures continue to increase (Hitz and Smith 2004; Tol 2009). Several researchers have argued that crop productivity is likely to follow a similar pattern for several major crops (Parry et al. 2004; Schlenker and Roberts 2009) or that changing conditions are already creating a drag on global crop yields (Lobell and Field 2007; Lobell et al. 2011). Easterling et al. (2007) projected that crop productivity would begin to decline, even in temperate regions, when temperature increases exceed 1 to 3° C (1.8-5.4° F). Burke et al. (2011) project that both corn yields and farm profits would decline under a large range of climate projections in the United States for time ranges in the mid- and late-21st Century. In a hedonic farmland value regression for dryland acreage in the United States, Schlenker et al. (2005) estimate annual losses of $5-5.3 billion under a 5°F increase in temperature and an 8% precipitation increase.

Uncertainty in climate projections is therefore a critical element of crop and economic impact uncertainty (Adams et al. 1995; Sands and Edmonds 2005; Burke et al. 2011; Malcolm et al. 2012). Nevertheless, both effects on crops and economic assessment efforts have been slow to develop the tools necessary to accommodate climate uncertainty. Burke et al. (2011) argue that although more than 20 climate models are regularly used by the climate change community, none of which have been determined to be more reliable than others for long-term climate projections, the median number of model projections used for economic, political, or social impact studies is generally two models. Furthermore, Adams et al. (2003) find that applying climatological projections at a finer spatial scale in determining effects on crop yield can substantially change estimates of a set of producer and consumer welfare measures, with finer scale data generally

leading to reduced benefits or greater damages relative to a coarser scale analysis. The failure of economic impact estimates to capture the uncertainty generated by choice of climate projections, as well as application scale and downscaling method, artificially constrains the uncertainty associated with the impact estimates themselves. Greater attention to methods of quantifying and tracking multiple sources of uncertainty is required in climate change studies (Lobell and Burke 2008; Challinor et al. 2009; Soussana et al. 2010; Winkler et al. 2010).

Sensitivity of Economic Impact Estimates to Scope of Analysis

One of the most significant limitations in U.S. economic impact assessment is confining the scope of analysis to a consideration of the effects of climate change on domestic yields. Effects on regional yields alone are generally a poor predictor of regional welfare effects because domestic markets are highly interconnected with international markets, which will also be responding to yield and production changes worldwide. International trade mediates a larger, global response to highly decentralized yield and production changes, causing adjustments in world prices and trade patterns that can be equally important in affecting the domestic welfare measures being calculated (Adams et al. 1995; Hertel et al. 2010a). Changes in *relative* productivity by region, and the price and trade effects arising in response, are therefore a critical determinant of the economic and welfare effects of climate change (Reilly et al. 2007; Hertel et al. 2010a; Winkler et al. 2010).

Assessments assuming a generally positive U.S. yield effect from climate change, for instance, when looking at U.S. yield effects in isolation from the rest of the world, suggest that domestic yield increases could stimulate supply and depress prices, with positive welfare implications for consumers and mixed implications for producers. The functioning of world markets, however, ensures that the actual effects of climate change on domestic consumers and producers will depend on what is happening to yields in the rest of the world and on the associated effects on world production and world price. If global yield effects are generally negative, they can drive global prices up despite domestic yield increases; the resulting price increases can benefit U.S. producers through increased return for their product, but U.S. consumer welfare is depressed by the global-market-mediated price increase (Reilly et al. 2003; Sands and Edmonds 2005). On the other hand, if net global yield effects for a given crop are also positive, then

world yield effects can further lower world and domestic prices and push benefits associated with price changes even more in favor of consumers.

In countries that experience yield declines, producer returns may therefore increase if rising global prices are sufficient to offset the adverse income effects of reduced yields (Reilly et al. 2007; Hertel et al. 2010a). Consumers, on the other hand, always suffer welfare losses from reduced availability of food and increased prices associated with declining yields (Hertel et al. 2010b); in developing countries, certain non-agricultural demographics, such as the urban labor strata and the non-agricultural self-employed, can be highly vulnerable to increased poverty arising from higher food prices (Hertel et al. 2010a; Hertel et al. 2010b).

The potential for climate change to alter the variability of production returns, as well as relative variability across crops or livestock enterprises, is also likely to affect farmers' risk management decisions, making "climate risk management" an increasingly important driver of production and adaptation decisions (Chen et al. 2004; Howden et al. 2007). A risk-averse farmer can be expected to allocate more acreage to crops with relatively low variability, for instance. Climate change adaptation behavior may therefore include shifts among crops based on differences in climate change effects on yield variability and co-variance of yields across crops (Isik and Devadoss 2006). Adams et al. (2003) found that adaptation behavior leads to greater welfare gains when climate change effects on yield variability are considered, in part because adaptation behaviors will help mitigate increases in yield variability due to climate change. Other risk management options that may play an increasingly important role in farmers' decisionmaking under conditions of increased yield variability include crop insurance, expansion of irrigation or other inputs, and adoption of moisture-conserving tillage operations and other best management practices (Knutson et al. 2011; Darwin 2004; Beach et al. 2010). Economic impacts of climate change will therefore also be sensitive to the availability, effectiveness, and costs of adaptation measures adopted in response to yield variability as well as to those adopted in response to changes in average yields (Adams et al. 2003). Because little information is available on projected changes in climate and yield variability, however, such considerations have not been integrated into economic impact analyses.

The scope of analysis is also defined by the number of sectors included in the impact analysis. Existing analyses of agricultural have focused on climate change's effects on cropland agriculture with some expansion, often in the case of simulation modeling efforts, to include the effects of changing feed prices or competition for pasture land on the livestock sector (Malcolm et al. 2012; Reilly et al. 2003). Although climate change will also have direct effects on both the productivity and management costs of the livestock and dairy sectors, through pathways such as lowered feed efficiency, reduced forage productivity, reduced reproduction rates, and costs associated with modifying livestock housing to reduce thermal stress, relatively few economic impact studies have estimated these costs. In the absence of such estimates, most system-wide economic impact assessments, with few exceptions (i.e., Adams et al. 1999; Reilly et al. 2003), do not account for the potential direct costs and productivity effects of climate change on livestock, forage, and rangeland production (Antle and Capalbo 2010; Izaurralde et al. 2011). Furthermore, crop sector studies have focused largely on the implications of climate change for commodity crop production, with less attention paid to market and revenue impacts in specialty crop sectors (such as fruits, vegetables, tree nuts, and nursery crops), though in 2010 such crops accounted for roughly 37% of all U.S. cash receipts for farm crops (Antle and Capalbo, 2010; ERS 2012).

Interactions within agricultural sectors and across other sectors of the economy will also be an important determinant of aggregate economic impact, as climate change will directly affect cropland, forestry, and livestock (as well as all other economic sectors) simultaneously. However, only a small subset of studies has looked at the effects of changing relative productivity across sectors on resource allocation decisions such as shifting land use among crop, livestock, and timber production (Darwin et al. 1995; Alig et al. 2002; Sands and Edmonds 2005; Reilly et al. 2007). Reilly et al. (2007) argue that because agricultural adaptation requires shifting resources into or out of the agricultural sector, the full economic impact of those changes can only be assessed using economy-wide measures of well-being that take into account aggregate consumption across all goods and services.

Other issues of scope are implicit in the methodology used to estimate changes in environmental indicators arising as a result of climate change. Economic impact estimates of climate change are often derived by comparing economic outcomes at some future date under climate change to outcomes under the current climate (Antle and Capalbo 2012). The regional costs (or benefits) of climate change are measured as the differences in welfare or revenue (or other economic indicator of interest) between these two points of comparison. This method, however, fails to take into account the costs associated with transitioning from the current to the future position – including, for instance, the costs of developing the transportation, distribution, and irrigation infrastructure necessary to support new patterns of agricultural production – or how such "adjustment" costs are affected by the rate and variability with which climate conditions change over time (Quiggin and Horowitz 2003; Patt et al. 2010). Quiggin and Horowitz (2003) argue that adjustment costs are likely to be the greatest element of cost in response to climate change, but they are largely ignored in sector-level economic impact studies (Patt et al. 2010, Antle and Capalbo 2010; Hertel et al. 2010).

Sensitivity of Economic Impact Estimates to Socioeconomic and Technology Projections and Treatment of Adaptation Constraints

Because likely future responses to climate change depend upon a wide array of uncertain variables, the climate change community has relied heavily on the development of future scenarios, or plausible narratives describing how the future might unfold with respect to characteristics such as socioeconomic variables, technological and environmental conditions, and greenhouse gas (GHG) emissions (Moss et al. 2010). The IPCC, for instance, created a storyline describing population, economic growth, technology and clean energy adoption, and agriculture and land use to inform each of the potential emissions scenarios used in its analyses (Nakicenovic et al. 2000).

No such narratives have historically been available for more sector-specific input assumptions such as future crop and livestock productivity, farm policy, farm size, and input and output prices, however, so economic impact assessments generally assume that historical conditions or trends continue into the future. Smooth continuity is unlikely given the magnitude of disturbance to the agricultural system expected under a changing climate and the extended time horizon of many such analyses, so recent research has focused on developing a set of "Representative Agricultural Pathways" (RAPs) that expands the coverage of global socioeconomic scenarios to include more region-specific agricultural and economic development conditions relevant to agricultural modeling efforts. Such RAPs include

assumptions about size of farm households, availability of agricultural labor, investments in transportation and communication infrastructure, price of fertilizer and seed inputs, and trade policy on agricultural exports (Claessen et al. 2012). Research studying the sensitivity of climate change effects to RAP specification in Kenya finds that estimates of climate change are highly sensitive to assumptions made about future socioeconomic and technological conditions, in part through differential implications for farm livelihood and technical and financial ability to adapt (Claessen et al. 2012).

While limitations of scope were described above as a problem for generating robust estimations of the economic impacts arising from effects to and adaptation within the agricultural sector, trade-offs may exist between expanding economic analyses (i.e., considering effects within other sectors of the economy or world markets) and the ability to represent adaptation options and constraints effectively at the producer level. For instance, studies that look at international trade in a rigorous way are generally based on an analysis of highly aggregated data and regions; such models and methodologies are unable to capture the dynamics of potential adaptive responses at the farm level, such as changes in harvest in planting days or changes in capital equipment or infrastructure as farms within a region find it necessary to change their production methods.

Few economic impact analyses in the United States have incorporated potential constraints to adaptation related to farm financing and credit availability, for instance, though research suggests that such constraints may be significant. Farmer members of a sustainable agriculture organization in Nebraska reported lack of capital as their largest perceived barrier to implementing drought risk-reduction practices (Knutson et al. 2011). In their analysis of dairy and specialty crop farms in the Northeastern United States, Wolfe et al. (2008) identify small family farms with little capital as those most vulnerable to climate change (Wolfe et al. 2008). In an analysis of adaptation capacity by production region, Antle et al. (2004) argue that areas with marginal financial and resource endowments, such as the Northern Plains, are especially vulnerable to climate change (Antle et al. 2004). In addition to technical and financial ability to adapt to changing average conditions, farm resilience to climate change is also a function of financial capacity to withstand increasing variability in production and returns, including catastrophic loss (Smit and Skinner 2002; Beach et al. 2010). Such farm-level analyses emphasize the importance of complementing and

In an analysis of adaptation capacity by production region, Antle et al. (2004) argue that areas with marginal financial and resource endowments, such as the Northern Plains, are especially vulnerable to climate change.

informing economy- and sector-wide impact studies with more detailed analyses of the implications of the heterogeneity of farms and farmers in determining farm viability, and the potential for adaptation under a changing climate both within the United States and internationally (Claessen et al. 2012).

Farmers' adaptation decisions may also be constrained by elements of "path dependency" within the agricultural system (Vanloqueren and Baret 2009; Chhetri et al. 2010). Examples of path dependence include technological lock-in (e.g., arising from sunk machinery costs), social, economic, and cultural reinforcement of prevailing development paths, and lags in institutional response that might otherwise enable more rapid adoption of innovative technologies (Chhetri et al. 2010). Little research has been done on the potential effects of such constraints on the speed and efficiency of agricultural adaptation or on the resulting economic implications of such obstacles. Differences in farm-scale ability to absorb the costs associated with adaptation may also have implications for existing agricultural trends toward large-scale agricultural production and vertical integration in U.S. agriculture.

The Changing Geography of Production

The migration of crop production in response to climate change has been recognized as a likely adaptation mechanism since the early days of integrated assessment modeling (Adams et al. 1995; Darwin et al. 1995). Regional capacity for expanding agriculture or irrigated production will depend on resource constraints such as the availability of land and water (Darwin et al. 1995; Schlenker et al. 2007). Large bands of uncertainty around future projections for regional precipitation change make it difficult to predict with precision regional changes in relative productivity, and estimates of net land brought into production as a result of climate change are mixed and highly sensitive to which models and climate assumptions or scenarios are used in the estimation (Zhang and Cai 2011; Malcolm et al. 2012). In general, however, studies estimate that arable land will increase at the higher latitudes, including Canada, Russia, Northern United States, and southern Argentina, and decrease in western Africa, Central America, western Asia, the South Central United States, and northern South America (Ramankutty et al. 2002; Zhang and Cai 2011). Fischer et al. (2005) estimate that by the 2080s, expansion of cropland in Southeast Asia will be particularly constrained due to land-use competition from other sectors combined with a lack of suitable agricultural land.

Sensitivity of Economic Impact Estimates to Estimation Methodology

Methods used for climate change assessment vary widely and have included expert opinion, statistical estimation using hedonic and production function approaches, and integrated assessment modeling (Schlenker et al. 2005). These assessment methodologies have differing capacities for reflecting adaptation options, allowing the adoption of adaptation technologies that don't yet exist, capturing the effects of market responses such as changes in the prices of inputs and outputs, and accommodating scope and scale considerations like those described above (Antle and Capalbo, 2010).

Statistical estimation methods, for instance, use observed data on agricultural production and climate between regions to parameterize functional relationships between climate variables and production (or production value, or value of land used for production) (Adams et al. 1998). Projected climate effects can then be inferred by changing the input climate variables and observing a production change based on the historically derived relationship. Estimating future effects based on relationships observed in past data, however, cannot take into account the possibility of future technological changes that might fundamentally change production decisions and adaptation options. Such estimation methods are also highly sensitive to model structure. In an exploration of the hedonic estimation method – a widely used statistical methodology for impact assessment – Schlenker et al. (2005) demonstrate that pooling dryland and irrigation acreage in a single statistical model, as other authors have done, can yield biased estimates of economic impact because different explanatory variables are required for the different types of production system (Schlenker et al. 2005). When they run a hedonic climate change impact estimate for dryland acreage only, they predict unambiguously negative effects on U.S. agriculture from climate change. The hedonic estimation method has also been criticized as highly sensitive to seemingly minor model structure choices related to weighting schemes, dummy variables, or control variables (Deschenes and Greenstone 2007).

A second major approach to economic impact assessment is the structural approach, which employs integrated assessment models to measure the economic consequences of climate change (Adams et al. 1998). Integrated assessment models have been broadly defined as "any model which combines scientific and socio-economic aspects of climate change primarily for the purpose of assessing

policy options for climate change control" (Kelly and Kolstad 1999). Over the past few decades, integrated assessment modeling efforts have used model ensembles from several different disciplines to tie together the dynamics of climate effects at various scales for a broader picture of projected agricultural system response and effects. These analyses allow for the introduction of a wide range of potential adaptation behaviors, though that flexibility is limited by the structure and scale of the component models, as well as by the need to specify for newly introduced adaptation options cost and benefit information that may be unknown or highly uncertain.

Integrated assessments of climate change effects must synthesize information on dynamics and relationships that occur at multiple scales, identifying and capturing relevant driving forces and feedbacks without getting bogged down in "unwarranted precision" at any point in the system (Challinor et al. 2009; Polsky and Easterling 2001). Finding the appropriate balance of generality and specificity in region and scale is challenging, as are the technical details associated with using information produced at one scale (i.e., field-scale crop dynamics) in an analysis at a completely different scale (i.e., international trade modeling) (Challinor et al. 2009).

These modeling efforts have increased the sophistication with which the dynamics of the climate-crop-international economy nexus can be represented, but many are now confronting a lack of reliable data in critical areas such as soil types, land use, and hydrological processes worldwide. Furthermore, determining the validity and robustness of results emerging from such efforts has been hampered by the lack of consistent data, model structure, and input assumptions across modeling

efforts (see sidebar below). Technical research needs related to the synthesis of information at multiple scales include improved understanding of the implications of different methods of downscaling and upscaling information in climate change assessments, and more sophisticated model linkages to more accurately reflect the effect of factors that influence adaptation options and behavior at multiple scales.

International Effects and Food Security Implications

Using a linked series of 34 national and regional agricultural economic models, Fischer et al. (2005) estimate that under a range of climate scenarios evaluated in 2080, agricultural gross domestic product (GDP) increases in most developed countries and decreases in most developing countries (with the exception of Latin America). In North America, gains to agricultural GDP range from 3 to 13%, depending on the climate scenario (Fischer et al. 2005), however the effects of climate change are generally projected to be more severe in poor developing countries (Winters et al. 1998; Parry et al. 2004; Mertz et al. 2009; Hertel et al. 2010a). Productivity may be more negatively affected because many developing countries are already at the upper end of their temperature ranges, and precipitation is not expected to increase as it is in many temperate regions (Easterling et al. 2007; Mertz et al. 2009). Overall economic impacts may be more severe because developing countries rely on agriculture for a much greater proportion of their national income and employment than do developed countries (Mertz et al. 2009).

Furthermore, relative capacity for adaptation varies by region, country, sector, and crop, and is therefore

The Agricultural Model Intercomparison and Improvement Project (AgMIP)

AgMIP is advancing the integrated assessment modeling effort by bringing together climate scientists, crop scientists, soil scientists, and social and behavioral scientists from around the world to create common protocols and compare climate projections, crop modeling projections, and production and trade results across research efforts. These comparisons will be used to better understand and isolate sources of variation and uncertainty across analysis tools and scales, as well as to improve the compatibility and availability of the spatially explicit climate, resource, and yield data necessary for such analyses. Through improved tools and data for characterizing world food security implications of climate change, AGMIP hopes to provide substantially improved inputs into international research efforts and decision-making processes about climate change impacts and risks.

itself a factor in determining how the economic impacts of climate change will be distributed across and within agricultural sectors worldwide. Tol (2009) suggests that "low-income countries are typically less able to adapt to climate change both because of a lack of resources and less capable institutions." Such differences in relative adaptive capacity, together with differential climate change effects on yields, may entrench and exacerbate existing production and consumption discrepancies between developed and developing countries (Fischer et al. 2005; Tubiello et al. 2007a; Parry et al. 2005). Even future climate scenarios with mild to inconsequential aggregate global effects on food production may result in severe implications for the food security of the world's poorest and most vulnerable populations.

Concerns about whether future food supply can meet the demands of a growing population have been raised independently of climate change issues, often citing issues related to increasing meat consumption and increasing use of grain for biofuel production (Edgerton 2009; Funk and Brown 2009). Questions about the evolution of agriculture under changing climate conditions, however has added several new levels of risk and uncertainty to those analyses. The food security implications of climate change vary significantly according to the assumptions made about level of development and population growth into the future that underlie emissions trajectories used in the climate scenarios, for instance (Fischer et al. 2005). As with economic impacts, the food security implications of climate change are also significantly different across regions (Funk and Brown 2009; Hertel et al. 2010a; Acevedo 2011).

Climate change in the near term is not expected to significantly affect aggregate global food production (Rosenzweig and Parry 1994; Darwin et al. 1995; Parry et al. 2004). Studies have consistently suggested that climate change is not a significant food security risk for the United States and other developed countries in the near to medium term (Adams et al. 1995; Cline 2007; UNDP 2007). This dynamic is only partly due to the yield assumptions associated with climate change assessments in the United States. Research suggests that production in the United States is much more variable across possible climate projections than is consumption; trade patterns adjust to keep U.S. consumption fairly stable despite effects on production (Sands and Edmonds 2005).

Concerns about food security are more acute for other regions of the world, however. Regional differences in yields and adaptation capacity

are expected to result in regional differences in vulnerability to effects of hunger and poverty, with particularly severe implications for tropical semi-arid developing countries (Parry et al. 2004; Fischer et al. 2005; Parry et al. 2005). Almost 90% of world hunger is concentrated in Asia, the Pacific, and Sub-Saharan Africa (Acevedo 2011). These regions are particularly vulnerable to climate change; by the end of the 21st century, there is a high probability (>90%) that normal growing season temperatures in the tropics and the subtropics will exceed the hottest temperatures on record for those regions from 1900 to 2006 (Battisti and Naylor 2009).

The production and calorie consumption implications for these regions are significant. Fischer et al. (2005) estimate a global increase in undernourishment of 15% by 2080 under a "worst case" high population development scenario (A2). Using a statistical analysis of historical relationships between harvest and temperature/precipitation for major crops in 12 "food-insecure" regions, Lobell et al. (2008) identified south Asia and southern Africa as two regions with a high probability of suffering production losses to crops important to large, food-insecure populations. These results are consistent with several other studies that project negative effects of climate change on productivity and food security in Africa and south Asia (Parry et al. 2005; Schlenker and Lobell 2010; Challinor et al. 2007; IPCC 2007a; Funk and Brown 2009). Hare et al. (2011) review a number of studies projecting significant risks to food security in south Asia, Sahelian and northern Africa, and Russia with mean global temperature increases of 2°C (Hare et al. 2011). Funk and Brown (2009) estimate that interactions between drought exacerbated by climate change and declining agricultural capacity (including the effects of population growth) could increase demand for World Food Program humanitarian assistance by 83% by 2030 in the absence of agricultural development that mitigates effects on yields.

Other climate-related market and political dynamics may further increase the vulnerability of poorer countries. Yu et al. (2011) report that trade policy changes resulting from increasing food prices (such as the export bans or export restrictions observed in 2007-2008) have served to further increase food prices, particularly for the poorer, food-deficit countries/regions, causing them to lower their imports of agricultural and food commodities. Such trade disruptions can increase the risk of food insecurity among vulnerable populations. Other aspects of the food supply system, including the distribution infrastructure, demand factors, and

> Studies have consistently suggested that climate change is not a significant food security risk for the United States and other developed countries in the near to medium term.

other elements related to access and utilization may also be affected by climate change, however little research has been done on elements of food supply and security beyond global food production (Schmidhuber and Tubiello 2007; Jarvis et al. 2011). Like economic impact assessments, food security assessments have also focused on the implications of shifts in mean climate conditions and excluded the effects on production of increased incidence of extreme events like drought and flooding (Shmidhuber and Tubiello 2007).

While developing countries may be particularly vulnerable to climate change effects, substantial gaps between crop yield potential and actual yields ("yield gaps") in those countries may represent an opportunity to offset negative climate change effects through investments that narrow yield gaps on existing croplands (Lobell et al. 2009; Schlenker and Lobell 2010). Furthermore, Tilman et al. (2011) suggest that "strategic intensification" of agriculture that targets yield gaps and elevates yields on existing croplands of under-yielding nations can significantly reduce the potential environmental effects associated with meeting 2050 global crop demands.

Climate Change Effects and the Environment

Meeting food demand in the future will involve multiple strategies, including intensification of production on existing land, expansion of agricultural land, and reduction of waste along the food supply chain (Pfister et al. 2011). Reliance on specific adaptation mechanisms will depend on regional patterns of climate change; however, intensification and expansion of agriculture can have significant environmental implications. A multitude of concerns are linked with climate change, including increased water stress and competition with downstream aquatic systems, increased GHG emissions associated with land clearing, increased pesticide use, increased nutrient loading, and loss of natural systems and the ecosystem services they provide (Malcolm et al. 2012; Reilly et al. 2003; Pfister et al. 2011; Tilman et al. 2011; Antle and Capalbo 2010).

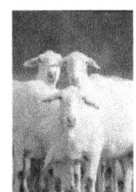

Economic impact studies generally do an inadequate job of addressing such environmental concerns. Antle and Capalbo (2010) concisely articulate the problem: "Due to both data and model limitations, ecosystem services have not been incorporated into integrated assessment studies and cannot be linked to reduced-form statistical studies that do not model

land use changes and other aspects of management decisions." However, some economic impact studies have explicitly linked climate change and adaptation with environmental effects. Malcolm et al. (2012) found that the changes in environmental indicators of erosion and nutrient loss associated with changing agricultural patterns in response to climate change were disproportionately larger than the increase in agricultural acreage experienced; climate changes were therefore projected to result in some combination of an intensification of agriculture (and environmental effects) on existing acreage and an expansion of production onto acres with higher-than-average environmental effects. Antle et al. (2004) also found that shifting production patterns in response to changing climate conditions within a dryland, grain-producing region in Montana had substantial effects on soil carbon stocks.

Potential environmental effects are associated with both intensification of agriculture and expansion of cropland. Identifying and incentivizing the adoption of environmentally friendly management practices that deal effectively with climate-change-related challenges, such as shifting diseases and pests and increased incidence of flooding and other extreme events, will be a critical and challenging element of a sustainable agricultural adaptation strategy for climate change. Environmental effects may also be reduced through adaptation and agronomic advancements that result in increased yields per acre (Burney, 2010; Tilman, 2011). Pfister et al. (2011) suggest that incorporating environmental affects into decisionmaking may fundamentally change agricultural systems by directing crop production toward areas where environmental effects from production are relatively low.

Climate Change, Economic Resilience, and Extreme Weather Events

Economic approaches to climate change impact assessment are just beginning to make use of concepts of system resilience and adaptability that have traditionally been more widely developed in the ecological literature (Antle et al. 2010; Chapter 7 of this volume). A farm's economic resilience to climate change refers to its ability to survive a large climate-related economic shock, such as those associated with sudden reductions in output or increases in input prices (Antle and Capalbo 2010). A farm's capacity for adaptation and its economic resilience are inextricably linked. Both are functions of a farm's access to the natural, physical, human, and financial resources necessary to absorb economic shocks in the short-term, while simultaneously responding to long-term shifts in growing conditions and market prices. Economic impact assessments focusing on the long-term adaptability of the agricultural sector have nevertheless been largely unable to address the challenges and implications associated with short-term resilience of farm production enterprises under a changing climate. There is little information available on the relationship between climate change and the incidence of critical economic thresholds related to profitability and financial sustainability for different farm sectors and types of operations. Climate change analysis has generally focused on the effects of mean changes in climate variables rather than on the effects of variability and extreme events, due in large part to a lack of data on the variability associated with the climate projections derived from general circulation models (GCMs). Nevertheless, as is the case for effects on crop productivity (Tubiello et al. 2007a; Gornall et al. 2010), farm financial vulnerabilty and resilience may be more sensitive to the magnitude and timing of extreme events than to the effects of mean growing season changes under a changing climate. Extreme events may directly affect crops at critical developmental stages, such as flowering, for instance, or may reduce the efficiency of farm inputs by reducing the flexibility of timing of farm operations and applications (Tubiello et al. 2007b; Hatfield et al. 20011). Livestock and dairy production may also be more affected by changes in number of days of extreme heat than by adjustments of average temperature. Catastrophic crop or livestock losses are likely to affect the financial viability of production enterprises in a fundamentally different way than moderate losses over longer periods of time.

Attention is increasingly turning in both biophysical and economic research arenas to likely changes in the timing and variability of climate conditions, with particular attention to the incidence of extreme events such as drought or flooding. Rosenzweig et al. (2002) estimated that, under climate change, losses to corn production in the United States from precipitation extremes would be expected to increase substantially and by 2030 could average $3 billion per year. Such events may represent critical economic thresholds for farming operations and compromise their ability – and the ability of the agriculture sector as a whole – to engage in long-term adaptation.

Extreme Events

Climate change projections suggest a likely increase in regional and seasonal variability of temperature and precipitation. There is a spatial and temporal component to these changes across the United States. Karl et al. (2009) showed that precipitation events would change in frequency and intensity with a projected increase in spring precipitation, particularly in the Northeast and Midwest, and a decline in the U.S. Southwest; summer precipitation is projected to decrease.

Temperature trends will likely be more uniform than those of precipitation; projections generally call for more occurrences of "heat events," or episodes that exceed the expected average temperatures by 3 to 5°C (Karl et al. 2009). In a recent analysis, Munasinghe et al. (2011) showed that the frequency of high temperature extremes increased 10-fold in the first three decades of the 20th century (1900-1929), and in the last decade (1999-2008). The change in frequency of high temperatures was greater in the tropics than in the higher latitudes, and the frequency of extremes was greater in the daily minimum temperatures compared to the daily maximum temperatures. Increases in temperature are also often associated with lack of precipitation, potentially leading to more drought occurrences.

Some evidence exists that the United States is already experiencing an increased incidence of extreme weather events. A compilation of the economic impact of extreme events with an economic impact in excess of $1 billion shows an increase in this extent of economic damage over the last 30 years, as shown in Table 6.2 (NOAA NCDC 2011). The regions affected by extreme events vary across the years both in economic impact and spatial extent. Across the United States from 1980 through 2011 there has been an increase in the number of events with significant economic impact. An increased occurrence of extreme events associated with climate change across the United States will likely lead to an increased incidence of weather events with significant economic impact.

Patterns already evident in crop insurance payments, workable field days, and soil erosion provide a glimpse into the implications for agriculture of an increased incidence of extreme events. The following analyses focus on Iowa as a case study but are typical of the upper Midwest in terms of the expected outcomes. Because of regional heterogeneity in the expected effects of climate change, the implications of climate change may be quite different for other regions within the United States.

....farm financial vulnerabilty and resilience may be more sensitive to the magnitude and timing of extreme events than to the effects of mean growing season changes under a changing climate.

Crop Insurance

Extreme events trigger claims for crop insurance and claims in the United States have been made in response to drought (40%), excess moisture (25%), hail (5-10%), hurricane (5%), excess heat (<5%), and other causes (20%) (OECD-INEA-FAO 2010). An evaluation of the changes in Iowa crop insurance indemnities show that indemnities in the first decade of the 2000s far exceed those occurring during 1971-1999; 2001-2010 indemnities are 3.5 times those

for the period 1971-2000 (Figure 6.1). Indemnities are paid out by crop insurance companies to farmers for losses occurring, for example, due to drought, flooding, or crop price declines. Several factors may have contributed to this trend, including changing frequency and intensity of weather extremes, increased acreage enrollment in crop insurance programs, and an accompanying reduction in reliance on ad hoc disaster payments. While future research is required to clarify the relative importance of these factors, insight into changing crop insurance

Table 6.2. Extreme event, location, and economic impact for the United States. Source: NCDC, 2011.

Year	Event	Location	Sector	Economic Impact (2011 $)
2011	Upper Missouri River Flooding	Upper Midwest (MT,ND,SD, IA, KS, MO)	Agriculture	2.0 Billion
2011	Mississippi River Flooding	Lower Mississippi River (AR, TN, LA, MS, MO)	Agriculture	1.9 Billion (3-4 Billion total)
2011	Heat/Drought	Southern Plains/Southwest	Agriculture	10.0 Billion
2009	Drought	Southwest/Great Plains (TX, OK, KS, AZ, NM, CA)	Agriculture	5.3 Billion
2008	Drought	South and West (CA, TX, GA, TN, NC, SC)	Agriculture	2.0 Billion
2008	Flooding	Upper Midwest (IA, IL, IN, MO, MN, NE, WI)	Agriculture	15.8 Billion Total
2007	Drought	Great Plains and Eastern U.S.	Agriculture	5.5 Billion
2007	Freeze	East and Midwest U.S.	Agriculture	2.2 Billion
2007	Freeze	California	Agriculture	1.5 Billion
2006	Drought	Central U.S.	Agriculture	6.7 Billion
2005	Drought	Central US (AR,IL,IN, MO, OH,WI)	Agriculture	1.2 Billion
2003	Storms and Hail	Southern Plains and lower MS valley	Agriculture	1.6 Billion
2002	Drought	30 states, western, Great Plains, and eastern U.S.	Agriculture	12.5 Billion
2000	Heat/Drought	South-Central and Southeastern U.S.	Agriculture	5.2 Billion
1999	Heat/Drought	Eastern U.S.	Agriculture	1.4 Billion
1998	Freeze	California	Agriculture	3.5 Billion
1998	Heat/Drought	TX/OK to the Carolinas	Agriculture	8.3-12.4 Billion
1995-1996	Drought	TX/OK	Agriculture	7.2 Billion
1993	Flooding	Upper Midwest	All Sectors	32.8 Billion
1993	Drought	Southeastern U.S.	Agriculture	1.6 Billion
1990	Freeze	California	Agriculture	5.9 Billion
1989	Drought	Upper Great Plains (ND,SD)	Agriculture	1.4 Billion
1988	Drought	Central and Eastern U.S.	Agriculture	76.4 Billion
1986	Heat/Drought	Southeastern U.S.	Agriculture	2.1-3.1 Billion
1985	Freeze	Florida	Agriculture	2.5 Billion
1983	Freeze	Florida	Agriculture	4.5 Billion
1980	Heat/Drought	Central to Eastern U.S.	Agriculture	54.8 Billion

programs is gained by examining the relative fraction of indemnities from different causes.

Crop indemnities and their causes are examined for Iowa for 1971-2010. A shift has occurred in primary causes of climate and weather extremes. In the 1970s and 1980s, indemnities for hail and drought accounted for 70-80% of total indemnity (Figures 6.2 and 6.3). Drought remains an important factor, but is exceeded by indemnity claims for excessive moisture and flood (Figures 6.4 and 6.5). The reduction in the fraction of indemnity from hail claims likely is an indicator of shifts in crop insurance policy and enrollment in crop insurance programs. Since little evidence exists of a decrease in hail frequency, the implication is that increased penetration of drought and excessive wetness/flood has increased exposure of the crop insurance industry to rain and drought events. If we assume farm bill changes to insurance coverage have not disproportionately increased the likelihood of payouts (or insurance payments) due to either drought or excessive wetness/flood, then the relative fractions of indemnity from excessive wetness/flood and drought may be an indicator of changes in frequency of extreme wet and dry periods.

The relationship between annual precipitation and fraction of indemnity paid out for drought, flood, and excessive wetness is examined in Figure 6.6. For precipitation levels below the annual average rainfall for Iowa (33.64 inches for the 1955-2010 period, which is the period of the indemnity database) there is generally a relatively high fraction of indemnity going to drought, with low fractions paid out due to flood and excess precipitation events. For precipitation levels above 1.8 – 1.3 inches/year, the fraction of indemnities paid out to drought and excess flood events increases markedly.

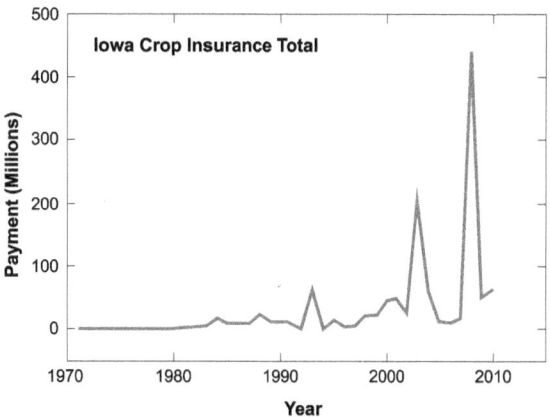

Fig. 6.1. Crop insurance indemnity (unadjusted dollars) in Iowa 1971–2010. Source: USDA-RMA.

Since the 1980s, many of the indemnity claims occur during individual years that can be described by a single climate extreme. During the 1980s, drought in the extreme years of 1988 and 1989 accounts for 30% and 19%, respectively, of the decade's total indemnity claims. In the 1990s, the 1993 flood accounts for 24% of the decade total. In the 2000s, the drought of 2003 and flood of 2008 account for 21% and 45%, respectively, of the decade total. However, in 2011 (which ranks third in crop indemnities for the years between 2001 and 2011, behind 2003 drought and 2008 flood), crop indemnity in Iowa includes claims for both excessive wetness in the spring and drought during the summer. Climate model projections for the mid-21st century project a continuing trend of wet springs, while also maintaining the potential for summer drought. Other hybrid years occurred prior to 2011 (Table 6.3), but the total indemnity in 2011 represents from 2 times to 24 times the indemnity of the 8 other hybrid years since 1971.

Table 6.3. For Iowa, Hybrid Drought: Excessive Wetness Years since 1971, Total Indemnity, and Ratio to 2011. Source: U.S. Department of Agriculture, Risk Management Agency. "Cause of Loss Historical Data Files." Accessed from http://www.rma.usda.gov/data/cause.html on 3/19/12.

Io⬚a ⬚⬚⬚ri⬚ ⬚ro⬚⬚⬚t In⬚e⬚			
⬚cc⬚rrence o⬚E⬚ce⬚⬚ive ⬚ etne⬚⬚⬚⬚ Year	Fraction of total indemnity for ⬚ro⬚⬚⬚t⬚ etne⬚⬚	Total Indemnity (⬚S$)	Ratio to 2011
1984	48⬚ ⬚25⬚	18003685	5.0
1985	40⬚ ⬚23⬚	12053131	15.6
1991	26⬚ ⬚46⬚	12053131	7.4
1992	28⬚ ⬚20⬚	12053131	7.4
1997	32⬚ ⬚30⬚	4881255	18.4
2001	31⬚ ⬚43⬚	48888822	1.8
2005	42⬚ ⬚16⬚	12803619	7.0
2007	20⬚ ⬚24⬚	16234329	5.5
2011	22⬚ ⬚24⬚	89782894	1.0

Figs. 6.2, 6.3, 6.4, and 6.5. Fraction of total indemnity by cause for 1971–1980, 1981–1990, 1991–2000, and 2001–2010. Source: U.S. Department of Agriculture, Risk Management Agency. "Cause of Loss Historical Data Files." Accessed from http://www.rma.usda.gov/data/cause.html on 3/19/12.

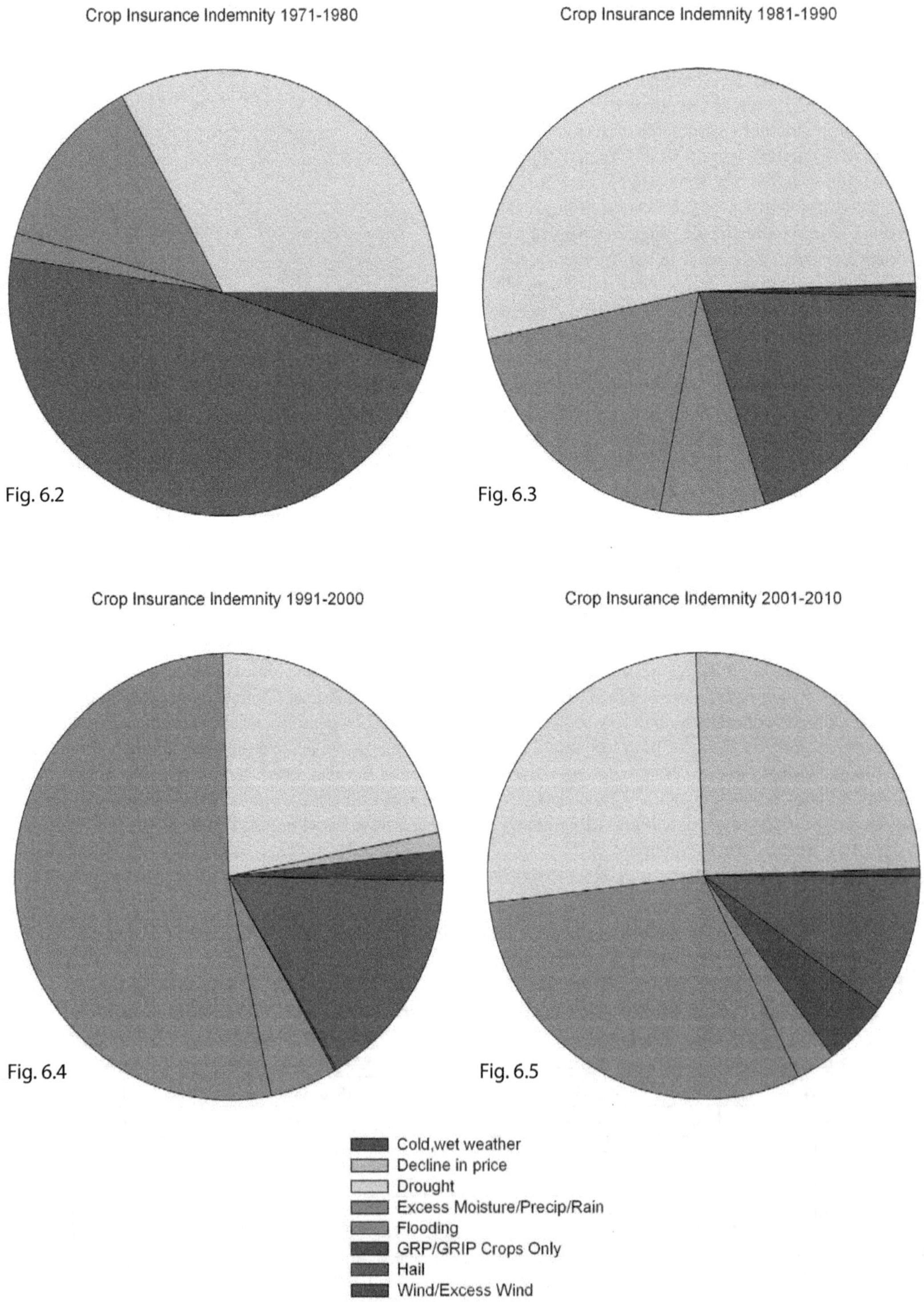

Crop Insurance Indemnity 1971-1980

Crop Insurance Indemnity 1981-1990

Fig. 6.2

Fig. 6.3

Crop Insurance Indemnity 1991-2000

Crop Insurance Indemnity 2001-2010

Fig. 6.4

Fig. 6.5

- Cold,wet weather
- Decline in price
- Drought
- Excess Moisture/Precip/Rain
- Flooding
- GRP/GRIP Crops Only
- Hail
- Wind/Excess Wind

Fig. 6.6. In Iowa during 1955–2010, fraction of total annual indemnity from drought, flood, and excessive wetness versus the annual rainfall. Source: U.S. Department of Agriculture, Risk Management Agency. "Cause of Loss Historical Data Files." Accessed from http://www.rma.usda.gov/data/cause.html on 3/19/12.

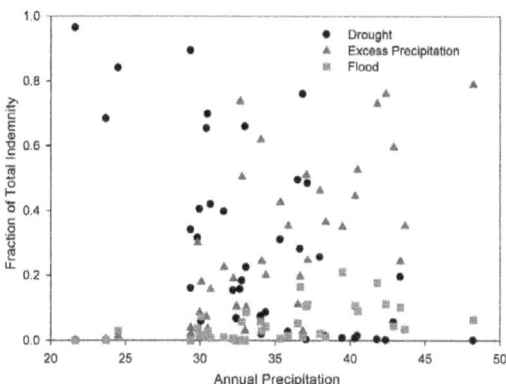

Projections of rainfall change are placed in context of changes since 1873 to provide an indication of how indemnities may be affected. Figure 6.7 shows the annual Iowa rainfall and 30-year average rainfall (computed in accordance with World Meteorological Organization (WMO) standards). The most recent 30-year periods (1981-2010, 1971-2000, 1961-1990) show an upward trend of 30-year average precipitation, increasing by 9% to 35.1 inches from 1981 to 2010, up from the 1951 to 1980 average of 32.1 inches. Prior to 1951-1990, fluctuations in the 30-year average were much smaller, ranging from 30.7 to 32.0 inches. Projections of change in 30-year average annual precipitation for Iowa show a range of percent change from -10% to +20%, with a large

fraction of projections experiencing between a 5% and 15% increase. Projections at the low end return Iowa's annual precipitation back to that experienced prior to the 1961-1990 period. Projections at the high end continue the recent trend in precipitation increase with roughly the same rate of change.

Since warmer temperature occurs in all projections, the lower end projections might result in a higher indemnity fraction from droughts while the high end projections could create mixed conditions of excessive wet, excessive dry, and hybrid years. If it is assumed that the standard deviation of annual precipitation in 30-year periods is not greatly changed in the future (historically it has ranged from 3.6 to 5.9 inches, but all but two 30-year periods fall within a range of 4.8 to 5.9 inches), the projection of a 5% to 15% increase in 30-year average precipitation suggests that excessive wetness/flood will become the dominant cause of indemnities, though adaptation mechanisms such as innovative drainage systems may help buffer this effect. In fact, a 15% increase of 30-year average precipitation is just over half the current standard deviation of annual precipitation. This implies a much increased frequency of having at least 30% indemnity from excessive wetness and flood, based upon the historical record (Figure 6.8).

Workable Field Days

Workable days in the field are a critical component of most field-crop operations. Producers require days in which field work can be conducted without causing problems that would negatively crop production. The

Fig. 6.7. Annual (solid line with black squares) and 30-year average precipitation (yellow squares) for Iowa 1873–2011. The 30-year average is computed based on the WMO definition of climate normal. Precipitation in 2011 is through October. Source: U.S. Department of Agriculture, Risk Management Agency. "Cause of Loss Historical Data Files." Accessed from http://www.rma.usda.gov/data/cause.html on 3/19/12.

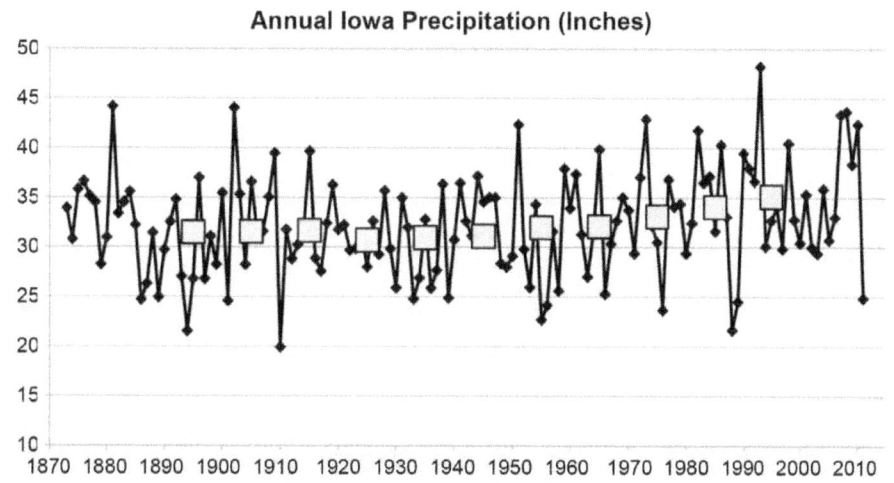

recent trend toward wetter springs has resulted in fewer workable field days during the planting season. The 50th percentile for 1976-2010 was 22.4 days. Prior to 1995 (the mid-point of the data record), the 50th percentile was exceeded 10 times; since 1995, it has been matched or exceeded 6 times.

Between 1976 and 1994, the average number of workable field days between April 2 and May 13 was 22.65 days. This declined to 19.12 days for the 1995-2010 period. The number of workable field days for April through mid-May is negatively correlated with April through May rainfall (-0.716) (Figure 6.9). The 1873-2010 Iowa statewide April-May rainfall has a mean of 7.9 inches, median of 7.04 inches, 75th percentile of 8.8 inches, and 95th percentile of 10.3 inches. When April-May rainfall is below the median, field work days are rarely less than 20; whereas above the 75th percentile they are rarely more than 20. One adaptation to reduced workable field days is to increase equipment size to allow more area to be managed in a shorter period of time; another adaptation would be to diversify crops to spread the operation times over a larger portion of the growing season, and, finally, selection of varieties with different maturity lengths would allow for a longer period for planting.

Estimates of workable field days should be calculated for all agricultural regions for the critical fall and spring field seasons to help producers understand the

effects of a changing climate on their decisionmaking process. Excessive precipitation during the fall can disrupt harvest operations and negatively affect product quality. Examples of untimely events affecting product quality have been reported for raisin harvest, hay harvest, and grain harvest, where excessive soil water causes delays in harvest and soil damage from equipment passing over saturated fields. Excessive moisture during harvest can also lead to disease outbreaks, which affect hay, grain, vegetable, or fruit quality.

Soil Erosion

Soil erosion rates are useful for illustrating the potential effects of extreme events on agriculture. Soil erosion is the result of inadequate infiltration rates and excessive rainfall that exceeds the soil's capacity to absorb water (see Chapter 4 of this report). Iowa State University (ISU) has conducted an extensive effort to provide estimates of statewide soil erosion using a soil-erosion model and the National Weather Service's 15-minute radar estimates of rainfall.

A spring 2010 aerial survey conducted by the Environmental Working Group (EWG) indicated that soil erosion and runoff (Cox et al. 2011) are likely far worse than the ISU predictions suggest, however. Part of this underestimate is the inability of current models to account for the effect of widespread

Fig. 6.8. Projected mid-21st Century change (2040–2069 minus 1970–1999) of 30-year average annual temperature and precipitation. Source: "Bias Corrected and Downscaled WCRP CMIP3 Climate Projections" archive at http://gdo-dcp.ucllnl.org/downscaled_cmip3_projections/. Maurer, E. P., L. Brekke, T. Pruitt, and P. B. Duffy (2007), 'Fine-resolution climate projections enhance regional climate change impact studies', Eos Trans. AGU, 88(47), 504.

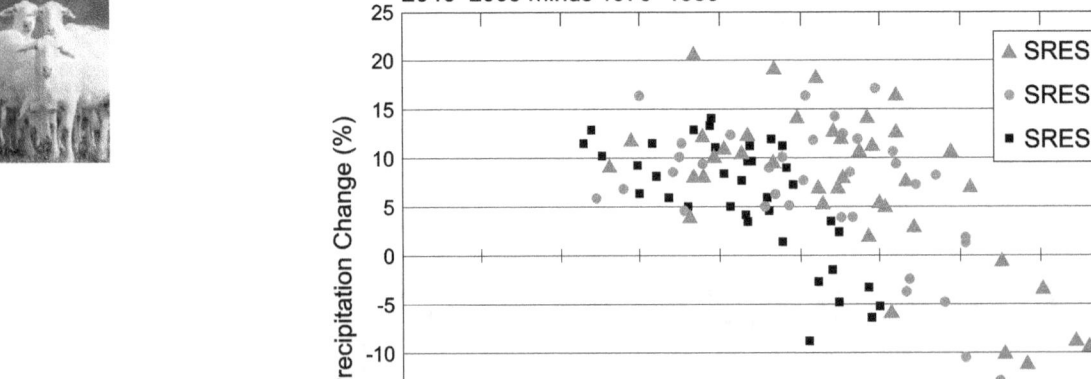

"ephemeral gullies." During heavy rains, these gullies reappear rapidly where farmers have tilled and planted over natural depressions in the land, forming "pipelines" that swiftly carry away the water the soil cannot absorb.

The ISU data and EWG's survey reinforce long-standing doubts about the current system used to determine sustainable levels of erosion for working cropland; sustainable here is defined as how much soil loss the land can tolerate before it loses its ability to support a healthy crop. There is substantial and growing evidence that these tolerable soil losses ("T values") greatly overstate the ability of cropland to remain fertile when experiencing soil erosion and water runoff. These concerns may become particularly relevant at a time when a warming climate is producing ever more frequent severe storms that can quickly exceed soil's ability to absorb water and can produce high levels of erosion over very short periods of time.

Over 3 days in 2007 (May 5-7), such a storm pummeled large portions of southwest Iowa. Data available from USDA's 2007 National Resources Inventory (USDA 2009) calculate the average erosion rates in Iowa at 5.2 tons per acre per year, only slightly higher than the "sustainable" T-value of 5 tons per acre per year for most Iowa soils (Figure 6.10). However, according to results from the Iowa daily erosion project (Iowa State University), average erosion exceeded sustainable rates in 198 townships

Fig. 6.9. Iowa average field work days during April through mid May versus April through May rainfall. Light blue line is 50% April-May rainfall (7.04 inches); dark blue is 75% (8.8 inches); black is 95% (10.3 inches). Source: Monthly Iowa Precipitation Data is from Iowa State Climatologists Office (http://www.iowaagriculture.gov/climatology.asp), who is responsible for quality control and quality assurance of the 33-station long-term climate reference network. After QC/QA, the data are submitted to the NOAA data archive at NCDC. U.S. Department of Agriculture, National Agricultural Statistics Service. "Field Work Days Data Files." Accessed from http://www.rma.usda.gov/data/cause.html on 3/19/12.

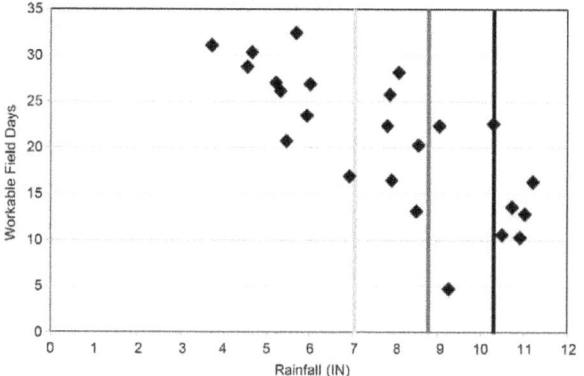

(4.6 million acres). On May 6, the worst day of the storm, 182 townships encompassing 4.2 million acres suffered erosion exceeding the sustainable rate for an entire year. In 69 townships (1.6 million acres), soil eroded at twice the sustainable rate, an average of 10 tons per acre. In 14 townships (323,000 acres), the rate was more than 20 tons per acre.

Fig. 6.10. Estimates of soil displacement per unit of daily rainfall obtained from Iowa data. Data Source: The data collected for this summary plot are archived under the Iowa Daily Erosion Project. http://wepp.mesonet.agron.iastate.edu/ (Verified Apr 2, 2012).

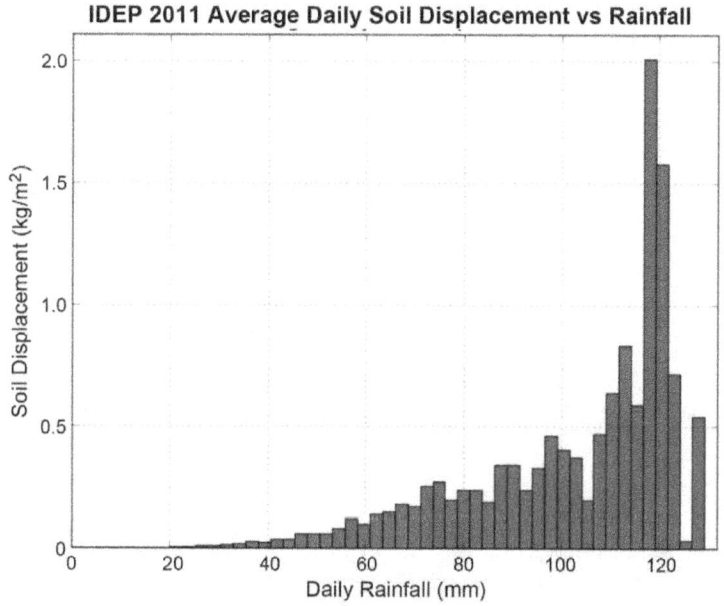

Observations of soil erosion and daily rainfall rates show a rapid increase in soil displacement as the daily rainfall exceeds 100 mm. These daily totals are not uncommon during extreme events for the upper Midwest. More intense rainfall events are likely to cause more erosion events unless improved conservation practices (e.g., residue cover, reduced tillage, installed waterway conduits) are adopted to reduce rain energy, protect soil, and reduce runoff.

Conclusions

There remains a high degree of uncertainty in estimating both the biophysical and the economic impacts of climate change. That uncertainty is due to limitations of data; inherent uncertainty in future projections of emissions, available technology, and socio-economic conditions within the agricultural sector, and more broadly; and due to limitations in the availability of estimation methods that capture system-wide interactions among sectors in climate effects and opportunities for adaptation. While several economic impact assessments have suggested that climate change may not substantially affect domestic producers and consumers in the short-term, such results are highly sensitive to the future climate scenarios selected for analysis and to boundaries placed on the scope of climate effects considered. Estimates of aggregate economic impacts of climate changes often mask considerable variability across demographics and regions, both within the U.S. and worldwide. Even in the short-term, climate change will likely increase the incidence of global hunger through effects on the world's poorest and most at-risk populations.

Chapter 7

Adapting to Climate Change

Climate change presents unprecedented challenges to the adaptive capacity of the U.S. agricultural sector. U. S. producers are adapting crop and livestock management practices in an effort to reduce new production risks associated with the increased weather variability accompanying climate change. Current climate change effects are challenging agricultural management and are likely to require major adjustments in production practices over the next 30 years. Projected changes over the next century have the potential to transform U.S. agriculture, particularly for production systems at their marginal climate ranges. Effective adaptive action across the multiple dimensions of the U.S. agricultural system offers the potential to capitalize on the opportunities presented by climate change and minimize the costs by reducing the severity or avoiding negative effects of a changing climate.

Understanding Agricultural Vulnerability

Vulnerability and adaptive capacity are characteristics of human and natural systems, are dynamic and multi-dimensional, and are influenced by complex interactions among social, economic, and environmental factors (Adger et al. 2007). The vulnerability of a system is a function of the exposure and the sensitivity of the system to hazardous conditions mediated by the ability of the system to cope, adapt, or recover from the effects of those conditions, i.e., the adaptive capacity or resilience of the system (Smit and Wandel 2006) (Figure 7.1). Because agricultural systems are human-dominated ecosystems, the vulnerability of agriculture to climatic change is strongly dependent not just on the biophysical effects of climate change but also on the responses taken by humans to moderate those effects (Marshall 2010). Adaptive decisions are shaped by the operating context within which decisionmaking occurs (for example, existing natural resource quality and non-climate stressors, and government policy and programs), access to effective adaptation options, and the individual capability to take adaptive action.

The concepts of vulnerability, adaptation, adaptive capacity, and resilience are well developed in the global change literature (synthesized in Smit and Wandel 2006; Adger et al. 2007; Nelson et al. 2007) (Table 7.1); however, the methodological development to apply these concepts to adaptation planning and assessment lags behind, particularly in developed countries (Moser et al. 2008; Kenny 2011). Efforts to

Fig 7.1. Linked human and biophysical factors that determine the ultimate vulnerability of agricultural systems to climate change (Marshall et al. 2010).

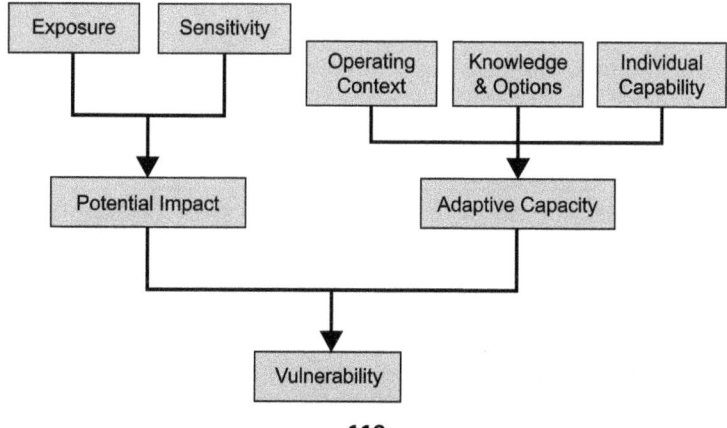

identify key factors that contribute to system vulnerability to climate change effects, that address issues of uncertainty, scale, and multidimensional system interactions, and that develop effective integrated indices of vulnerability or adaptive capacity typify methodological research (e.g., Adger and Vincent 2005; Brooks et al. 2005; Alberini et al. 2006; Eakin and Bojórquez-Tapia 2008) and participatory research methods are increasingly employed (e.g., Petheram et al. 2010; Krishnamurthy et al. 2011). Agricultural researchers have contributed to this literature, though the primary focus of this work has been on small-holder agriculture in the developing world (e.g., Vincent 2007; Simões et al. 2010; Below et al. 2012).

Adaptation Drivers

Agricultural productivity is determined by a diverse set of biophysical, social, economic, and technological drivers operating across multiple dimensions of time and space. These drivers create opportunity and present risk to agricultural production. In particular, agriculture is highly sensitive to weather effects with climate variations, soil type, biotic stressors, and management being the factors linked to production variability across many regions (Howden et al. 2007; Hatfield et al. 2011; Lal 2011). As climate change intensifies, Howden et al. (2007) suggests that "climate risk" is likely to be added to the production, finance, and marketing risks already commonly managed by producers (Harwood

Table 7.1. Definitions of adaptation concepts.

Concept	Definition
Adaptation	Adjustment in natural or human systems in response to actual or expected climatic stimuli or their effects, which moderates harm or exploits beneficial opportunities. Various types of adaptation can be distinguished, including anticipatory and reactive adaptation, private and public adaptation, and autonomous and planned adaptation (Parry et al 2007, p 869).
Adaptive capacity	The ability of a system to adjust to climate change including climate variability and extremes) to moderate potential damages, to take advantage of opportunities, or to cope with the consequences (Parry et al 2007, p 869).
Coping capacity	The ability of a system to deal with the impacts of present-day weather extremes or climate variability (Luers and Moser 2006).
Exposure	The nature and degree to which a system is exposed to significant climatic variations (Parry et al 2007, p 987).
Maladaptation	Any changes in natural or human systems that inadvertently increase vulnerability to climatic stimuli; an adaptation that does not succeed in reducing vulnerability but increases it instead (McCarthy et al 2001, pg 990).
Mitigation	An anthropogenic intervention to reduce the sources or enhance the sinks of greenhouse gases. (Parry et al 2007, p 878)
Vulnerability	The degree to which a system is susceptible to, or unable to cope with, adverse effects of climate change, including climate variability and extremes. Vulnerability is a function of the character, magnitude, and rate of climate variation to which a system is exposed, its sensitivity, and its adaptive capacity (Parry et al 2007, p 883).
Resilience	Resilience is the capacity of a system to absorb disturbance and reorganize while undergoing change so as to still retain essentially the same function, structure, identity, and feedbacks. (Walker et al 2004).
Sensitivity	Sensitivity is the degree to which a system is affected, either adversely or beneficially, by climate-related stimuli. The effect may be direct (e.g., a change in crop yield in response to a change in the mean, range, or variability of temperature) or indirect (e.g., damages caused by an increase in the frequency of coastal flooding due to *sea level rise*) (Parry et al 2007, p 881).
Social-Ecological systems	A set of critical resources (natural, socioeconomic, and cultural) whose flow and use is regulated by a combination of ecological and social systems (Holling and Gunderson 2002).
Vulnerability	The degree to which a system is susceptible to, or unable to cope with, adverse effects of climate change, including climate variability and extremes. Vulnerability is a function of the character, magnitude, and rate of climate variation to which a system is exposed, its sensitivity, and its adaptive capacity (Parry et al 2007, p 883).

et al. 1999). Climate risk will add complexity and increase uncertainty in agricultural decision environments throughout the multiple dimensions of the U.S. agricultural system.

Key drivers that shape adaptive responses to climate change at the agricultural enterprise scale include: experience of indirect and direct effects of climate change (Field et al. 2007; Spence et al. 2011; Knutson et al. 2011), market signals (Antle 2009), current and proposed climate change policies (Batie 2009), institutional strategies (Preston et al. 2011), farmer perceptions and preferences (Blackstock et al. 2010; Nelson et al. 2010a; Arbuckle 2011; Weber and Stern 2011), issues awareness (e.g., food security) (Godfray et al. 2010), and information sources and types and how they are interpreted (Malka et al. 2009; Blackstock et al. 2010; Tarnoczi and Berkes

2010). University and industry research priorities that focus on science agendas and technologies are also drivers of adaptation investments because they influence the options available to those considering or undertaking adaptation (McDaniels et al. 1997; Cabrera et al. 2008). Taken together, these drivers inform both short- and long-term cost/benefit considerations by producers seeking to maintain profitability in the face of climate variability and change (Antle and Capalbo 2010).

A Typology of Adaptation

Adaptation strategies can be categorized according to a variety of central attributes: by temporal or spatial scope; by intentionality (e.g., spontaneous versus deliberate strategies); by specific aim (e.g.,

Table 7.2. **A Typology of Climate Change Adaptation Strategies for Agriculture.** *This table presents examples of climate change adaptation strategies to key biophysical and social drivers of adaptation (Iglesias, et al. 2007, Smit and Skinner, 2002). The adaptation strategies are grouped according to the actors involved and the form the adaptation takes. The first three categories mainly involve enterprise-scale decision-making by producers. The last two are typically the responsibility of public agencies and agribusiness. Adaptations included in these categories could be thought of as system-wide.*

Key Adaptation Drivers	Adaptation Strategies				
	Farm Production Practices	Farm Financial Management	Farm Infrastructure	Technological Developments	Government Programs and Insurance
Increased variability in growing conditions (changes in seasonal temperature and precipitation patterns)	Change crop variety and breed, change timing of farm operations, use season extension and irrigation, Build soil health	Purchase crop insurance, invest in crop shares/future, participate in income stabilization programs, diversify household income	Install water management (eg, catchment, swales), irrigation systems, weather protection systems, data collection/analysis systems	Drought/cold/heat tolerant crop varieties, Efficient irrigation, Weather and climate information systems and decision-support tools, Farm-level resource mgt practices to improve resilience	Modify gov. insurance, subsidy, support and incentive programs to influence farm-level risk management strategies, provide technical support for risk mgt. Modify land and water resource management policies and programs to improve resilience to climate change
Increased soil degradation (increased erosion reduces soil quality)	Soil conservation practices (eg, no-till, mulch), Build soil health	Participate in soil conservation cost share and easement programs	Install soil conservation structures (eg, terraces, grassed waterways, riparian areas)	Farm-level soil conservation practices, Soil building amendments (eg. biochar, stabilizing agents)	Modify land and water resource management policies and programs to promote soil conservation and soil health mgt.
Increased pest pressure, novel pests	IPM practices, Resistant crop varieties and breeds, Farmscaping	Participate in insurance programs	Purchase improved application technologies, Pest protection structures	Pest resistant crop varieties, IPM options and early warning information systems, Decision-support tools, Pest suppression technologies	Insurance programs, Risk analysis, IPM and weather-based decision-making, Technical advice
Increased number, length and/or intensity of drought events	Resistant varieties/breeds, adjust crop/livestock development, Build soil health	Participate in insurance programs	Install water management systems (eg, catchment, swales), Install irrigation systems	Drought resistant crop varieties and breeds, Alternative crops/livestock, Efficient irrigation, Farm-level water management decision-support tools	Insurance programs, Weather-based decision-making, Farm-level and regional contingency planning and water use priority planning, Technical advice
Increased number and/or intensity of flood events	Avoid high risk locations/time periods	Participate in insurance programs	Increase drainage capacity, Build defense structures, Restore/create wetlands, Floodplain mgt. plan	Flood tolerant varieties, Excess water management technologies	Insurance programs, Weather-based decision-making, Farm-level and regional contingency planning, Technical advice
Shift in optimum zones for current production systems	Change in crop/livestock systems	Participate in insurance programs	Adapt existing infrastructure to new crop/livestock systems	New climate control technologies, Adapt existing equipment to new crop/livestock systems	Create transition insurance and cost-share programs, Develop technical advice for transitioning to alternative resilient farming systems
Government climate change policy	Use GHG emissions reduction practices	Participate in financial incentives programs	Install GHG reduction measures	GHG Monitoring/reduction and decision-tools	Agricultural GHG management policies and programs
Economic (eg. carbon markets)	Adjust crop/livestock mix appropriate to new market	Participate in new market	Alter tillage and water management regimes and storage and use of livestock waste, invest in necessary equipment, re-train staff.	Develop capabilities to manage GHG emissions.	Develop and provide advice and guidance on BMP
Consumer behavior (eg. diet change)	Adjust crop/livestock mix to meet demand	Participate in new market	Develop flexibility to respond to changes in consumer behavior.	Utilisation of web resources to stay informed and make informed decisions.	Provision of information and advice on trends, preferences and market conditions.
Perception of climate risk	Short-term vs. long-term adjustments	Participate in insurance programs	Develop flexibility to enable rapid responses.	Utilisation of most appropriate sources of information for decision making.	Seasonal and decadal forecasts with associated probabilities of error.

to modify effect versus reduce vulnerability); by sector (agriculture, tourism, public health); by specific process or outcome (e.g., to increase drought resistance versus maintain profit); by actor (individuals, local community, private sector, or government); by type of action (physical, technological, behavioral, regulatory or market); or by some combination of these and other attributes (Smit et al. 1999; Smit et al. 2000; Adger et al. 2007). Based on a comprehensive analysis of Canadian agriculture, Smit and Skinner (2002) recommend organizing adaptation options by actor (producers, agro-industry, and government) and the type of action (farm management, infrastructure, technology, and government programs).

Organizing agricultural adaptation options by actor and type of action facilitates adaptation planning and assessment because it identifies stakeholder agency and clarifies potential cross-scale interactions that may influence adaptive capacity. For example, adaptations to increased variability in growing conditions (the adaptation driver) can be made by producers (change crop cultivar), agribusiness (develop new crop cultivars), and government (provide climate risk insurance, cost-share installation of conservation practices that increase resilience to climatic variability). This typology also

facilitates identification of the range of potential adaptive actions available to any particular actor and adaptive actions that address multiple drivers, such as managing for high quality soils. Table 7.3 presents general examples of adaptation options for key adaptation drivers acting on the U.S. agricultural system organized, according to Smit and Skinner's (2002) typology. This table can be viewed as a menu of potential adaptation responses to specific drivers by producers, agribusiness, and government actors operating in the U.S. agricultural system.

Recently, a new typology has emerged to classify adaptation options consistent with the systems perspective (e.g., Nelson et al. 2007; Millar et al. 2007; Easterling 2009; van Apeldoorn et al. 2011; Pelling 2011). This typology classifies adaptation options along a spectrum of intention and action – resistance, resilience and transformation – that describe successively greater change in the adaptive capacity of the agricultural system.

Millar et al. (2007) explain the differences in management intention along the resistance-resilience-transformation spectrum in the management of forests under climate change. Resistance strategies seek to maintain the status quo over the near term through management actions that

Table 7.3. The components of agricultural adaptation (Bryant et al 2000).

Component	Elements or Examples
Characteristic Stresses	Climatic change and variability
	Government policies
	Consumer pressure
	Economic conditions
	Non-climatic Environmental Factors
Multiple Dimensions of the Agricultural System	Cultural
	Economic
	Institutional
	Political
	Social
	Technological
Scales of System Vulnerabilities and Responses	International
	National
	Agricultural Sector
	Region
	Community, Locality
	Farm, Field, Plant
Responses	Producer: crop choice, diversification, irrigation, crop insurance
	Public and Institutional: information, research and development, infrastructure, taxes and subsidies

resist climate change disturbance. These strategies are typically reactive, site-specific interventions that do not enhance the adaptive capacity of the ecosystem, but operate to defend the existing ecosystem from climate change effects through more intensive management intervention. Resistance strategies can be costly, will likely increase in cost and difficulty over time, and may ultimately fail as climate change effects intensify. Resilience strategies are typically proactive actions that increase the adaptive capacity of the ecosystem by improving its ability to self-organize so as to moderate effects of climate-related disturbances and return to a healthy condition after a disturbance, either naturally or with minimal management intervention. Transformation strategies increase adaptive capacity by facilitating transition of the existing ecosystem to a new ecosystem with a different structure and function that are better suited to sustained production under rapidly changing climate conditions.

The resistance-resilience-transformation framework that is applied to climate risk management in U.S. national forests (USDA Forest Service 2010; Spies et al. 2010) defines adaption options in the National Fish, Wildlife, and Plants Climate Adaptation Strategy (NFWP Climate Adaptation Strategy 2012) and is recommended for use in an ecosystem-based approach to agricultural adaptation (Easterling 2009). Research and development efforts to better understand and manage agricultural ecosystem resilience and stability in the face of climate change are explicit in the strategic goals of climate change adaptation programs administered by the USDA Global Change Program, the Agricultural Research Service, and the Natural Resources Conservation Service (GAO 2009).

Resistance strategies in use today by U.S. farmers coping with current changes in weather variability may include changes in management practices such as adjustments in cultivar selection and the timing of field operations, increased use of pesticides to control higher pest pressures, and the purchase of crop insurance. Adjustments in management practices include the use of multiple cultivars within monocultures (Newton et al. 2011) and diversifying crop rotations (Lin 2011) to manage pest populations, integrating livestock with crop production systems to manage resource cycles (Tomich et al. 2011), building soil quality to manage water cycles, and other practices typically associated with sustainable agriculture. Such actions may increase the capacity of the agricultural system to self-organize in response to climate change effects and avoid loss of productivity with minimal reactive management intervention (Wall and Smit 2005; Easterling 2009; Lin 2011;

Tomich et al. 2011). Transformation adaptations are those that might include the northward migration of existing production systems and the shift of cultivated row crops into forest, perennial grasslands, or wetlands. Based on projected climate change effects, over the next century it is likely that agricultural systems in some areas of the United States will have to undergo a transformation to remain productive and profitable (Easterling, 2009).

Enhancing the Adaptive Capacity of Agriculture

The U.S. agricultural system has demonstrated a remarkable adaptive capacity over the last 150 years as crop and livestock production systems spread across the diverse American landscape and successfully responded to variations in climate and other natural resources, as well as to dynamic changes in agricultural knowledge, technology, markets, and, most recently, public demands for the sustainable production of agricultural products (Reilly and Blanc 2009; NRC 2010). This adaptive capacity has been driven in large part by public sector investment in agricultural research, development, and extension activities (Antle 2009) made during a period of climatic stability and abundant technical, financial, and natural resource availability.

Agricultural Sustainability

Sustainability has been described as the ability to meet core societal needs in a way that can be maintained indefinitely without significant negative effects. Authors of the National Academies of Science (NAS 2010) study on sustainable agricultural systems in the 21st century identified four generally agreed-upon goals that help define sustainable agriculture:

- Satisfy human food, feed, and fiber needs, and contribute to biofuel needs.

- Enhance environmental quality and the resource base.

- Sustain the economic viability of agriculture.

- Enhance the quality of life for farmers, farm workers, and society as a whole.

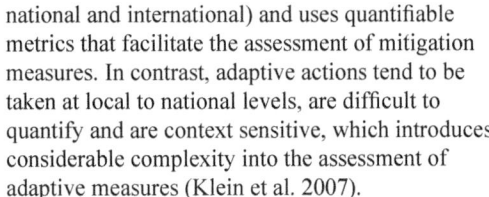

Government policy and programs will be crucial to effective adaptation efforts as the agricultural system responds to projected increases in temperature and precipitation variability and extremes accompanying climate change that likely will be outside the range of individual, community, and institutional experience (Adger et al. 2007, Antle and Capalbo 2010). Government efforts to enhance the adaptive capacity of the agricultural system will be complicated by the dynamic nature and complexity of interactions between the climate system, the agricultural system (Adger et al. 2007), and a scarcity of crucial agricultural resources such as land, water, energy, and ecosystem services (NRC 2010). The potential for effective adaptive action is dynamic, involves social, economic, and ecological processes, and is driven by decisionmaking at multiple scales and by multiple actors (Adger et al. 2007; Marshall et al. 2010).

Howden et al. (2007) and others (e.g., Smit et al. 1999; Moser et al. 2008; Wolfe et al. 2008) discuss some of the long-term opportunities that an improved understanding of adaptation presents to society. Adaptation has the potential to reduce the risks of climate change by improving planning, preventing maladaptation, and informing investment and management of resources such as perennial crops, major infrastructure projects, and capacity building programs. This section discusses some key influences that shape the operating context for adaptive responses by the U.S. agricultural system: climate policy, economic perspectives, finding the balance between mitigation and adaptation, and the limits to adaptation. Choosing among options for adaptive action and the influence of individual decision making on adaptation responses within this context are presented in the next section.

Climate Policy

Climate policy is a dynamic area of public policy-making under active investigation by the global change research community. Initial climate policy development and analysis focused on managing the mitigation of climate change through production and use of energy (Klein et al. 2007), but as awareness of the need to adapt to climate change has grown in this century, governments world-wide have initiated adaptation planning, even though many questions remain about effective adaptation strategies (Adger et al. 2007). Several crucial differences between mitigation and adaptation processes influence the nature of policy development efforts (Klein et al. 2007). Climate change mitigation involves international cooperation to manage greenhouse gas (GHG) emissions on a global scale (typically

national and international) and uses quantifiable metrics that facilitate the assessment of mitigation measures. In contrast, adaptive actions tend to be taken at local to national levels, are difficult to quantify and are context sensitive, which introduces considerable complexity into the assessment of adaptive measures (Klein et al. 2007).

The local nature of adaptation complicates the implementation of government support for adaptation efforts because of the potential for complex cross-scale interactions between top-down policy decisions made at national or international scales, and bottom-up adaptive responses, for example, see sidebar) Belliveau et al. 2006; Klein et al. 2007; Urwin and Jordan 2008). As experience with government adaptation planning for climate change grows, the potential for synergies and tradeoffs created by interactions between adaptation and mitigation policies (Klein et al. 2007) and non-climate policies (Belliveau et al. 2006; Urwin and Jordan 2008) add additional complexities to those imposed by other social, economic, and ecological conditions (Adger et al. 2007). Researchers and policy analysts have given little attention to addressing the critical challenges to adaptation governance, such as building support for action, identifying effective policy strategies, or addressing institutional barriers to adaptive action (Moser et al. 2008; Smith et al. 2009).

In an analysis of climate change adaptation policymaking by U.S. municipalities, States, and the Federal Government, Smith et al. (2009) identified an "adaptation architecture" fundamental to facilitating successful governance of adaptive action. Components of the architecture include governance processes that provide clear leadership, enable coordination between agencies and departments, incorporate mainstream climate considerations into daily decision making, integrate new funding for adaptation into baseline support for climate-sensitive sectors, address institutional and policy barriers to adaptation efforts, and involve stakeholders in policy development and implementation. In addition, government decisionmakers employ decision tools that are robust under uncertainty and informed by accurate, timely, and scale-appropriate climate change information. Finally, government programs invest in adaptation research to understand conditions that promote or impede adaptation decisions and in technology development and diffusion to expand adaptation options. Smith et al. (2009) suggest that in order for adaptation planning to be fully effective at the local level, these components of the adaptation architecture must be integrated within an adaptive management framework.

National Climate Change Adaptation Strategies

Over the last decade, national climate change adaptation policies and programs have been adopted or are under development in Australia, Canada, the European Union, and the United States (Biesbroek et al. 2010). Agriculture is commonly recognized as a key climate-sensitive sector in these national plans, along with related sectors such as water resources, energy, finance and insurance, and natural resources. Adaptation measures are typically planned as an element within broader sectoral initiatives, such as water-resource or disaster-risk planning (Adger et al. 2007).

National adaptation planning got underway in the United States in 2009, when the Interagency Climate Change Adaptation Task Force (Task Force) was established to provide Federal support and coordination for adaptation planning at Federal, State, local, and tribal levels of government (ICC Adaptation Task Force 2011). Senior representatives from more than 20 departments and agencies participate in the work of the Task Force, which is co-chaired by representatives of the Council on Environmental Quality, the Office of Science and Technology Policy, and the National Oceanic and Atmospheric Administration. All Task Force work is being conducted in accordance with a set of goals and guiding principles that foster locally focused, participatory, ecosystem-based approaches to planning, integrated assessment and effective decisionmaking, and international collaboration (see sidebar, page 126). Initial work has involved agency assessment of climate change effects on operations and services, the preparation of climate adaptation plans for each participating agency, and development of cross-cutting adaptation plans for fresh and ocean waters, and fish, wildlife, and plant resources.

Agricultural Adaptation Policy

Agricultural climate policy is an active area of research, development, and analysis in the global change community because of the climate-sensitive nature of agricultural production, the critical importance of agricultural production to human well-being, the dependence of agriculture on natural resources and ecosystem services, and the unique relationship between agriculture and climate, relative to other sectors of the economy. In developed economies, agricultural policy has focused on climate change mitigation, with agriculture being viewed as both a significant source of GHG emissions and a significant sink for sequestration of carbon (Adger et al. 2007; Klein et al. 2007). In contrast, according to Yohe et al. (2002), agricultural climate policy for developing economies features adaptation because food security and sustainable rural livelihoods are primary aims. Much of this work has taken place within a sustainable development framework, and there is considerable research and policy literature addressing the issues of agricultural adaptation to climate change particularly for small-holder farmers, the rural poor, and other resource-dependent social groups in developing countries.

The Local Nature of Adaptation Complicates National Adaptation Efforts

Cross-scale interactions between top-down, non-climate policies and bottom-up adaptive responses by producers can increase agricultural vulnerability to climate change.

Adaptive actions taken by wine producers in British Columbia, Canada demonstrate the perverse outcomes that sometimes occur through interactions between "top down" policy and "bottom-up" place-based adaptations. In response to increased competition from foreign wine imports following the North American Free Trade Agreement, grape producers responded to increased competition from high quality foreign wine imports by replacing existing low-quality, but winter-hardy, grape varieties with more cold-sensitive but higher quality varieties in an adaptation that was facilitated by government aid. This change enhanced the wine industry's domestic and international competitiveness, thereby reducing market risks, but simultaneously increased its susceptibility to winter injury. Producers must irrigate to prevent frost damage in winter, an adaptation that decreases market competitiveness because it reduces the quality of the grapes. Winter irrigation also increases production costs, disease risks and producers' vulnerability to water shortages (Belliveau et al. 2006).

National agricultural adaptation planning has only recently begun, so documentation of the process is sparse and most of the relevant literature presents policy goals and updates of progress on initial implementation efforts; however, general recommendations for agricultural adaptation policy measures are sometimes offered by researchers or analysts exploring climate change effects at the national scale. The comprehensive assessment of existing or proposed agricultural policies addressing issues such as climate change mitigation, subsidy and trade, insurance and disaster assistance, soil and water conservation, environmental quality, and the production of biofuels to identify potential synergies and tradeoffs with proposed adaptation policies is often recommended (e.g., Antle and Capbalo 2010; Olesen et al. 2011). Because of the uncertainties associated with climate change effects and the complexity of adaptation processes, adaptive governance strategies are recommended to implement, evaluate, and revise adaptation strategies (e.g., Olesen and Bindi 2002; Howden et al. 2003; Biesbroek et al. 2010). Enhancing the resilience of agriculture to climate change through adaptation strategies that promote development of sustainable agriculture is a common "no-regrets" recommendation (e.g., Howden et al. 2003; Howden et al. 2007; Olesen et al. 2011). Broad policy measures that may enhance the adaptive capacity of agriculture include strengthening climate-sensitive assets, promoting adaptive governance approaches that encourage climate-learning and adaptive management, integrating adaptation into all relevant government policies, and addressing non-climate stressors that degrade adaptive capacity (Marshall et al. 2010).

Mitigation and Adaptation: Complement or Tradeoff?

According to Klein et al. (2007), only recently have policymakers begun to understand that effective climate policy will involve a balanced portfolio of mitigation and adaptation actions that rely on interactions between the two as a means to enhance adaptation. Recognizing the need for both and the need to explore tradeoffs and synergies between the two, policymakers are faced with an array

U.S. National Adaptation Strategy (ICC Adaptation Task Force 2011)

Integrating Adaptation into Federal Government Planning and Activities: Agencies are taking steps to manage climate impacts to Federal agency missions, programs, and operations to ensure that resources are invested wisely and Federal services remain effective for the American people. Agencies are developing climate adaptation plans to identify their vulnerabilities and prioritize activities that reduce climate risk.

Building Resilience to Climate Change in Communities: Recognizing that most adaptation occurs at the local level, Federal agencies are working with diverse stakeholders in communities to prepare for a range of extreme weather and climate impacts (e.g., flooding, drought, and wildfire) that put people, property, local economies, and ecosystems at risk.

Improving Accessibility and Coordination of Science for Decision Making: To advance understanding and management of climate risks, the federal government is working to develop strong partnerships, enhance regional coordination of climate science and services, and provide accessible information and tools to help decision makers develop strategies to reduce extreme weather impacts and climate risks.

Developing Strategies to Safeguard Natural Resources in a Changing Climate: Recognizing that American communities depend on natural resources and the valuable ecosystem services they provide, agencies are working with key partners to create a coordinated set of national strategies to help safeguard the Nation's valuable freshwater, ocean, fish, wildlife, and plant resources in a changing climate.

Enhancing Efforts to Lead and Support International Adaptation: To promote economic development, regional stability, and U.S. security interests around the world, the Federal Government is supporting a range of bilateral and multilateral climate change adaptation activities and coordinating defense, development and diplomacy policies to take into account growing climate risks.

of complex questions that cannot be answered with any certainty (Klein et al. 2005). Studies of the agricultural sector show the importance of identifying potential synergies between adaptation and mitigation strategies, and the synergies possible with the use of coherent climate policy frameworks that link issues such as carbon sequestration, GHG emissions, land-use change, regional water management, and the long-term sustainability of production systems (Easterling et al. 2007; Jones et al. 2007 Rosenzweig and Tubiello 2007).

Mitigation and adaptation have some similarities and several crucial differences that interact to complicate adaptation efforts (Klein et al. 2007). Both mitigation and adaptation responses involve technological, institutional, and behavioral options, and are driven by a similar set of factors that determine the capacity to act. Both are implemented on a local or regional scale, and may be motivated by local and regional interests, as well as global concerns. Differences between mitigation and adaptation actions arise as a result of differences in scale and intention of effect. Mitigation actions engage the global climate system, involve long lag times between action and response, produce global benefits through local investment, and aim to reduce all potential climate effects by slowing global warming. In contrast, adaptation actions typically engage the local agricultural system, yield immediate benefits by reducing climate vulnerability, produce local benefits through local investments, and can selectively manage agricultural system response to climate to reduce or avoid negative effects and take advantage of positive effects. Mitigation and adaptation can be complementary, because each addresses a different aspect of climate risk (Jones et al. 2007); however, interactions between mitigation and adaptation responses, both within and across scales, complicate the management of both (Klein et al. 2007). For example, intensive livestock producers may respond to increased average temperatures by making adaptations that enhance the cooling and ventilation of animal housing, or they may respond by reducing stocking densities. The former adaptation would likely increase energy use and interfere with mitigation efforts, while the latter would contribute to mitigation efforts through the reduction of GHG emissions (Rosenzweig and Tubiello 2007).

As a result of these differences, mitigation action is often driven by international initiatives managed by national governments, while adaptive action is usually initiated by the individuals, communities and regions experiencing the damaging effects of climate variability and change (Klein et al. 2007). In addition, the research exploring mitigation and adaptation involve different communities of scholars with very different analytical approaches; mitigation research and policy focuses on technological and economic issues, utilizes quantitative metrics and relies on top-down aggregate modeling for studying mitigation tradeoffs, while adaptation research focuses on qualitative, place-based, systems analysis (Klein et al. 2005; Wilbanks et al. 2007). These differences create barriers to the integrated analysis of mitigation and adaptation synergies and tradeoffs even though adaptation is now recognized as a necessary complement to mitigation efforts (Klein et al. 2007).

Integrated Assessment of Mitigation and Adaptation Responses

The relationships between mitigation and adaptation are being explored through conceptual and policy analysis using approaches ranging from complex quantitative simulation modeling to participatory case studies designed to elucidate interactions and their implications for specific locales or sectors within the context of broader development objectives (Klein et al. 2007). Integrated assessment models (IAMs) offer a quantitative approach to the integrated analysis of the societal costs and benefits of climate change and climate policy; however, in current IAMs, climate policy is dominated by mitigation (Klein et al. 2007) and the models have other limitations that call into question their usefulness as an analytical tool in adaptation planning and policy analysis (see Chapter 6). For example, IAMs are likely to underestimate the negative effects of climate change on crop production because the models use average weather data and do not simulate pests and disease effects (Antle and Capalbo 2010).

As discussed in greater detail below, recent efforts to develop methods to support the comprehensive assessment of mitigation and adaptation options recommend seeking out win-win and no-regrets solutions within a robust decision framework, rather than using optimization approaches, because of the uncertainties involved in the analysis (Klein et al. 2005; Jones et al. 2007; Wilbanks et al. 2007). Sustainable development has emerged as a potentially powerful integrator of mitigation and adaptation options that support the development of resilient communities and sectors in the developed and the developing world (Yohe et al. 2002). Sustainable development, mitigation and adaptation share many determinants (Yohe et al. 2007; Goklany 2007), and the interactions between the global climate system and socioeconomic development patterns are increasingly recognized (Klein et al. 2005).

Sustainable development has emerged as a potentially powerful integrator of mitigation and adaptation options that support the development of resilient communities and sectors in the developed and the developing world.

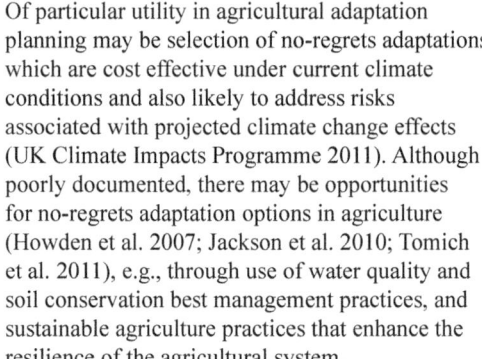

Adaptation Costs and Benefits

The complexity of adaptation processes and the uncertainty in projections of climate change effects challenge efforts to estimate the costs of adaptation and the benefits gained from taking adaptive action. The literature addressing the costs and benefits of adaptation to climate change is sparse and fragmented in scope (Adger et al. 2007). Because of uncertainties about the potential damages that would be avoided by adaptation and the scarcity of information on adaptation processes and associated costs, the methodologies used to estimate adaptation costs are widely acknowledged to be speculative at best (Adger et al. 2007) and likely underestimate the costs of adaptation (Parry et al. 2009). This is because of the bias toward hard adaptations (e.g., publically financed structural measures like expansion of water supply) over soft adaptations (e.g., change in behavior to more efficient use of existing water supplies) that exist in national and international assessments.

The economic response of the agricultural sector to climate change effects has received relatively extensive attention by researchers considering the benefits of farm-level adaptations and adjustments made through markets and international trade (Adger et al. 2007). This work suggests that, at larger scales, the benefits of adaptation will be sufficient to offset the costs of climate change effects in temperate regions, but there are likely to be large variations across and within regions, including the United States (Adger et al. 2007; Chapter 6).

The recognition of the inherent uncertainties associated with adaptation cost-benefit analysis coupled with the desire to move ahead with adaptation planning despite these uncertainties have driven research to develop robust adaptation-decision strategies (e.g., Wilbanks et al. 2007; Parry et al. 2009; World Bank 2010; Moser and Ekstrom 2010; Jones and Preston 2011). Such strategies support robust adaptation planning through use of adaptive management practices, case studies, hedging mechanisms, methods that prioritize and sequence adaptation investments, and methods that support a consideration of the social, institutional, and cultural factors that influence adaptation efforts (e.g., Council of Australia Governments 2007; Parry et al. 2009; World Bank 2010; Brown et al. 2011). Australia (Hills and Bennett 2010) and Canada (BC Agriculture 2012) currently use robust adaptation planning frameworks in regional agricultural adaptation planning.

Of particular utility in agricultural adaptation planning may be selection of no-regrets adaptations, which are cost effective under current climate conditions and also likely to address risks associated with projected climate change effects (UK Climate Impacts Programme 2011). Although poorly documented, there may be opportunities for no-regrets adaptation options in agriculture (Howden et al. 2007; Jackson et al. 2010; Tomich et al. 2011), e.g., through use of water quality and soil conservation best management practices, and sustainable agriculture practices that enhance the resilience of the agricultural system.

Limits to Adaptation

High adaptive capacity does not guarantee successful adaptation to climate change. Adaptation assessment and planning efforts routinely encounter conditions that serve to limit adaptive action regardless of the adaptive capacity of the system under study (Adger et al. 2007; Moser and Ekstrom 2010). Limits to adaptation are "conditions or factors that render adaptation ineffective as a response to climate change and are largely insurmountable" (Adger et al. 2007, pg. 733), while barriers to adaptation are "obstacles that can be overcome with concerted effort" (Moser and Ekstrom 2010, pg. 2). These terms are sometimes used interchangeably because the perception of an obstacle as a barrier or a limit to adaptation depends on social perspectives such as cultural norms or level of technological development (Adger et al. 2009).

The limits and barriers to adaptation – some ecological, and others arising from economic or social considerations that complicate adaptation efforts – add uncertainty to the adaptation process and raise ethical questions about adaptation as a response to climate change. As a result of these limits, the capacity for adaptation and the processes by which it occurs vary greatly within and across economic sectors, communities, regions, and countries. Successful adaptation planning processes will include an assessment of the limits to adaptation in the design and implementation of policy and programs.

Ecological Limits to Adaptation

Increasing evidence exists that the resilience of agricultural systems to global change is dependent on a wide range of ecosystem processes that provide services to agriculture, such as the regulation of water quality and quantity, waste processing, climate protection, and the suppression of pest populations (Peterson 2009; Jackson et al. 2010; Tomich et al. 2011). The ability of ecosystems to provide these services to agriculture is increasingly compromised by multiple stressors such as pollution, agricultural intensification, overgrazing, ecological simplification, and the effects of climate variability and change (Folke et al. 2004; Falkenmark et al. 2007; Peterson 2009; Jackson et al. 2010). In some cases, the combined effects of these stressors have pushed ecosystems past critical ecological thresholds (tipping points), resulting in a sudden shift from a productive to an unproductive state (Folke et al. 2005; Adger et al. 2007; Peterson 2009). Ecosystem shifts represent a significant challenge to resource management that is often outside human experience (Folke et al. 2004). Climate change effects that surpass critical ecosystem thresholds or result in dramatic transformations of the physical environment of a system may present limits to adaptation.

Social Barriers to Adaptation

Agricultural producers routinely plan for and manage seasonal weather and weather-related events, but projected increases in weather variability, and frequency and intensity of weather events associated with climate change present novel risk-management challenges. The increased complexity and projected changes in the variability and intensity of temperature and precipitation are likely to challenge both the structure and function of individual enterprises and possibly the U.S. agricultural system as a whole; however, adaptation efforts may be hampered if producers do not recognize the value of taking adaptive action to prevent loss.

Social and cultural limits to adaptation are determined by individual and group experience, shared values, beliefs, and world views (Adger et al. 2007). Social adaptation barriers represent a significant challenge to climate change adaptation in U.S. agriculture. The perception of the need for adaptation is influenced by access to finance (Knutson et al. 2011), political norms and values (Roser-Renouf and Nisbet 2008; Malka et al. 2009; Borick et al. 2011; McCright and Dunlap 2011), and culture and religious ideologies (Wardekker et al. 2009; Kahan 2012). For example, a recent survey of 1,276 Iowa farmers revealed that 32% believed there was insufficient evidence of climate change or that climate change is not occurring (Arbuckle 2011). A substantial portion of the U.S. public does not perceive that solid evidence exists to support global warming or are unsure of the evidence (either immediately or in the long term), with the level varying over time and

with the availability of new or different information (Leiserowitz 2006; Borick et al. 2011; Leiserowitz et al. 2011).

Research on producer adoption rates of conservation best-management practices (BMPs) (Hua et al. 2004; Valentin et al. 2004; McCown 2005; Smith et al. 2007; Prokopy et al. 2008) offers valuable insights on the willingness and capacities of producers to put in place adaptive management in response to changing climate conditions. For example, McCown (2005) reports that learning and adoption of conservation practices is not simply a function of knowledge transfer from scientists to farmers. Many interventions assume that objective knowledge is sufficient to convince farmers to adopt new practices. This assumption overlooks how farmers (and humans in general) make decisions, which integrate objective "fact" science, and subjective personal knowledge and experiences, though seldom with equal weight (Slovic 2010). Producers making risk management decisions use a combination of analytical knowledge (facts they know) and experiential-affect heuristics (the positive or negative feelings, consciously or subconsciously, associated with the view of the task) in assessing the need for adaptation (Slovic 2010).

Prokopy (2008) reported that *no* factors consistently determine BMP adoption, based on a meta-analysis of 55 studies conducted over 25 years. However in some studies, several variables have been found to be significant and positively associated with BMP adoption rates, including presence of younger farmers, higher education levels, more income and capital, diverse operations, larger area under management, more access to labor, access to information, positive environmental attitudes, environmental awareness, and utilization of social networks. Decisions to adapt to climate change are likely to have similar complexity and uncertainties. A National Research Council report (NRC 2009) found that human dimensions research lags behind research investigating the natural climate system, stating that "the preparation of [climate change] knowledge for use in decisionmaking, as well as the effective communication of scientific insights to stakeholders, lags significantly behind or is entirely inadequate."

Decisionmakers managing adaptation in an agricultural system must make choices within an adaptive operating space shaped by government policy, economics, and the limits to adaptation. Within this context, decisionmakers must consider the alternative adaptive actions available, selecting the best option and implementing it. The next section presents options for assessing the vulnerability

Agricultural producers routinely plan for and manage seasonal weather and weather-related events, but projected increases in weather variability, and frequency and intensity of weather events associated with climate change present novel risk-management challenges.

and adaptive capacity of an agricultural system, reviews research efforts to develop the educational programs and decision tools that producers need to effectively manage climate risk, and describes specific adaptation options recommended for U.S. agriculture.

Assessing Options, Taking Action

Taking adaptive action requires stakeholders throughout the U.S. agricultural system to make decisions about the system under their management despite the multidimensional uncertainties created by climate change. The place-based nature of adaptation adds additional complexities to adaptive responses and drives the development of flexible management strategies to identify and assess context-specific adaptive options rather than prescriptive solutions. Adaptation will be more effective if decisionmakers have the knowledge, information, and tools they need to manage climate risk effectively.

Research and development efforts to support adaptation planning in agriculture are underway. These efforts aim to provide agricultural decisionmakers with effective methods to assess the vulnerability and adaptive capacity of the systems under their management, to guide the selection and implementation of adaptation options, and to manage dynamic systems in a complex decision environment. To date, much of this work addresses the regional, local, or enterprise scale, and focuses on policy and technical support for decisionmaking by producers at the field and farm scale.

Vulnerability Assessment

Vulnerability assessment aims to estimate the exposure, sensitivity, and adaptive capacity of the agricultural system of interest in order to quantify vulnerability to climate change effects for a specific geographic location. Vulnerability assessment is typically integrated across multiple scales; national or regional climate projections are integrated with individual, community, or regional estimates of adaptive capacity. Understanding the vulnerability of an agricultural system may aid adaptation decisions through the identification of system elements at the greatest risk of exposure and those most sensitive to projected climatic effects, and by clarifying the most effective response options to enhance the adaptive capacity of the system.

Bryant et al. (2000) report on a synthesis of research investigating the vulnerability of Canadian agriculture to climate variability and change

conducted during the last two decades of the 20th century. This early work identified the components of agricultural adaptation and elucidated the multidimensional, multi-scale, and context-specific complexity of the agricultural adaptation decision environment (see Table 7.2), highlighted the critical role of human agency in agricultural adaptation, and established the need for participatory research to understand agricultural adaptation to climate change.

Some key producer perceptions of adaptation to climate change emerged from Bryant et al.'s (2000) work. Many producers were skeptical about the reality of projected rates of climate change, but they did respond to climatic risks specific to local landscape associated with seasonal variability and the potential for extreme events during critical crop development stages. Most producers expressed a high level of confidence in their ability to manage climatic variability with available technologies and so were not concerned about projected changes in climate, although the level of confidence varied depending on region and type of enterprise. Bryant et al. (2000) challenge producer perceptions that the Canadian agricultural sector is well adapted to uncertainties in climate, pointing to frequent widespread losses and economic hardships associated with unexpected weather events and the ongoing need for public relief (with disaster payments, crop insurance, and the like). They suggest that weather variability has declined in importance relative to other factors in farm decisionmaking as result of improvements in agricultural technology in combination with programs that reduce production vulnerability and conclude that institutional and policy programs that reduce producer vulnerability to weather variability may actually serve as a barrier to effective adaptive action over the long term (i.e., resistance adaptation that inhibits resilience and transformation adaptation).

This early participatory research with Canadian producers generated a wealth of knowledge about the multidimensional complexity and dynamic nature of agricultural vulnerability at the enterprise level, which has since been confirmed by more recent research in Canada, Australia, New Zealand, Europe, and the United States (e.g., Reid et al. 2007; Belliveau et al. 2006; Wolfe et al. 2008; Marshall 2010; Nelson et al. 2010a; Nelson et al. 2010b; Reidsma et al. 2010; Kenny 2011; Olesen et al. 2011).

Agricultural vulnerability to climate change in California's Central Valley (Jackson et al. 2012), Washington State (Miles et al. 2010), and the Northeastern U.S. (Wolfe et al. 2008) has been evaluated with interdisciplinary case study

approaches using various combinations of emissions scenarios to estimate exposure, landscape planning, simulation modeling, qualitative analysis of adaptive capacity, and adaptation scenarios. The results of these studies suggest that agriculture is vulnerable to climate change and that sensitivities to climatic exposures vary substantially by region and enterprise type; however, potential adaptive capacity is high and near-term productivity can be maintained through a combination of adjustments in agricultural practices and government support.

The vulnerability of agriculture in California's Central Valley was assessed in a detailed case study utilizing scenario analysis to explore planning issues at the farm and landscape levels over the next 50 years (Jackson et al. 2009; Jackson et al. 2011). Researchers concluded that increased temperatures and a more uncertain water supply leave the Central Valley highly vulnerable to climate change. A comprehensive analysis of three distinct agricultural adaptation scenarios suggests measures that integrate changes in crop mix, irrigation methods, fertilization practices, tillage practices, and land management may be the most effective approach to managing climate risk.

In an assessment of regional climate change effects and adaptation strategies for Washington State (Miles et al. 2010), eight climate-sensitive sectors of the State economy, including agriculture, were evaluated. Climate change effects on the agriculture sector were explored by model simulations of apple, potato, and wheat production under different climate change scenarios. Results suggest that Washington agriculture is vulnerable to climate change effects, but productivity can be maintained over the short term (i.e., 10-20 years) by adjusting production practices and adopting new technologies, by improvements in agricultural water management, and by State-wide monitoring to gather and interpret data on climate change effects (Miles et al. 2010; Stöckle et al. 2010).

Research exploring the vulnerability of agriculture in the Northeastern U.S. through the end of this century suggests that producers, government agencies, and others in the region will benefit from strategic adaptive actions that anticipate projected climate change effects (Wolfe et al. 2008). This research found that some producers will likely benefit from climate change effects, such as those currently producing or willing to shift to better adapted crops, those with multi-regional production options, those who guess correctly about climate and market trends, and those who have the financial resources to implement adaptation strategies in

a timely manner. Wolfe et al. (2008) concluded that farmers most vulnerable to climate change effects are those without the financial resources to adapt, those unwilling or unable to exit from their current production system, and those who make poor decisions regarding the type and/or timing of adaptations. Subsequent research conducted in other regions (e.g., Chhetri et al. 2010; Knutson et al. 2011) report farmer vulnerabilities similar to those found by Wolfe et al. (2008).

Using participatory research methods to explore the vulnerability of the grape industry in Canada's Okanagan Valley, Belliveau et al. (2006) identified multiple climatic and non-climatic factors that influence the region's vulnerability to climate change (Figure 7.2). They concluded that reducing the region's vulnerability to climate change will likely require action beyond the control of individual producers, and present a diverse mix of technological developments, educational programs, and economic incentives designed to enhance the capacity of Okanagan Valley producers to manage a greater range of climatic conditions in the present and reduce future vulnerabilities to projected climate changes. An Australian study provides an example of an innovation in hazards/impact modeling that allows the explicit consideration of adaptive capacity in the assessment of rural community vulnerability to climate change (Nelson et al. 2010 a, b). This new method was developed to address the limitations presented by standard hazard/impact modeling to adaptation planning. For example, because such modeling is dominated by a focus on technical strategies to reduce exposure and sensitivity to climate change, the diverse potential for regional and local adaptive capacity is often overlooked. Nelson et al. (2010 a) propose a new method that broadens the quantitative estimates of rural community exposure and sensitivity developed using standard hazard/impact modeling through the addition of an integrated index of adaptive capacity based on a rural-livelihoods framework. This framework views decisionmaking as a dynamic response in a decision environment shaped by changing access to five broadly defined types of capital (human, social, natural, physical, and financial). In an application of this new method, Nelson et al. (2010b) report that Australian rural communities are vulnerable to climate variability and change and discuss a complex set of interacting environmental, economic, and social factors that contribute to this vulnerability. In addition, this new assessment approach informed the selection of specific adaptive responses likely to reduce the vulnerability of rural communities through local and regional actions that enhance adaptive capacity.

Fig. 7.2. A schematic framework of the multiscale factors influencing production-level vulnerability to climate change of grape growers and winery operators in the Okanagan Valley of Canada. This framework provides specific examples of the socio-economic forces interacting with local climate, landscape and enterprise level factors to drive adaptation responses in the agricultural SES (Belliveau et al 2006).

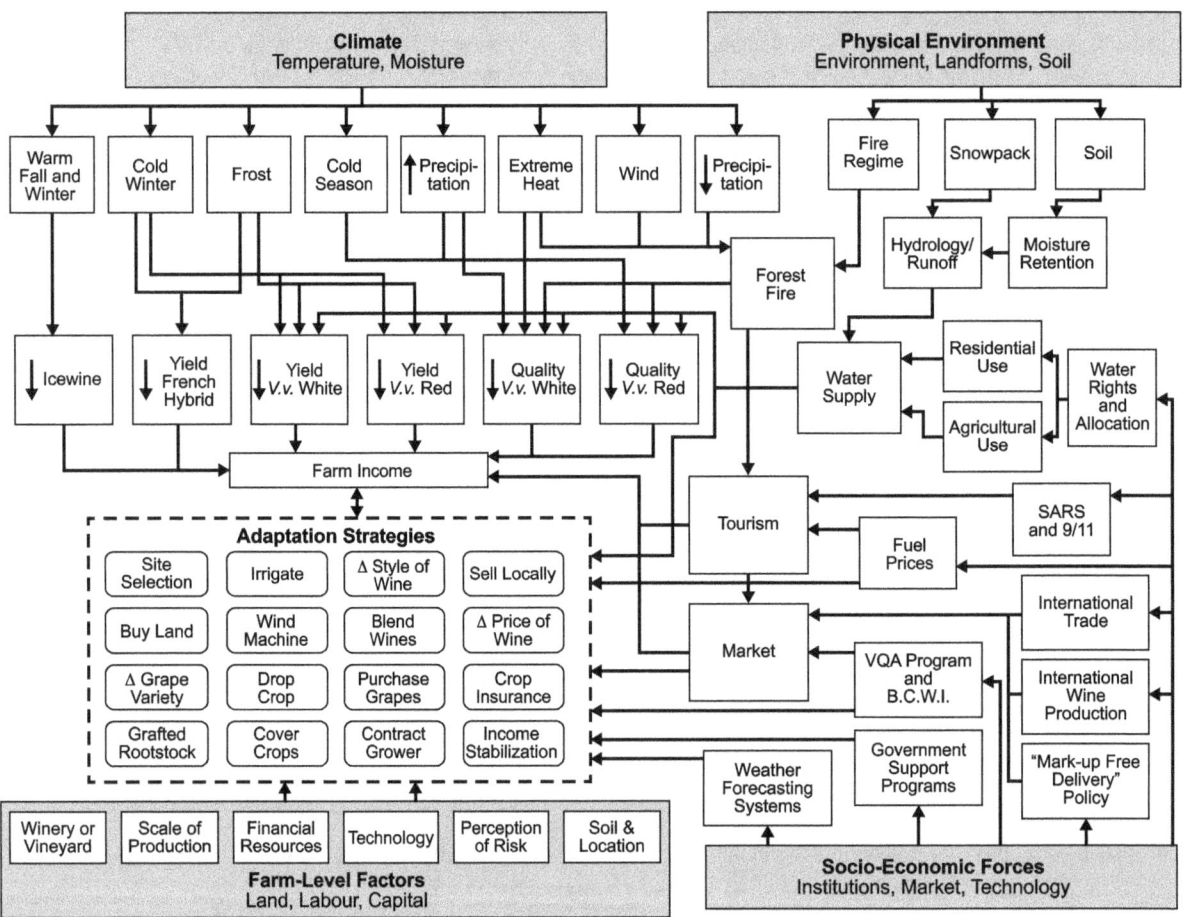

Vulnerability assessment is emerging as a potentially powerful tool for decisionmakers seeking to understand and manage agricultural adaptation processes. Integrated research approaches, like those described above, can provide valuable qualitative and quantitative information about system exposures and sensitivities to climate change effects and the capacity for adaptive response.

Assessing Adaptive Capacity

Agricultural vulnerability to climate change effects can be reduced by enhancing the adaptive capacity of the agricultural system. A better understanding of the key determinants of adaptive capacity in agricultural systems would aid efforts to sustain agricultural production and productivity in the face of projected increases in the frequency and intensity of climatic events (NRC 2010).

Research and development is underway to understand the determinants of agricultural adaptive capacity in all its dimensions and to develop assessment methods useful to decisionmakers operating within an agricultural system. Key to the utility of adaptive capacity concepts in decisionmaking is to be able to identify critical determinants and their links to potential adaptive responses in the system of interest (Moser 2008). Consistent with the results of research to assess the vulnerability of agriculture, this emerging body of work suggests that the adaptive capacity of agricultural systems is dynamic and determined by a complex mix of economic, ecological, and social factors that interact with climatic effects across multiple dimensions of space and time.

In an assessment of the adaptive capacity of the northeastern United States economy, Moser et al.

(2008) present the dairy industry as an example of the contribution that access to financial resources plays in the adaptive capacity of a system. Their assessment found that projected climatic changes impose additional uncertainty on an already fragile dairy industry because of projected changes in productivity and production costs. They suggest that farmers with the financial resources to respond to these uncertainties are in the best position to benefit from climate change effects, while others on the margin of economic viability may not be able to afford to continue in agriculture. Based on this assessment, they predict a northward shift of dairy production in the Northeast, and a gain in market share by larger corporate dairy operations.

The adaptive capacity of Canada's Prairie agricultural system was explored in a project using mixed methods to identify and map a quantitative index of adaptive capacity useful to regional adaptation planning (Swanson et al. 2009). An integrated index of 20 quantitative indicators representing Smit et al.'s (2001) six determinants of adaptive capacity was developed, with existing data from Statistics Canada sources to represent a top-down, conceptually-based approach to the

assessment of adaptive capacity in the region (Figure 7.3). Field interviews with producers and producer organizations in the region were also conducted to develop a set of indicators representing a bottom-up, producer-driven assessment of adaptive capacity. A comparison of the two approaches confirmed the context-specific nature of adaptive capacity and the need for site-specific, participatory research when exploring adaptation processes; producers in the study identified only nine of the 24 indicators included in the index as meaningful determinants of adaptive capacity.

The assessment of the adaptive capacity of rural communities in Australia provides an example of an emerging quantitative approach to the integrated geographic analysis of adaptive capacity (Nelson et al. 2010 a, b). This method estimates the adaptive capacity of the rural agricultural system using farm survey data to create an integrated quantitative index based on a rural livelihoods framework. This framework views decisionmaking as a dynamic response within a decision environment shaped by changing access to the five types of capital (human, social, natural, physical, and financial). This integrated analysis of adaptive capacity

Fig. 7.3. *Determinants of the Adaptive Capacity of Agriculture to Climate Change in the Canadian Prairie Region:* *The adaptive capacity of the Prairie region agricultural SES was estimated using an integrated framework of indicators selected to represent the 6 determinants of adaptive capacity (Swanson et al. 2009).*

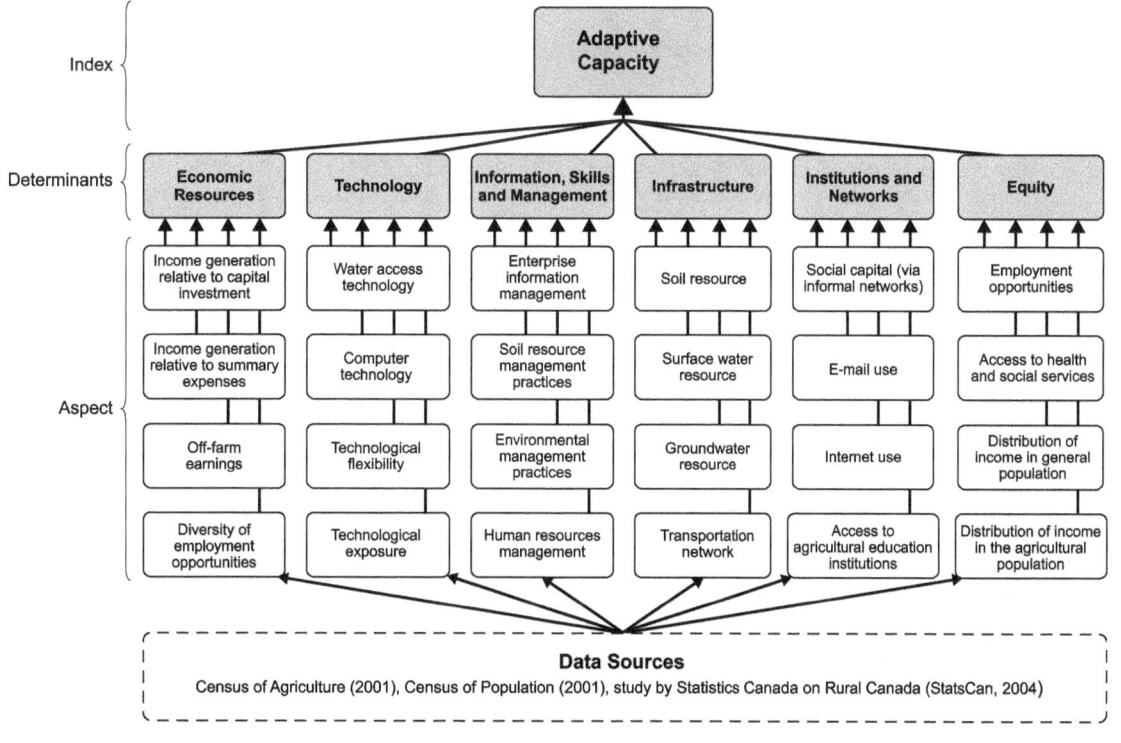

enhanced adaptation planning by identifying specific community-based factors that limit adaptive capacity in rural areas. The identification of specific local factors facilitated the collaborative exploration by community groups, industry and local governments of the tradeoffs between building adaptive capacity and attaining other goals associated with specific adaptive actions and was useful to regional natural resource management plann ing.

In a comprehensive analysis of the adaptive capacity of Australian agriculture, Howden et al. (2003) identified three key strategies that governments might use to reduce risk and capture opportunities presented by climate change. These include use of participatory methods to develop improved cost/benefit analyses of adaptation options; develop socioeconomic and cultural/institutional structures to support more resilient agricultural systems; and use adaptive management strategies to cope with the inherent uncertainties in adaptation efforts.

Adaptive capacity assessment holds much promise as a useful approach to managing adaptation to climate change in agricultural systems. To take effective adaptive action, decisionmakers need specific information about how adjustments in resources are likely to influence the adaptive capacity of the system under management, both now and in the future. The next section turns to adaptive actions that have been taken or are likely to be effective at the enterprise-level in the U.S. agricultural system.

Incremental Adaptation: Extending Existing Production Practices

Agriculture has a long history of successful adaptation to varying environmental conditions through adjustments in crop and livestock management practices that prevent losses of productivity. While producer adaptations to climate variability and change are not well documented in the United States, during the last 25 years agricultural producers in Canada, Europe, Australia, and New Zealand report success using existing management practices to cope with increases in seasonal variability of temperature and precipitation, and extreme weather events (Bryant et al. 2000; Harrington and Lu 2002; Reid et al. 2007; Marshall 2010; Kenny 2011; Olesen et al. 2011, Rogovska and Cruse 2011). These researchers suggest that this past success in managing climatic variability has contributed to the confidence expressed by these producers in their ability to manage future climatic variability with available agricultural technologies, policies and programs. Table 7.2 presents examples of potential actions that could be taken by producers,

agribusiness, and government to address specific climate change adaptation drivers. Many of these options are extensions of existing practices that serve to increase enterprise resistance or resilience to climate change effects.

Research and development efforts to address climate change effects on U.S. agriculture have largely focused on investigating single-factor climate change effects (e.g., increased CO_2, ozone, temperature, or water availability) at cell (metabolic) and whole-plant and animal scales (e.g., Hatfield et al. 2011; Izaurralde et al. 2011). Although adaptation is not directly investigated in these studies, the results are often extended to include recommendations for agricultural adaptation options at the enterprise scale.

Crop scientists routinely recommend a combination of changes in management practices (e.g., timing of field operations, cultivar selection, or irrigation management), and development of CO_2 responsive and stress tolerant germplasm as resistance and resilience adaptation strategies for crop production; however, crop-specific constraints are often recognized (see Chapter 5).

More transformative adaptive options involve altering species of crops grown at a given location, and the northward migration of crops (see Chapter 5). New strategies are under development to manage the rapid evolution of "climate ready" cropping systems including: a shift from fixed technological packages to the participatory design of cropping systems for specific locales; the integration of environmental and social goals into cropping system design criteria; a shift from field scale to multi-scale design; and the addition of resilience design-objectives to accommodate increased climatic variability and change (Wery and Langeveld, 2010). These strategies typically involve the use of cropping systems modeling as a design and assessment tool, coupled with multi-criteria, indicator-based analysis at the field, farm and landscape scales, risk analysis to account for variable climate conditions and the use of participatory approaches that utilize local knowledge of agriculture and natural resources and facilitate the community-based exploration of the trade-offs between the multiple functions of agriculture.

Adaptation options for managing novel crop pest and disease management challenges may involve increased use of pesticides, new strategies for preventing rapid evolution of pest resistance to chemical control agents, development of new pesticide products, and improved pest and disease forecasting. Adaptation options that may increase resilience of agricultural systems to changes in pest pressures include crop diversification and management of biodiversity at both field and landscape scale to suppress pest outbreaks and pathogen transmission (see Chapter 4).

Recommended adaptive responses by livestock producers managing for resistance or resilience to climate change focus on obtaining the appropriate education and training required to understand and manage animal needs, potential stress levels, and options for reducing stress, as well as developing an adaptive management plan, and selecting animals and management strategies compatible with the production enterprise. Transformative adaptations may include a transition to livestock species or breeds that have greater tolerance of relatively high temperatures and that are more capable of utilizing existing vegetation and more resistant to novel pests and diseases (see Chapter 4). Wolfe et al. (2008) recommend a variety of low-, medium-, and high-cost adaptation measures for dairy production systems in the Northeast, including practices that reduce heat stress, changes to feed composition and feeding schedules, and planning for adequate water supplies. In a recent review of climate change effects on forage and rangeland production, Izaurralde et al. (2011) recommend a conversion to integrated crop/livestock farming systems as a transformative strategy to reduce detrimental environmental impacts, improve profitability and sustainability, and enhance ecological resilience to climate change in U.S. livestock production systems.

Sustainable natural-resource management strategies inform effective adaptation options for U.S. agriculture. The ability of healthy soils to regulate water resource dynamics at farm and watershed scales is widely recognized and particularly critical for the maintenance of crop and livestock productivity under conditions of variable and extreme weather events. Soil conservation practices like cover cropping, diversifying annual cropping systems, inclusion of perennial crops in rotations, changing from annual to perennial crops, organic soil amendments, grazing management, conversion of cropland to pasture, agroforestry and natural areas, and wetland restoration may enhance the resilience of the U.S. agricultural system to climate change effects (see Chapter 5).

In recent years, a number of reports suggest that U.S. agricultural systems may be able to enhance resilience and reduce climate risk by adopting sustainable agriculture practices (Hendrickson et al. 2008; Jackson et al. 2010; NRC 2010; Izaurralde et al. 2011; Lin 2011; Merrill et al. 2011; Tomich

et al. 2011). A transition to knowledge-intensive, low-input, resilient production systems has been identified as an effective means of managing climate risk in New Zealand agriculture (Kenny 2011). International development programs routinely recommend development of sustainable agriculture systems as a proactive, cost-effective climate risk management approach in less developed economies (IAASTD 2009; FAO 2010; World Bank 2011). For example, the United Nations Food and Agriculture Organization (FAO 2010) recommends a number of sustainable agricultural practices that reduce climate risk and may have application to U.S. agricultural systems. These include intercropping within a crop rotation or in agro-forestry systems; improved water harvesting and retention (e.g., in ponds, behind dams, through construction of retaining ridges, etc.) and water use efficiency (e.g., in irrigation systems); and ecosystem-based management of biodiversity to provide pest and disease management, regulate microclimate and nutrient cycles, decompose wastes, and crop pollination.

In an exhaustive review of more than 300 recent publications addressing climate change issues in European agriculture, Iglesias et al. (2011) linked specific agricultural adaptation actions to key risks and opportunities associated with projected climate change effects in five agroclimatic zones across Europe. Adaptation options at the enterprise scale are consistent with those recommended for U.S. production systems, such as changes in crop management (e.g., cultivar selection, timing of field operations, management of landscape biodiversity, and increased use of pesticides), improved water management (e.g., floodplain and wetlands restoration, efficient irrigation, and water harvesting), and changes in livestock management (e.g., shelter and heat protection, breed selection, grazing regime, and timing of breeding).

Adaptation strategies that extend existing practices at the farm level can be very effective over the near term; however, if the intent is to resist climate change effects in order to maintain an existing farming system adapted to previous and more stable climate conditions, then this strategy is likely to become increasingly costly and may ultimately fail as climate change effects intensify. This pattern of investing in resistance strategies at the expense of resilience or transformation strategies may become maladaptive over the longer term and could lead the agricultural system into an adaptation trap (Allison and Hobbs 2004; van Apeldoorn et al. 2011). There is a critical need for adaptation assessment tools that inform the selection of enterprise-level options along the resistance-resilience-transformation spectrum.

Managing Climate Risk: New Strategies for Novel Uncertainty

Decisionmakers in the U.S. agricultural system face novel uncertainties if climatic variability and change intensify as projected over the next 30 years. Adaptation efforts will be enhanced by availability of effective climate risk management tools, the use of adaptive management strategies, and the mainstreaming of climate knowledge throughout the multiple dimensions of agriculture decisionmaking.

Efforts to develop and extend new climate risk management tools to enhance the adaptive capacity of U.S. agriculture must take into account the complex decision environments encountered by producers managing climate risk (Reid et al. 2007; Brown et al. 2010). Community-based research and education led by innovative producers to develop strategic planning skills, increase climate awareness, improve financial security, and adopt climate tools such as seasonal climate forecasts may be effective methods to promote farm, landscape, and regional adaptive capacity (Marshall 2010; Kenny 2011). Collaborative training and transformative learning promote flexible decisionmaking and autonomous thinking that is most advantageous for managing changing environmental conditions (Tarnoczi 2011).

New decision tools utilizing adaptive management practices are being developed to address the novel uncertainly and complexity that climate risk presents to agricultural management decisions. Natural resource managers in Australia can take advantage of self-assessment processes to select and monitor the effect of sustainable practices on the adaptive capacity of systems under their management (Brown et al. 2010). A simple on-farm tool is available to producers making drought-adaptation decisions in rain-fed, field-based livestock production systems in Australia (Reid 2009). New planning tools to aid Canadian farmers managing climate change and climate variability have been developed (Bryant et al. 2008). Notable examples of technical efforts underway in the United States to encourage use of seasonal climate information in enterprise management decisions include Agroclimate, a project of the Southeast Climate Consortium (Agroclimate 2011), and Adapt-N, a web-based decision support tool that provides field-specific, locally adjusted N fertilizer recommendations for corn production based on the effects of local early-season weather (Moebius-Clune et al. 2011).

A key strategy to enhance the adaptive capacity of the U.S. agricultural system may be the use of adaptive management practices to support learning by doing (Howden et al. 2007; Easterling et al. 2007; Carpenter et al. 2009). Adaptive management is particularly useful in decision environments characterized by high uncertainty and complexity. Performance-based management decision tools and methods for evaluating adaptive responses are needed so as to avoid lost opportunities for learning (Howden et al. 2007; Morton 2008; Winsten 2009). By tracking the successes and failures of different adaptation actions, individuals, businesses, and institutions can identify effective, efficient, and equitable policies and measures. Adaptive management promotes the development of more robust adaptation strategies over time (Howden et al. 2007; Preston et al. 2011).

Mainstreaming climate knowledge to improve climate risk management has been proposed as a core adaptation strategy in agriculture, as well as in many other economic sectors (Howden et al. 2007; Adger et al. 2007; Field et al. 2007). Mainstreaming climate knowledge improves decisionmaking by ensuring that land managers, technical advisors, researchers,

Agriculture and Food Research Initiative

Agriculture and natural resources science for climate variability and change adaptation strategies for U.S. agriculture are among the topics being supported by USDA funding. Several projects just underway in 2011 are actively researching adaptation strategies for cereal and legume crops, livestock, and forestry production systems. This work encompasses a number of specific disciplines, notably crop breeding, climate forecasting, atmospheric dynamics, soil science, hydrology, entomology, agricultural engineering, sociology, economics, forestry, weed sciences, and landscape ecology with significant efforts to conduct field trials and build extensive data bases that enable life-cycle analyses and other systems analysis methodologies. Several research projects are explicitly connected to the development of national agricultural policy, adaptive management tools for farmers, and education to develop the next generation of scientists to address the dynamic and complex problems of climate and agriculture.

private businesspeople, and government program managers and policymakers are aware of current and projected climate effects, and can access best management practices to reduce risks and capture opportunities. Taking such a comprehensive, climate risk management approach to agricultural adaptation offers great potential to promote effective adaptive action by decisionmakers throughout the multiple dimensions of U.S. agriculture.

Conclusions

The increasing pace of climatic change, the complex interactions between the global climate system, ecosystems and social systems, and the complexity of climate change adaptation processes presents a novel challenge to the sustainability of U.S. agricultural system. Current climate change effects are challenging agricultural management and are likely to require major adjustments in production practices over the next 30 years and projected climate changes over the next century have the potential to transform U.S. agriculture. Taking adaptive action to avoid the damages and capitalize on the opportunities presented by climate change requires stakeholders throughout the U.S. agricultural system to make decisions about the system under their management despite the multidimensional uncertainties associated with a changing climate. The place-based nature of adaptation adds additional complexities to adaptation planning and assessment and drives the development of flexible management strategies to identify and assess context-specific adaptive options rather than prescriptive solutions. A climate-ready U.S. agricultural system will depend on easy access to useable climate knowledge, improved climate risk management strategies, effective adaptation planning and assessment methods, and the development of more resilient production systems.

Risk Assessment and Climate Change: An Overview

There is uncertainty associated with many of the steps necessary to assess the effects of climate change on agriculture. Some of that uncertainty arises because the science of estimating climate change is complex and continuously evolving. Other sources of uncertainty arise from an incomplete understanding of the effects of a multitude of climate variables and conditions on crop and livestock growth and development. The practice of anticipating human adaptation behavior in the future is inherently uncertain; observations of past behavior provide a good starting point, but advances in communication,

information, and technology may fundamentally alter future conditions, and decisionmaking options, in ways that are not easy to predict.

The lack of certainty about the expected effects of climate change complicates decisionmaking about how, and when, to develop adaptive strategies or invest in mitigating technologies. Nevertheless, decisions are made under uncertainty on a daily basis; a thunderhead on the horizon does not render us paralyzed with indecision about whether to carry an umbrella, for instance. Without consciously realizing it, we weigh the likelihood that it will rain, together with the costs of carrying an umbrella and our aversion to getting wet, and reach a decision. While that particular decision is relatively trivial, the same tools and processes can be applied to dissect much more complex problems, arriving at decisions of far greater importance in a systematic way despite the presence of uncertainty.

Risk management is the field of decisionmaking that refines the tools and processes used in situations with risky or uncertain outcomes in order to allow decisionmakers to manage the risk associated with a full suite of potential outcomes. A number of generalities emerge from the field of risk management that can be useful in climate and adaptation planning:

Risk management weighs outcomes as a function of both likelihood and consequence. As a result, outcomes with a low probability (or likelihood) of occurrence but a very high negative effect can be of as much import in the decisionmaking process as outcomes with a high probability of occurrence but a low negative effect. In the context of climate change, design and incorporation into decisionmaking of low-probability, high-impact potential outcomes, based on expert assessment of the literature, is not "fear mongering" but a necessary part of representing the range of outcomes under management consideration. Such outcomes may or may not influence the management path ultimately chosen, depending on the relative effects, probabilities of occurrence, and management options available.

Because there are many possible outcomes in risky situations, a management path chosen now may not be the optimal one given the outcome that actually materializes. When faced with a number of possible outcomes, it may be preferable to design management strategies based on a consideration of how they perform over a range of those outcomes, rather than selecting strategies based on a single "most likely" or "high impact" outcome.

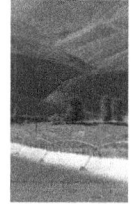

When there are irreversible costs associated with taking action now, there may be a value associated with waiting and acquiring additional information before accepting irreversible costs when the outcome is uncertain (Antle and Capalbo 2010). This "option value" captures the irreversible costs that can be avoided if a management decision is postponed until more information is acquired and uncertainty in the management decision is reduced.

Even in the absence of irreversible management investments and costs, there is a value to information that reduces uncertainty and enables improved management decisionmaking over the remaining uncertainty in outcomes.

Comprehensive risk management in the context of climate change would allow a subjective examination of the "risk-weighted" costs and benefits of launching various adaptation strategies, including potential investments in early-response systems, adaptation technologies, communication and research infrastructure, capacity building, etc., given uncertainty about which climate and impact scenario will ultimately emerge. The approach requires the quantification of an enormous amount of information about potential climate outcomes, their probability of occurrence, and their effects, however. Few efforts have been made to develop such comprehensive quantification efforts in the context of climate change.

Chapter 8

Conclusions and Research Needs

Agriculture in the United States has followed a path of continual adaptation to a wide range of factors driving change both from within and outside of agricultural systems. Agriculture is a social-ecological system (SES). The complex relationships among different commodities, production practices, institutions, and stakeholders have enabled successful adaptation to past levels of climate variability and gradual changes in climate, as well as to other environmental, economic, and policy-environment changes and consumer purchase behavior. As a result, agriculture in the United States over the past century has steadily increased its productivity and integration into world markets.

The expected increases in frequency, duration, and intensity of weather events driven by changing climate present novel and unprecedented challenges to the sustainability of U.S. agriculture. Past experience with agricultural production under a relatively stable climate has created a sense of confidence in management decisions; faced with a problem, a producer or land manager could turn to a time-tested response. With few exceptions, climate was expected to follow well-established boundaries of temperature and precipitation. The response was formulated by the producer and resulted in changes of land use and/or management decisions. With increasing uncertainty about weather and future climate projections, a novel sense of uncertainty is being introduced into agriculture. Producers are faced with new types of climate-driven problems; and there is a lack of knowledge and experience-derived responses from which to formulate new management strategies. As a result, climate is expected to become a more significant factor in future decisionmaking by producers and other land managers, scientists and technical advisors, agribusiness, and policymakers. While past management decisions have largely focused on adjusting to mean values of precipitation and temperature, future decisions will likely require a greater emphasis on managing high levels of

Fig. 8.1. The hills are alive with the sounds of pollinating insects, and that's exactly what technicians Rebekah Andrus (left) and Olivia Messinger are netting in a field near the Wellsville Mountains (Utah). Image courtesy ARS.

uncertainty, and planning for and adjusting to the extremes.

Furthermore, climate change effects on U.S. agriculture cannot be fully examined or understood without consideration of their global context. Climate change is a global phenomenon that is affecting the global agricultural system of which U.S. agriculture is a part. Climate-driven yield reductions and, in some cases, enhancements, in different regions will affect world markets, sometimes to the benefit, and sometimes to the deficit of other countries with competing production. Less developed countries are expected to have less capacity for adapting to climate change and thus, even in the short-term, there are likely to be significant effects on global hunger and well-being. Additionally, risk exists of relocating pests and pathogens in agricultural products exported to world markets from their native habitats, thus creating a demand for increased vigilance by inspectors at the Nation's entry ports.

With increasing uncertainty about weather and future climate projections, a novel sense of uncertainty is being introduced into agriculture. Producers are faced with new types of climate-driven problems; and there is a lack of knowledge and experience-derived responses from which to formulate new management strategies.

For the near future, adaptation by agriculture to changing climate will likely continue the existing fluid, gradual pace driven primarily by producer-level adaptive actions to manage increased variability and extreme weather events. The effectiveness of near-term, producer-level adaptations to the novel risks associated with climate change will be enhanced by new knowledge, technology, policies, and programs that both contribute to managing climate risk at the enterprise level and assist with avoiding actions that might reduce future adaptive capacity. These climate risk management strategies will likely involve both short-term and longer term adaptation planning that takes into account the projected exposures and specific sensitivities of different production systems. Many short-term adaptive actions at the producer level will likely involve extensions of existing management strategies to reduce the risk of weather variability to agricultural productivity.

The concept of vulnerability provides a useful framework from which to manage the complexities of adapting agriculture to climate change. Effective adaptation requires understanding and adjusting one (or more) of the three determinants of agricultural system vulnerability to climate change. These include agricultural system: (1) exposure to changing climate conditions, (2) sensitivity to changing climate conditions, and (3) capacity for effective adaptive action. The sections that follow summarize the findings and conclusions of this report with respect to these three elements (exposure, sensitivity, and adaptive capacity) of agricultural vulnerability to a changing climate.

Exposure to Changing Climate Conditions

Agriculture's exposure to climate change will depend on the trajectory of global greenhouse gas emissions in the coming decades and on how those emissions ultimately translate into a changing climate. Efforts to estimate such projections have a high degree of uncertainty, related to uncertain assumptions about factors such as population increases and extent of emissions mitigation efforts, as well as to uncertainty in the science of climate change. Nevertheless, it is very likely that U.S. climate conditions will continue to change throughout the 21st century, largely driven by overall emissions of GHGs and aerosols, as well as due to the strength of feedbacks in the climate system. Looking ahead to 2100, a low-emissions scenario is likely to produce summer-time warming of 3°C to 4°C degrees in much of the Interior West

(excluding coastal areas), with warming of 2°C to 3°C almost everywhere else. A high-emissions scenario is likely to result in warming of 5°C to 6°C in much of the Interior West and Midwest, with warming of 3°C to 5°C degrees in the Southeast and far western regions, and significant increases in hot nights during the summer.

Projected changes in precipitation for North America are more uncertain because they are sensitive to both local conditions, as well as to shifts in large-scale circulation patterns. The seasonality of precipitation is an important factor for agriculture, particularly in western regions that rely on winter accumulation of snow and gradual release of water stored in snowpack throughout the spring and summer. Most regions of the northern and central United States are projected to see an increase of 5% to 15% in winter precipitation over the next 30-40 years; areas along the southern border will likely see decreases of 5% to 10%, with southern Texas possibly experiencing decreases of up to 15% to 20%. Projections of change in summer precipitation over the next 30 to 40 years show that the Northwest is likely to become noticeably drier, with reductions of 15% to 25% in summertime precipitation. Much of the central South will likely sees decreases of about 5%, while some northern central and eastern U.S. regions are projected to experience increases of 5% to 15%. Over the Midwest, springtime precipitation is expected to increase with the potential for more intense storms. It is important to note, however, that increased precipitation does not necessarily translate into more available moisture for agriculture at the time when the water is needed; changes in timing and distribution of precipitation will be critical determinants of water availability and management options under changing climate conditions.

Sensitivity to Changing Climate Conditions

The effects of climate change on agricultural production can be classified as either direct or indirect. Direct effects refer to the biophysical effects of changing abiotic climate conditions on crop and livestock growth, development, and conditions. Indirect effects include biotic effects – effects arising from changing agro-ecosystem conditions related to insect, disease and weed pressure – as well as induced effects on input resources (land, water, soil) and market-mediated effects on input and output prices.

Plant response to climate change is dictated by a complex set of interactions to CO_2, temperature, solar radiation, and precipitation. To date, research has focused on single factors in controlled environments, creating considerable uncertainty about climate change effects on crop production. Changes in average climate conditions are important, as are changes in the timing and incidence of extreme climate events. Each crop species has a given set of temperature thresholds that define the upper and lower boundaries for growth along with an optimum temperature, with critical periods of exposure to temperatures such as the pollination stage when pollen is released to fertilize the plant and trigger development of reproductive organs, for fruit, grain, or fiber (Hatfield et al. 2011). The effects of higher temperatures on the quality of crop production is not well understood, but of particular concern to specialty crops; the value of specialty crops is derived not just by tonnage but also by the quality of the harvested product, such as the size of a peach, the red blush on an apple, or the bouquet of a red wine produced from a particular vineyard. Extreme events may also reduce the efficiency of farm inputs by reducing the flexibility of timing of farm operations and applications (Tubiello et al. 2007).

While increasing CO_2 in the atmosphere has a positive effect on plant growth and decreases soil water use rates (Kimball 2011), the magnitude of influence of increasing atmospheric CO_2 on crop yields also depends on the status of other constraints such as nutrient and water limitations, and timing of crop exposure to temperature and water extremes. Further, the overall effects on crop production, depends on the relative response of the crop versus the response of most weeds and other competitors for resources; changes in climate will affect both the crop and the pathogen, and understanding these changes will be critical to avoid increased losses in crop productivity. Quality of crop may also be affected; in forage and grain crops, exposure to increased CO_2 causes a reduction in grain and forage quality (Morgan et al. 2004).

Livestock agriculture is similarly affected through direct climate impacts on the animals, the resources that they rely on, and their management costs. Abiotic climate impacts on animals are directly related to the ability to maintain a body temperature within the optimum range for growth and reproduction. Furthermore, conception rates decline with increasing levels of the Thermal Heat Index (THI) (Hahn 1995; Amundson et al. 2006). While a portion (estimated to be about 50%) of the declines of domestic livestock

production during hotter summers can be offset by increased production due to milder winter conditions, loss of productivity, as reflected in increased time to slaughter weight or decreased dairy milk production, has been estimated to represent significant costs to producers (Frank et al. 2001). Positive changes to winter THI levels will not offset summer declines in conception rates, particularly in cattle that breed primarily in spring and summer.

While climate change impact analysis has primarily focused on such direct, abiotic impacts, the limited amount of research on indirect effects of climate change suggests indirect impacts are likely to be significant in both the crop and livestock sectors. The most common projections for pest insects, pathogens, and viral diseases are expanded or shifted ranges with increasing temperature (Gutierrez et al. 2006; Diffenbaugh et al. 2008; Mika et al. 2008; Gutierrez et al. 2009; Savary, 2011; Canto et al. 2009; Navas-Castillo et al. 2011). Warming temperatures may lead to additional insect generations in a single season, resulting in increased insect abundance and faster development of pesticide resistance.

Weeds are also likely to thrive under changing climate conditions. The habitable zone of many weed species is largely determined by temperature, and weed scientists have long recognized the potential for northward expansion of weed species' ranges as the climate changes (Patterson et al. 1999). Furthermore, many weeds respond more positively to increasing CO_2 than most cash crops. To date, for all weed/crop competition studies where the photosynthetic pathway is the same, weed growth is favored as CO_2 increases (Ziska and Teasdale 2000; Ziska and Bunce 1998). Recent research suggests that glyphosate, the most widely used herbicide in the United States, loses its efficacy on weeds grown at CO_2 levels that likely will occur in the coming decades (Ziska et al. 1999).

Indirect impacts of climate change on animal agriculture will play out through forage and feed markets as well as through biotic impacts of disease and other pests. Regional warming and changes in rainfall distribution may lead to changes in the spatial or temporal distributions of diseases sensitive to moisture, such as anthrax, blackleg, hemorrhagic septicemia, and vector-borne diseases (Baylis and Githeko 2006). Climate change also may influence the abundance and/or distribution of the competitors, predators, and parasites of vectors themselves (Thornton 2010). Hotter weather may increase the incidence of ketosis, mastitis, and lameness in dairy cows, and enhance

The effects of higher temperatures on the quality of crop production is not well understood, but of particular concern to specialty crops; the value of specialty crops is derived not just by tonnage but also by the quality of the harvested product, such as the size of a peach, the red blush on an apple, or the bouquet of a red wine produced from a particular vineyard.

growth of mycotoxin-producing fungi, particularly if moisture conditions are favorable (Gaughan et al. 2009).

The impacts of climate change on forage and feed price, availability, and quality are mixed and sensitive to climate projection. It is likely that rising CO_2 concentrations over the last 150 years have increased productivity of pastures (Polley et al. 2003; Izaurralde et al. 2011) and that future climatic conditions will enhance productivity on most rangelands over the next 30 years (Izaurralde et al. 2011). However, increased CO_2 concentrations may also affect forage crop quality (Morgan et al. 2004) and the projected impacts of climate change on feed price are highly sensitive to uncertain climate projections and crop yield assumptions.

Climate change will also indirectly affect agriculture through its impacts on soil and water resources as well as on the ecosystem services upon which agricultural productivity relies. Future changes in the climatic drivers of soil erosion (e.g., changes in the intensity of rainfall) and enterprise-level management adaptations to a changing climate (e.g., crop selection and dates of planting, harvest, and tillage), for instance, have the potential to greatly influence soil erosion rates, with a general trend in the United States toward higher rates of erosion. Increased rates of erosion can decrease soil productivity through increased loss of soil organic carbon and other essential nutrients, as well as reduced soil water storage capacity.

An improved understanding of potential biotic and other indirect and induced impacts is therefore a critical element of a comprehensive climate risk assessment for agriculture. A notable element of the indirect effects of climate change is an expected increase of input costs for the management of insects, weeds, and pathogens.

Capacity of the Agricultural System to Adapt to Changing Climate Conditions

For the short term, the dynamics of the agricultural system will likely enable it to respond to climate changes in ways that partially offset the negative direct and indirect effects of climate change, while taking advantage of new opportunities that may arise through changing climate. Such adaptive behaviors can occur at multiple levels of the agricultural system, for example at the enterprise level (through shifts, expansion, or intensification of production),

the market level (through changing patterns of trade and consumption), and/or at the policy level (through programs that spread risk and support adaptive responses).

Adaptation by the plant sector will encompass potential strategies ranging from altering planting dates, selecting cultivars with different maturity ratings, utilizing more water-efficient crops and supplemental irrigation to offset precipitation deficits, or changing crop types and cropping patterns for a given location. Annual crops have more flexibility in adaptation strategies than perennial crops. The economic investment into perennial vine and tree crops and the expected lifetime of perennial crops will prove to be more challenging for adaptation because of the length of time required to develop new cultivars and/ or to introduce more adapted perennial plants into a region. Some adaptation strategies for perennial trees that have a specific chilling requirement may require development of chemical methods to mimic chilling hours. Adaptation strategies to cope with the direct impacts of abiotic stress will be different than strategies to address biotic stresses from insects, diseases, and weeds. One of the first approaches to offset the biotic stresses will be increased surveillance of emerging pest populations.

Adaptation of the animal production sector could involve a shift to livestock types with greater tolerance of relatively high temperatures which better utilize existing vegetation, and are more resistant to livestock pests (Morgan 2005). Preparing for climate change will require appropriate education and training, development of strategic plans for adjusting to changing conditions, recognition of animal needs and potential stress levels, adopting strategies to minimize and/or mitigate the stress, and selection of animals and management strategies that are compatible with the production enterprise (Gaughan 2009). Livestock managers will need to be proactive and consider resource availability (feed, water, health care, economic factors, the land base, human capital, and the animals) when selecting climate change adaptation strategies.

Economic impact research suggesting that domestic agricultural markets and producer and consumer welfare will remain relatively stable in the short-term despite changing climate conditions usually assumes that producers take successful adaptive actions, such as those described above. While such studies can be interpreted to indicate that the United States has a couple of decades – a buffer period – before the impacts of climate change will be sufficiently intense

to create large disruptions in the agricultural system, such results must be interpreted in the context of recognized limits to their analyses. Most existing climate change impact studies, for instance, are limited in scope, relying on an assessment of only one or two direct yield impacts, while excluding indirect impacts and interactions between impacts, such as changes in pest and disease pressures that can significantly decrease productivity and increase management costs in crops and livestock.

In addition, integrated economic analyses have focused on the impacts of average changes in temperature and precipitation patterns; however, the sensitivity of an agricultural system to climate change is a function of financial capacity to withstand increasing variability in production and returns, including catastrophic loss (Smit and Skinner 2002; Beach et al. 2010). A failure to consider the impacts of variability and extreme weather events on crop yields and farm returns, as well as potential credit and other resource constraints that limit a system's technical and financial ability to adapt, may underestimate the system's financial viability in the face of changing climate conditions. On the other hand, ongoing research and technology investment, such as breeding for drought-tolerant crops, management to improve ecosystem resilience to climate impacts and the use of adaptive management strategies may produce additional opportunities for adaptation that will increase the capacity of the system to respond to regional changes in climate conditions.

In the longer term, continuing changes in climate conditions are likely to overwhelm the ability of the agricultural system to adapt using existing technologies without significant disruptions to elements of the agricultural system such as producer welfare, consumer welfare, or the ecosystem services that support, and are impacted by, agricultural production.

Agricultural adaptation to climate change is challenged by the increasing pace of change, the complex interactions between the global climate system and the agricultural system, and the complexity of adaptation processes (Easterling et al. 2007). To date, U.S. agricultural research and development efforts have focused on improving adaptive capacity at the enterprise-level; however, strengthening adaptive capacity solely at this level may not be sufficient to successfully address the challenge of climate change (Burton and Lim 2005; Howden et al. 2007).

Mainstreaming climate knowledge has been proposed as a core adaptation strategy in agriculture as well as many other economic sectors (Howden et al. 2007; Adger et al. 2007; Field et al. 2007). Mainstreaming climate knowledge improves adaptive capacity of the agricultural system by ensuring that land managers, technical advisors, researchers, private businesspeople, government program managers, and policymakers are aware of current and projected climate impacts and can access best management practices to reduce risks and capture opportunities. Taking such a comprehensive, climate risk management approach to agricultural adaptation offers great potential to promote effective adaptive action by decisionmakers throughout the multiple dimensions of U.S. agriculture. Building a climate-ready U.S. agricultural system will require easy access to useable climate knowledge, improved climate risk management strategies, new processes to support effective adaptive actions, and development of resilient production systems (Howden et al. 2007).

Research Needs

Agricultural research, especially publically funded agricultural research, is a well-documented contributor to the success of U.S. agriculture (Fuglie and Heisey, 2007). Publically and privately funded research will provide innovations needed for agriculture to adapt to changing climate.

The research needs identified in this report are categorized below within a vulnerability framework and address specific actions that would serve to improve understanding and management of the exposure, sensitivity, and adaptive capacity of U.S. agriculture to climate change. Attention to these research needs would enhance the ability of the U.S. agriculture sector to anticipate and respond to the challenges presented by changing climate conditions.

Some overarching research needs include the following:

- Improve projections of future climate conditions for time scales of seasons to multiple decades, including more precise information about changes of average and extreme temperatures, precipitation, and related variables (e.g., evapotranspiration, soil moisture).

- Evaluate the sensitivity of diverse plant and animal production systems to key direct and indirect climate change effects and their interactions.

In the longer term, continuing changes in climate conditions are likely to overwhelm the ability of the agricultural system to adapt using existing technologies without significant disruptions to elements of the agricultural system such as producer welfare, consumer welfare, or the ecosystem services that support, and are impacted by, agricultural production.

- Develop and extend the knowledge, management strategies and tools needed by U.S. agricultural stakeholders to enhance the adaptive capacity of plant and animal production systems to climate variability and extremes. While existing management and agronomic options have demonstrated significant capacity for expanding adaptation opportunities, new adaptive management strategies, robust risk management approaches, and breeding and genetic advances offer much potential, but have yet to be evaluated.

Understanding Exposure

The vulnerability of an agricultural system to climate change is dependent in part on the character, magnitude and rate of climate variation to which a system is exposed. Effective adaptation will be enhanced by research to:

- Improve projections of future climate conditions for time scales of seasons to multiple decades, including more precise information about changes of average and extreme temperatures, precipitation, and related variables (e.g., evapotranspiration, soil moisture). Such projections are needed to better understand exposure to climate risks, and support effective assessment, planning, and decision-making across the multiple dimensions of the U.S. agricultural system.

- Enable projection of future climate conditions at finer temporal scales (hourly and daily versus weekly, monthly, or annual averages) and spatial scales (1-10 km, as opposed to 50-100 km). This finer scale information would permit decision-makers from many parts of the agricultural system to examine the potential effects of climate change on specific crop and livestock production systems in specific regions. There is also a need to include more precise decadal-scale projections to integrate climate information into longer term planning and improved information about the probability of potential changes to effectively manage climate risks.

- Develop the modeling systems that produce climate and impact projections through the use of standard socioeconomic scenarios and access to more accurate, complete, and integrated observations of climate change and its effects on agricultural systems to improve process-level understanding and validate model simulations.

- Improve the accuracy and range of weather predictions (as opposed to longer-term, scenario-dependent climate projections) and seasonal forecasts. Better forecasts are needed to understand near-term exposure and support tactical decision-making at all levels of the agricultural system. Improved forecasting is particularly critical given the expected increases in the variability of weather and the incidence of extreme conditions.

Understanding Sensitivity

The nature and degree of response to key climate change drivers determines the sensitivity of the agricultural system to climate change effects. Critical thresholds, feedbacks, and synergies operating at multiple temporal and spatial scales complicate efforts to assess agricultural system sensitivity to climate change. Effective adaptation to climate change effects will be enhanced by research to:

- Improve understanding of both direct and indirect climate change effects and their interactions on plant and animal production systems, together with new tools for exploring their dynamic interactions throughout the multiple dimensions of the U.S. agricultural sector;

- Enhance capabilities to quantify and screen plant and animal response to water and temperature extremes;

- Improve understanding of climate change effects on the natural and biological resources upon which agricultural productivity depends, particularly soil and water resources;

- Improve understanding of climate change effects on existing agricultural landscape patterns and production practices;

- Improve understanding of the economic impacts of climate change and how those impacts are distributed.

- Develop improved integrated assessment models and establish ecosystem manipulation sites to enable experiments that examine the impacts of simultaneous interacting multiple stresses on plant and animal production systems.

Enhancing Adaptive Capacity

Because agricultural systems are human-dominated ecosystems, the vulnerability of agriculture to climate change is strongly dependent on the responses taken by humans to adapt to climate change effects. The adaptive capacity of U.S. agriculture will be enhanced by research to:

- Improve understanding of the key determinants (social, economic, and ecological) of adaptive capacity and resilience in agricultural systems;

- Develop effective methods for the assessment of adaptive capacity;

- Identify and extend information about existing best management practices that offer "no-regrets" and "low regrets" adaptation options;

- Develop resilient crop and livestock production systems and the socio-economic and cultural/institutional structures needed to support them; Develop and extend adaptive management strategies and climate risk management tools to improve decision-making throughout the U.S. agricultural sector;

- Improve understanding of the social limits to adaptation, including the effects of cost/benefit considerations, technological feasibility, beliefs, values and attitudes, and resource constraints on adaptive response.

- Develop effective adaptation planning and assessment strategies useful to decision makers operating throughout the multiple dimensions of the U.S. agricultural system.

Understanding Basic Processes

Agricultural systems are notable for the complex interactions between the physical, biological, and chemical environment; climate plays a major role in affecting these basic processes. With additional research, we could advance our understanding of effects that climate has on agriculture. Some outstanding research needs are sketched out below. While not exhaustive, this list indicates some of the types of information that could advance our foundational knowledge of how the agroecosystem works, and also provide insight on the impacts of changing climate on this system.

- One critical environmental service is pollination. Lacking basic knowledge in this area makes it difficult to assess the potential response to climate change.

- Future crop yield increases will depend largely on our abilities to increase yield potential. To increase yield potential, an evaluation must be made as to the necessity of more targeted varietal selection designed with specific global change factors in mind (e.g., high temperatures, or deficit soil water). Alternatively, it may be that the best crop yield choice is to opt for a more generalized yield selection based on existing conditions, doing so with the expectation that enough new cultivars will be produced approximately every 7 years that will allow producers to manage in the face of a changing environment.

- Scientific understanding of crop response to changes in CO_2, temperature, ozone water and other environmental factors affected by climate change is far from complete; understanding these responses will guide genetic improvement.

- Linking physiological responses to genomic traits in plants will provide a level of understanding for potential resilience mechanisms to climate stress.

- More rapid generation turnover methods and effective selection criteria will be needed to make progress in developing perennial cultivars that can best adapt to changing climate. Marker-selected breeding techniques and other molecular tools will be needed for climate adaptation, as well as improved water use efficiency.

- Development of more robust methods of quantifying environmental stress on animals and differences among animal production systems will be required to define the parameters of adaptive systems capable of avoiding stresses caused by climate change.

- Understanding the fundamental role of environment variables on pest population dynamics will be necessary to define the potential role of climate change on the indirect impacts caused by pests.

Climate Change and Sustainable Agriculture

The direct and indirect effects of climate change on agriculture will challenge the Nation's ability to attain the four goals of agriculture sustainability as described by a National Academies of Science report (2010):

• Satisfy human food, feed, and fiber needs, and contribute to biofuel needs;

• Enhance environmental quality and the resources base;

• Sustain economic viability of agriculture; and

• Enhance the quality of life for farmers, farm workers, and society as a whole.

Effective adaptation to climate change will be necessary for U.S. agriculture to achieve these goals during the 21st century. Successful adaptation planning requires both an improved understanding of the potential system-wide impacts of climate change and opportunities for adaptation, as well as more effective methods of managing the novel uncertainty associated with the management of the U.S. agricultural system under a changing climate.

Appendix A

References Cited (by Chapter)

Chapter 1 References

CCSP, 2008. *The effects of climate change on agriculture, land resources, water resources, and biodiversity in the United States. A report by the U.S. Climate Change Science Program and the Sub-committee on Global Change Research.* P. Backlund, A. Janetos, S. Schimel, J. Hatfield, K. Boote, P. Fay, L. Hahn, C. Izaurralde, B.A. Kimball, T. Mader, J. Morgan, D. Ort, W. Polley, A. Thomson, D. Wolfe, M. Ryan, S. Archer, R. Birdsey, C. Dahm, L. Heath, J. Hicke, D. Hollinger, T. Huxman, G. Okin, R. Oren, J. Randerson, W. Schlesinger, D. Lettenmaier, D. Major, L. Poff, S. Running, L. Hansen, D. Inouye, B.P. Kelly, L. Meyerson, B. Peterson, R. Shaw 362 pp. U.S. Department of Agriculture, Washington, DC, USA.

IPCC, 2007a. Summary for policymakers, *in* S. Solomon, et al. (ed.), *Climate change 2007: The physical science basis. Contribution of Working Group I to the Fourth Assessment Report of the Intergovernmental Panel on Climate Change.* Cambridge University Press, Cambridge, United Kingdom, and New York, USA.

Chapter 2 References

Allison, H.E., and R.J. Hobbs., 2004. Resilience, adaptive capacity, and the "lock-in trap" of the Western Australian agricultural region. *Ecology and Society*, 9(1): 3.

Arbuckle, J.G.J., 2011. *Farmer perspectives on climate change and agriculture*, paper presented at Climate Change Conference, November 1-3, 2011. Heartland Regional Water Coordination Initiative, Lied Conference Center, Nebraska City, NE.

Atwell, R.C., L.A. Schulte, and L.M. Westphal, 2008. Linking resilience theory and diffusion of innovations theory to understand the potential for perennials in the U.S. corn belt. *Ecology and Society*, 14(1): 30.

Dessai, S., M. Hulme, R. Lempert, and R. Pielke, 2009. Climate prediction: A limit to adaptation?, *in* W.N. Adger, I. Lorenzoni and K.L. O'Brien (eds.), *Adapting to climate change: Thresholds, values, governance.* Cambridge University Press, Cambridge; New York.

Dimitri, C., A. Effland, and N. Conklin, 2005. *The 20th century transformation of U.S. agriculture and farm policy.* U.S. Department of Agriculture, Economic Research Service, Economic Information Bulletin No. 14.

Easterling, W.E., 2009. Guidelines for adapting agriculture to climate change, *in* D. Hillel and C. Rosenzweig (eds.), *Handbook of climate change and agroecosystems: Impacts, adaptation, and mitigation.* Imperial College Press; Distributed by World Scientific Publishing Co., London; Singapore; Hackensack, NJ.

Easterling, W.E., P.K. Aggarwal, P. Batima, K.M. Brander, L. Erda, S.M. Howden, A. Kirilenko, J. Morton, J.-F. Soussana, J. Schmidhuber, and F.N. Tubiello, 2007. Food, fibre and forest products, *in* M.L. Parry, O.F. Canziani, J.P. Palutikof, P.J. Van Der Linden and C.E. Hanson (eds.), *Climate change 2007: Impacts, adaptation and vulnerability: Working group II contribution to the Fourth Assessment Report of the IPCC Intergovernmental Panel on Climate Change*, pp. 273-313. Cambridge University Press, Cambridge, UK.

Gunderson, L.H., and C.S. Holling, 2002. *Panarchy: Understanding transformations in human and natural systems.* Island Press, Washington, DC.

Hatfield, J.L., K.J. Boote, B.A. Kimball, L.H. Ziska, R.C. Izaurralde, D. Ort, A.M. Thomson, and D. Wolfe, 2011. Climate impacts on agriculture: Implications for crop production. *Agronomy Journal*, 103(2): 351-370.

Hatfield, J.L., and J.H. Prueger, 2011. Spatial and temporal variation in evapotranspiration, *in* G. Gerosa (ed.) *Evapotranspiration - from measurements to agricultural and environmental applications*, p. 410. InTech.

Hendrickson, J., G.F. Sassenrath, D. Archer, J. Hanson, and J. Halloran, 2008. Interactions in integrated U.S. agricultural systems: The past, present and future. *Renewable Agriculture and Food Systems*, 23 (Special Issue 04): 314-324.

Hoppe, R.A., P. Korb, E. O'Donoghue, and D.E. Banker, 2007. *Structure and finances of U.S. farms: Family farm report, 2007*, U.S. Department of Agriculture, Economic Research Service Economic Information Bulletin No. 24.

Hoppe, R.A., J.M. MacDonald, and P. Korb, 2010. *Small farms in the United States: Persistence under pressure.* U.S. Department of Agriculture, Economic Research Service, Economic Information Bulletin No. 63.

Howden, M., A. Ash, S. Barlow, T. Booth, S. Charles, B. Cechet, S. Crimp, R. Gifford, K. Hennessy, R. Jones, M. Kirschbaum, G. McKeon, H. Meinke, S. Park, B. Sutherst, L. Webb, and P. Whetton, 2003. *An overview of the adaptive capacity of the Australian agricultural sector to climate change – options, costs and benefits.* Scientific and Industrial Research Organisation, Collingwood, Victoria http://www.cse.csiro.au/publications/2003/AGOAgClimateAdaptationReport.pdf.

Howden, S.M., J.F. Soussana, F.N. Tubiello, N. Chhetri, M. Dunlop, and H. Meinke, 2007. Adapting agriculture to climate change. *Proceedings of the National Academy of Sciences*, 104(50): 19691-19696.

Jackson, L., M. van Noordwijk, J. Bengtsson, W. Foster, L. Lipper, M. Pulleman, M. Said, J. Snaddon, and R. Vodouhe, 2010. Biodiversity and agricultural sustainability: From assessment to adaptive management. *Current Opinion in Environmental Sustainability*, 2(1-2): 80-87.

Janssen, M.A., Ö. Bodin, J.M. Anderies, T. Elmqvist, H. Ernstson, R.R.J. McAllister, P. Olsson, and P. Ryan, 2006. Toward a network perspective of the study of resilience in social-ecological systems. *Ecology and Society*, 11(1): 15.

Lin, B.B., 2011. Resilience in agriculture through crop diversification: Adaptive management for environmental change. *BioScience*, 61(3): 183-193.

Littell, J.S., D. McKenzie, B.K. Kerns, S. Cushman, and C.G. Shaw, 2011. Managing uncertainty in climate-driven ecological models to inform adaptation to climate change. *Ecosphere*, 2(9): art102.

Liu, J., R. Munang, M. Rivington, J. Thompson, P.V. Gardingen, and D. Dinesh, 2011. *Restoring the natural foundation to sustain a green economy: A century-long journey for ecosystem management.* International Ecosystem Management Partnership, United Nations Environmental Program Policy Brief 6. http://www.unep.org/ecosystemmanagement/Portals/7/Documents/policy%20series%206.pdf.

McLeman, R., D. Mayo, E. Strebeck, and B. Smit, 2008. Drought adaptation in rural eastern Oklahoma in the 1930s: Lessons for climate change adaptation research. *Mitigation and Adaptation Strategies for Global Change*, 13(4): 379-400.

McLeman, R., and B. Smit, 2006. Vulnerability to climate change hazards and risks: Crop and flood insurance. *The Canadian Geographer*, 50(2): 217+.

Medvigy, D., and C. Beaulieu, 2011. Trends in daily solar radiation and precipitation coefficients of variation since 1984. *Journal of Climate*, 25(4): 1330-1339.

Melhim, A., E.J. O'Donoghue, and C.R. Shumway, 2009. Do the largest firms grow and diversify the fastest? The case of U.S. dairies. *Review of Agricultural Economics*, 31(2): 284-302.

Melhim, A., E.J. O'Donoghue, and C.R. Shumway, 2009. What does initial farm size imply about growth and diversification? *Journal of Agricultural and Applied Economics*, 41(1): 193-206.

Moriondo, M., C. Pacini, G. Trombi, C. Vazzana, and M. Bindi, 2010. Sustainability of dairy farming system in Tuscany in a changing climate. *European Journal of Agronomy*, 32(1): 80-90.

Munang, R.T., I. Thiaw, and M. Rivington, 2011. Ecosystem management: Tomorrow's approach to enhancing food security under a changing climate. *Sustainability*, 3(7): 937-954.

Nelson, D.R., W.N. Adger, and K. Brown, 2007. Adaptation to environmental change: Contributions of a resilience framework. *Annual Review of Environment and Resources*, 32: 395-419.

Nelson, R., P. Kokic, S. Crimp, P. Martin, H. Meinke, S.M. Howden, P. de Voil, and U. Nidumolu, 2010. The vulnerability of Australian rural communities to climate variability and change: Part II—integrating impacts with adaptive capacity. *Environmental Science & Policy*, 13(1): 18-27.

New, M., A. Lopez, S. Dessai, and R. Wilby, 2007. Challenges in using probabilistic climate change information for impact assessments: An example from the water sector. *Philosophical Transactions of the Royal Society A: Mathematical, Physical and Engineering Sciences*, 365(1857): 2117-2131.

Newton, A., S. Johnson, and P. Gregory, 2011. Implications of climate change for diseases, crop yields and food security. *Euphytica*, 179(1): 3-18.

NRC, 2010. *Toward sustainable agricultural systems in the 21st century.*. National Research Council Committee on Twenty-First Century Systems, Agriculture, National Academies Press, Washington, D.C.

O'Donoghue, E.J., R.A. Hoppe, D.E. Banker, R. Ebel, K. Fuglie, P. Korb, M. Livingston, C. Nickerson, and C. Sandretto, 2011. *The changing organization of U.S. farming*, 77 pp. U.S. Department of Agriculture, Economic Research Service, Economic Information Bulletin No. 88.

Ostrom, E., 2009. A general framework for analyzing sustainability of social-ecological systems. *Science*, 325(5939): 419-422.

Pan, Z., M. Segal, R.W. Arritt, and E.S. Takle, 2004. On the potential change in solar radiation over the U.S. due to increases of atmospheric greenhouse gases. *Renewable Energy*, 29(11): 1923-1928.

Qian, T., A. Dai, and K.E. Trenberth, 2007. Hydroclimatic trends in the Mississippi River Basin from 1948 to 2004. *Journal of Climate*, 20(18): 4599-4614.

Reidsma, P., F. Ewert, A.O. Lansink, and R. Leemans, 2010. Adaptation to climate change and climate variability in European agriculture: The importance of farm level responses. *European Journal of Agronomy*, 32(1): 91-102.

Rivington, M., K.B. Matthews, G. Bellocchi, K. Buchan, C.O. Stöckle, and M. Donatelli, 2007. An integrated assessment approach to conduct analyses of climate change impacts on whole-farm systems. *Environmental Modelling & Software*, 22(2): 202-210.

Robertson, G.P., and S.M. Swinton, 2005. Reconciling agricultural productivity and environmental integrity: A grand challenge for agriculture. *Frontiers in Ecology and the Environment*, 3(1): 38-46.

Rockstrom, J., W. Steffen, K. Noone, A. Persson, F.S. Chapin, E.F. Lambin, T.M. Lenton, M. Scheffer, C. Folke, H.J. Schellnhuber, B. Nykvist, C.A. de Wit, T. Hughes, S. van der Leeuw, H. Rodhe, S. Sorlin, P.K. Snyder, R. Costanza, U. Svedin, M. Falkenmark, L. Karlberg, R.W. Corell, V.J. Fabry, J. Hansen, B. Walker, D. Liverman, K. Richardson, P. Crutzen, and J.A. Foley, 2009. A safe operating space for humanity. *Nature*, 461(7263): 472-475.

Romero, C., and A. Agrawal, 2011. Building interdisciplinary frameworks: The importance of institutions, scale, and politics. *Proceedings of the National Academy of Sciences*, 108(23): E196.

Rounsevell, M.D.A., D.T. Robinson, and D. Murray-Rust, 2012. From actors to agents in socio-ecological systems models. *Philosophical Transactions of the Royal Society B: Biological Sciences*, 367(1586): 259-269.

Smith, P., and J.E. Olesen, 2010. Synergies between the mitigation of, and adaptation to, climate change in agriculture. *The Journal of Agricultural Science*, 148(05): 543-552.

Stanhill, G., and S. Cohen, 2001. Global dimming: A review of the evidence for a widespread and significant reduction in global radiation with discussion of its probable causes and possible agricultural consequences. *Agricultural and Forest Meteorology*, 107(4): 255-278.

Stanhill, G., and S. Cohen, 2005. Solar radiation changes in the United States during the twentieth century: Evidence from sunshine duration measurements. *Journal of Climate*, 18(10): 1503-1512.

Tomich, T.P., S. Brodt, H. Ferris, R. Galt, W.R. Horwath, E. Kebreab, J.H.J. Leveau, D. Liptzin, M. Lubell, P. Merel, R. Michelmore, T. Rosenstock, K. Scow, J. Six, N. Williams, and L. Yang, 2011. Agroecology: A review from a global-change perspective. *Annual Review of Environment and Resources*, 36(1): 193-222.

Trostle, R., D. Marti, S. Rosen, and P. Westcott, 2011. *Why have food commodity prices risen again?* U.S. Department of Agriculture, Economic Research Service.

van Apeldoorn, D.F., K. Kok, M.P.W. Sonneveld, and T. Veldkamp, 2011. Panarchy rules: Rethinking resilience of agroecosystems, evidence from Dutch dairy-farming. *Ecology and Society*, 16(1).

Walker, B., S. Carpenter, J. Anderies, N. Abel, G. Cumming, M. Janssen, L. Lebel, J. Norberg, G.D. Peterson, and R. Pritchard, 2002. Resilience management in social-ecological systems: A working hypothesis for a participatory approach. *Conservation Ecology*, 6: 14.

Wolfe, D., L.H. Ziska, C. Petzoldt, A. Seaman, L. Chase, and K. Hayhoe, 2008. Projected change in climate thresholds in the northeastern U.S.: Implications for crops, pests, livestock, and farmers. *Mitigation and Adaptation Strategies for Global Change*, 13(5): 555-575.

USDA-National Agriculture Statistics Service 2007. Census of Agriculture. (www.agcensus.usda.gov) accessed 29-Jun-2012

Young, O.R., F. Berkhout, G.C. Gallopin, M.A. Janssen, E. Ostrom, and S. van der Leeuw, 2006. The globalization of socio-ecological systems: An agenda for scientific research. *Global Environmental Change*, 16(3): 304-316.

Chapter 3 References

Alexander, L.V., X. Zhang, T.C. Peterson, J. Caesar, B. Gleason, A.M.G. Klein Tank, M. Haylock, D. Collins, B. Trewin, F. Rahimzadeh, A. Tagipour, K. Rupa Kumar, J. Revadekar, G. Griffiths, L. Vincent, D.B. Stephenson, J. Burn, E. Aguilar, M. Brunet, M. Taylor, M. New, P. Zhai, M. Rusticucci, and J.L. Vazquez-Aguirre, 2006. Global observed changes in daily climate extremes of temperature and precipitation. *Journal of Geophysical Research*, 111(D5).

Allison, I., N.L. Bindoff, R.A. Bindschadler, P.M. Cox, N. de Noblet, M.H. England, J.A. Francis, N. Gruber, A.M. Haywood, D.J. Karoly, G. Kaser, C. Le Quéré, T.M. Lenton, M.E. Mann, B.I. McNeil, A.J. Pitman, S. Rahmstorf, E. Rignot, H.J. Schellnhuber, S.H. Schneider, S.C. Sherwood, R.C.J. Somerville, K. Steffen, E.J. Steig, M. Visbeck, and A.J. Weaver, 2009. *The Copenhagen diagnosis, 2009: Updating the world on the latest climate science.*, 60 pp. University of New South Wales Climate Change Research Centre (CCRC).

Booker, F.L., R. Muntifering, M.T. McGrath, K.O. Burkey, D.R. Decoteau, E.L. Fiscus, W. Manning, S.V. Krupa, A. Chappelka, and D.A. Grantz, 2009. The ozone component of global change: Potential effects on agricultural and horticultural plant yield, product quality and interactions with invasive species. *Journal of Integrative Plant Biology*, 51: 337-351.

Dentener, F., D. Stevenson, K. Ellingsen, T. Van Noije, and M. Schultz, 2010. *Hemispheric transport of air pollution 2010. Part A: Ozone and particulate matter.* United Nations, New York.

Dentener, F., D. Stevenson, K. Ellingsen, T. Van Noije, M. Schultz, M. Amann, C. Atherton, N. Bell, D. Bergmann, I. Bey, L. Bouwman, T. Butler, J. Cofala, B. Collins, J. Drevet, O.R. Doherty, B. Eickhout, H. Eskes, A. Fiore, M. Gauss, D. Hauglustaine, L. Horowitz, I. Isaksen, B. Josse, M. Lawrence, M. Krol, J.F. Lamarque, V. Montanaro, J.F. Muller, V.H. Peuch, G. Pitari, J. Pyle, J. Rodriguez, M. Sanderson, N.H. Savage, D. Shindell, S. Strahan, S. Szopa, K. Sudo, R. Vandingenen, O. Wild, and G. Zeng, 2006. The global atmospheric environment for the next generation. *Environmental Science and Technology*, 40: 3586-3594.

Emmons, L.K., S. Walters, P.G. Hess, J.F. Lamarque, G.G. Pfister, D. Fillmore, C. Granier, A. Guenther, D. Kinnison, T. Laepple, J. Orlando, X. Tie, G. Tyndall, C. Wiedinmyer, S.L. Baughcum, and S. Kloster, 2010. Description and evaluation of the model for ozone and related chemical tracers, version 4 (MOZART-4). *Geosci. Model Dev.*, 3(1): 43-67.

Groisman, P.Y., R.W. Knight, D.R. Easterling, T.R. Karl, G.C. Hegerl, and V.N. Razuvaev, 2005. Trends in intense precipitation in the climate record. *Journal of Climate*, 18(9): 1326-1350.

Hansen, J.E., R. Ruedy, M. Sato, and K. Lo, 2012. NASA GISS surface temperature (GISTEMP) analysis, *in Trends: A compendium of data on global change.* Carbon Dioxide Information Analysis Center, Oak Ridge National Laboratory, U.S. Department of Energy, Oak Ridge, Tenn., U.S.A.

IPCC, 2007a. Summary for policymakers, *in S. Solomon, et al. (eds.), Climate change 2007: The physical science basis. Contribution of Working Group 1 to the Fourth Assessment Report of the Intergovernmental Panel on Climate Change.* Cambridge University Press, Cambridge, United Kingdom, and New York, USA.

IPCC, 2007b. *Climate change 2007: Impacts, adaptation and vulnerability: Contribution of Working Group II to the Fourth Assessment Report of the Intergovernmental Panel on Climate Change.* Cambridge University Press, Cambridge, U.K.; New York.

Jones, P.D., D.H. Lister, T.J. Osborn, C. Harpham, M. Salmon, and C.P. Morice, 2012. Hemispheric and large-scale land-surface air temperature variations: An extensive revision and an update to 2010. *J. Geophys. Res.*, 117(D5): D05127.

Karl, T.R., and R.W. Knight, 1998. Secular trends of precipitation amount, frequency, and intensity in the United States. *Bulletin of the American Meteorological Society*, 79(2): 231-241.

Lamarque, J.-F., G.P. Kyle, M. Meinshausen, K. Riahi, S.J. Smith, D.P. Vuuren, A.J. Conley, and F. Vitt, 2011. Global and regional evolution of short-lived radiatively-active gases and aerosols in the representative concentration pathways. *Climatic Change*, 109(1-2): 191-212.

Matsuura, and Willmott, 2009. Data courtesy of University of Delaware, based on augmented global historical climatology network, version 2.

Maurer, E.P., L. Brekke, T. Pruitt, and P.B. Duffy, 2007. Fine-resolution climate projections enhance regional climate change impact studies. *EOS*, 88(47): 504.

Meehl, G.A., C. Tebaldi, G. Walton, D. Easterling, and L. McDaniel, 2009. Relative increase of record high maximum temperatures compared to record low minimum temperatures in the U.S. *Geophysical Research Letters*, 36(23).

Meehl, G.A., T.F. Stocker, W.D. Collins, P. Friedlingstein, A.T. Gaye, J.M. Gregory, A. Kitoh, R. Knutti, J.M. Murphy, A. Noda, S.C.B. Raper, I.G. Watterson, A.J. Weaver, and Z.-C. Zhao, 2007. Global climate projections, *in S. Solomon, et al. (eds.), Climate change 2007: The physical science basis. Contribution of working group 1 to the fourth assessment report of the Intergovernmental Panel on Climate Change*, pp. 747-845. Cambridge University Press, Cambridge.

Nakicenovic et al. 2000. *Special report on emissions scenarios: A special report of Working Group III of the Intergovernmental Panel on Climate Change.* Cambridge University Press, Cambridge; New York.

NOAA NCDC, 2011. *State of the climate: Global analysis for annual 2011.* NOAA National Climatic Data Center http://www.ncdc.noaa.gov/sotc/global/2011/13.

Prather, M., G. Gauss, T. Berntsen, I. Isaksen, J. Sundet, I. Bey, G. Brasseur, F. Dentener, R. Derwent, D. Stevenson, L. Grenfell, D. Hauglustaine, L. Horowitz, D. Jacob, L. Mickley, M. Lawrence, R. von Kuhlmann, J.-F. Muller, G. Pitari, H. Rogers, M. Johnson, J. Pyle, K. Law, M. van Weele, and O. Wild, 2003. Fresh air in the 21st century? *Geophysical Research Letters*, 30: 1100.

Soil and Water Conservation Society, 2003. *Soil erosion and runoff from cropland report from the USA*, 63 pp. Soil and Water Conservation Society, Ankeny, IA.

Solomon, S., G.K. Plattner, R. Knutti, and P. Friedlingstein, 2009. Irreversible climate change due to carbon dioxide emissions. *Proceedings of the National Academy of Sciences*, 106(6): 1704-1709.

Stevenson, D.S., F.J. Dentener, M.G. Schultz, K. Ellingsen, T.P.C. van Noije, O. Wild, G. Zeng, M. Amann, C.S. Atherton, N. Bell, D.J. Bergmann, I. Bey, T. Butler, J. Cofala, W.J. Collins, R.G. Derwent, R.M. Doherty, J. Drevet, H.J. Eskes, A.M. Fiore, M. Gauss, D.A. Hauglustaine, L.W. Horowitz, I.S.A. Isaksen, M.C. Krol, J.F. Lamarque, M.G. Lawrence, V. Montanaro, J.F. Müller, G. Pitari, M.J. Prather, J.A. Pyle, S. Rast, J.M. Rodriguez, M.G. Sanderson, N.H. Savage, D.T. Shindell, S.E. Strahan, K. Sudo, and S. Szopa, 2006. Multimodel ensemble simulations of present-day and near-future tropospheric ozone. *J. Geophys. Res.*, 111(D8): D08301.

US EPA, 2010. *Climate change indicators in the United States. EPA 430-r-10-007* www.epa.gov/climatechange/science/indicators/.

Zhang, X., F.W. Zwiers, G.C. Hegerl, F.H. Lambert, N.P. Gillett, S. Solomon, P.A. Stott, and T. Nozawa, 2007. Detection of human influence on twentieth-century precipitation trends. *Nature*, 448(7152): 461-465.

Chapter 4 References

Adkins, S., C.G. Webster, C.S. Kousik, S.E. Webb, P.D. Roberts, P.A. Stansly, and W.W. Turechek, 2011. Ecology and management of whitefly-transmitted viruses of vegetable crops in Florida. *Virus Research*, 159(2): 110-114.

Agrios, G.N., 2005. *Plant pathology.* Elsevier Academic Press, Amsterdam; Boston, MA.

Ahsan, N., Y. Nanjo, H. Sawada, Y. Kohno, and S. Komatsu, 2010. Ozone stress-induced proteomic changes in leaf total soluble and chloroplast proteins of soybean reveal that carbon allocation is involved in adaptation in the early developmental stage. *Proteomics*, 10(14): 2605-2619.

Ainsworth, E.A., P.A. Davey, C.J. Bernacchi, O.C. Dermody, E.A. Heaton, D.J. Moore, P.B. Morgan, S.L. Naidu, H.-s. Yoo Ra, X.-g. Zhu, P.S. Curtis, and S.P. Long, 2002. A meta-analysis of elevated [CO$_2$] effects on soybean (glycine max) physiology, growth and yield. *Global Change Biology*, 8(8): 695-709.

Ainsworth, E.A., and S.P. Long, 2005. What have we learned from 15 years of free-air CO$_2$ enrichment (FACE)? A meta-analytic review of the responses of photosynthesis, canopy properties and plant production to rising CO$_2$. *New Phytologist*, 165(2): 351-372.

Akey, D.H., B.A. Kimball, and J.R. Mauney, 1988. Growth and development of the pink bollworm, *Pectinophora gossypiella* (lepidoptera: Gelechiidae), on bolls of cotton grown in enriched carbon dioxide atmospheres. *Environmental Entomology*, 17(3): 452-455.

Alberto, A., L. Ziska, C. Cervancia, and P. Manalo, 1996. The influence of increasing carbon dioxide and temperature on competitive interactions between a C3 crop, rice *(Oryza sativa)* and a C4 weed *(Echinochloa glabrescens)*. *Functional Plant Biology*, 23(6): 795-802.

Alfaro, E.J., A. Gershunov, and D. Cayan, 2006. Prediction of summer maximum and minimum temperature over the central and western United States: The roles of soil moisture and sea surface temperature. *Journal of Climate*, 19(8): 1407-1421.

Altermatt, F., 2010. Climatic warming increases voltinism in European butterflies and moths. *Proceedings of the Royal Society B: Biological Sciences*, 277(1685): 1281-1287.

Andersen, C., 2003. Source-sink balance and carbon allocation below ground in plants exposed to ozone. *New Phytologist*, 157: 213-228.

Anderson, P.K., A.A. Cunningham, N.G. Patel, F.J. Morales, P.R. Epstein, and P. Daszak, 2004. Emerging infectious diseases of plants: Pathogen pollution, climate change and agrotechnology drivers. *Trends in Ecology & Evolution*, 19(10): 535-544.

Awmack, C., R. Harrington, and S. Leather, 1997. Host plant effects on the performance of the aphid *Aulacorthum solani* (kalt.) (homoptera: Aphididae) at ambient and elevated CO$_2$. *Global Change Biology*, 3(6): 545-549.

Bale, J.S., and S.A.L. Hayward, 2010. Insect overwintering in a changing climate. *Journal of Experimental Biology*, 213(6): 980-994.

Baylis, M., and A.K. Githeko, 2006. *T7.3: The effects of climate change on infectious diseases of animals*. Foresight http://www. bis.gov.uk/foresight/our-work/projects/published-projects/ infectious-diseases/reports-and-publications.

Belote, R.T., J.F. Weltzin, and R.J. Norby, 2004. Response of an understory plant community to elevated CO$_2$ depends on differential responses of dominant invasive species and is mediated by soil water availability. *New Phytologist*, 161(3): 827-835.

Bergant, K., S. Trdan, D. Žnidarčič, Z. Črepinšek, and L. Kajfež-Bogataj, 2005. Impact of climate change on developmental dynamics of *Thrips tabaci* (thysanoptera: Thripidae): Can it be quantified? *Environmental Entomology*, 34(4): 755-766.

Betts, R.A., P.D. Falloon, K.K. Goldewijk, and N. Ramankutty, 2007. Biogeophysical effects of land use on climate: Model simulations of radiative forcing and large-scale temperature change. *Agricultural and Forest Meteorology*, 142(2–4): 216-233.

Bezemer, T.M., and T.H. Jones, 1998. Plant-insect herbivore interactions in elevated atmospheric CO$_2$: Quantitative analyses and guild effects. *Oikos*, 82(2): 212-222.

Bijoor, N.S., C.I. Czimczik, D.E. Pataki, and S.A. Billings, 2008. Effects of temperature and fertilization on nitrogen cycling and community composition of an urban lawn. *Global Change Biology*, 14(9): 2119-2131.

Blumenthal, D., R.A. Chimner, J.M. Welker, and J.A. Morgan, 2008. Increased snow facilitates plant invasion in mixedgrass prairie. *New Phytologist*, 179(2): 440-448.

Booker, F., K. Burkey, P. Morgan, E. Fiscus, and A. Jones, 2012. Minimal influence of G-protein null mutations on ozone-induced changes in gene expression, foliar injury, gas exchange and peroxidase activity in *Arabidopsis thaliana l. Plant, Cell & Environment*, 35(4): 668-681.

Booker, F.L., and E.L. Fiscus, 2005. The role of ozone flux and antioxidants in the suppression of ozone injury by elevated CO$_2$ in soybean. *Journal of Experimental Botany*, 56(418): 2139-2151.

Booker, F.L., R. Muntifering, M.T. McGrath, K.O. Burkey, D.R. Decoteau, E.L. Fiscus, W. Manning, S.V. Krupa, A. Chappelka, and D.A. Grantz, 2009. The ozone component of global change: Potential effects on agricultural and horticultural plant yield, product quality and interactions with invasive species. *Journal of Integrative Plant Biology*, 51: 337-351.

Bradley, B.A., 2009. Regional analysis of the impacts of climate change on cheatgrass invasion shows potential risk and opportunity. *Global Change Biology*, 15(1): 196-208.

Bradley, B.A., D.M. Blumenthal, D.S. Wilcove, and L.H.H. Ziska, 2010. Predicting plant invasions in an era of global change. *Trends in Ecology and Evolution*, 25(5): 310-318.

Bradshaw, W.E., and C.M. Holzapfel, 2001. Genetic shift in photoperiodic response correlated with global warming. *Proceedings of the National Academy of Sciences*, 98(25): 14509-14511.

Bradshaw, W.E., and C.M. Holzapfel, 2006. Evolutionary response to rapid climate change. *Science*, 312(5779): 1477-1478.

Bradshaw, W.E., and C.M. Holzapfel, 2010. Insects at not so low temperature: Climate change in the temperate zone and its biotic consequences, *in* D.L. Denlinger and R.E. Lee (eds.), *Low temperature biology of insects*. Cambridge University Press, Cambridge; New York.

Bridges, D.C., and Weed Science Society of America, 1992. *Crop losses due to weeds in the United States, 1992*. Weed Science Society of America, Champaign, Ill., USA.

Bunce, J.A., 1995. Long-term growth of alfalfa and orchard grass plots at elevated carbon dioxide. *Journal of Biogeography*, 22(2/3): 341-348.

Bunce, J.A., 2001. *Weeds in a changing climate*, paper presented at International Symposium: World's Worst Weeds. British Crop Protection Council, Hilton Brighton Metropole Hotel, UK.

Bunce, J.A., 2004. Carbon dioxide effects on stomatal responses to the environment and water use by crops under field conditions. *Oecologia*, 140(1): 1-10.

Bunce, J.A., 2007. Effects of elevated carbon dioxide on photosynthesis and productivity of alfalfa in relation to seasonal changes in temperature. *Physiology and Molecular Biology of Plants*, 13(243-252).

Butler, G.D., B.A. Kimball, and J.R. Mauney, 1986. Populations of *Bemisia tabaci* (homoptera: Aleyrodidae) on cotton grown in open-top field chambers enriched with CO$_2$. *Environmental Entomology*, 15(1): 61-63.

Campbell, C.L., and L.V. Madden, 1990. *Introduction to plant disease epidemiology*. Wiley, New York.

Canto, T., M.A. Aranda, and A. Fereres, 2009. Climate change effects on physiology and population processes of hosts and vectors that influence the spread of hemipteran-borne plant viruses. *Global Change Biology*, 15(8): 1884-1894.

Carter, D.R., and K.M. Peterson, 1983. Effects of a CO$_2$-enriched atmosphere on the growth and competitive interaction of a C3 and a C4 grass. *Oecologia*, 58(2): 188-193.

Cavaleri, M.A., and L. Sack, 2010. Comparative water use of native and invasive plants at multiple scales: A global meta-analysis. *Ecology*, 91(9): 2705-2715.

Chellemi, D.O., C.G. Webster, C.A. Baker, M. Annamalai, D. Achor, and S. Adkins, 2011. Widespread occurrence and low genetic diversity of Colombian datura virus in brugmansia suggest an anthropogenic role in virus selection and spread. *Plant Disease*, 95(6): 755-761.

Chen, F., G. Wu, F. Ge, M.N. Parajulee, and R.B. Shrestha, 2005. Effects of elevated CO_2 and transgenic Bt cotton on plant chemistry, performance, and feeding of an insect herbivore, the cotton bollworm. *Entomologia Experimentalis et Applicata*, 115(2): 341-350.

Chen, F.J., G. Wu, and F. Ge, 2004. Impacts of elevated CO_2 on the population abundance and reproductive activity of aphid *Sitobion avenae* fabricius feeding on spring wheat. *Journal of Applied Entomology*, 128(9-10): 723-730.

Cheng, L., F.L. Booker, K.O. Burkey, C. Tu, H.D. Shew, T.W. Rufty, E.L. Fiscus, J.L. Deforest, and S. Hu, 2011. Soil microbial responses to elevated CO_2 and O_3 in a nitrogen-aggrading agroecosystem. *PLoS ONE*, 6(6): e21377.

Cho, K., S. Tiwari, S.B. Agrawal, N.L. Torres, M. Agrawal, A. Sarkar, J. Shibato, G.K. Agrawal, A. Kubo, and R. Rakwal, 2011. Tropospheric ozone and plants: Absorption, responses, and consequences. *Reviews of Environmental Contamination and Toxicology*, 212(61-111).

Clements, D.R., and A. Ditommaso, 2011. Climate change and weed adaptation: Can evolution of invasive plants lead to greater range expansion than forecasted? *Weed Research*, 51(3): 227-240.

Coakley, S.M., H. Scherm, and S. Chakraborty, 1999. Climate change and plant disease management. *Annual Review of Phytopathology*, 37: 399-426.

Craine, J.M., A.J. Elmore, K.C. Olson, and D. Tolleson, 2010. Climate change and cattle nutritional stress. *Global Change Biology*, 16(10): 2901-2911.

Crozier, L., 2004. Warmer winters drive butterfly range expansion by increasing survivorship. *Ecology*, 85(1): 231-241.

Daly, C., M.P. Widrlechner, M.D. Halbleib, J.I. Smith, and W.P. Gibson, 2010. Development of a new USDA plant hardiness zone map for the United States. *Journal of Applied Meteorology and Climatology*, 51(2): 242-264.

DeLucia, E.H., C.L. Casteel, P.D. Nabity, and B.F. O'Neill, 2008. Insects take a bigger bite out of plants in a warmer, higher carbon dioxide world. *Proceedings of the National Academy of Sciences*, 105(6): 1781-1782.

Dentener, F., D. Stevenson, K. Ellingsen, T. Van Noije, M. Schultz, M. Amann, C. Atherton, N. Bell, D. Bergmann, I. Bey, L. Bouwman, T. Butler, J. Cofala, B. Collins, J. Drevet, O.R. Doherty, B. Eickhout, H. Eskes, M. Gauss, D. Hauglustaine, L. Horowitz, I. Isaksen, B. Josse, M. Lawrence, M. Krol, J.F. Lamarque, V. Montanaro, J.F. Muller, V.H. Peuch, G. Pitari, J. Pyle, J. Rodriguez, M. Sanderson, N.H. Savage, D. Shindell, S. Strahan, S. Szopa, K. Sudo, R. Vandingenen, O. Wild, and G. Zeng, 2006. The global atmospheric environment for the next generation. *Environmental Science and Technology*, 40: 3586-3594.

Dermody, O., B. O'Neill, A. Zangerl, M. Berenbaum, and E. DeLucia, 2008. Effects of elevated CO_2 and O_3 on leaf damage and insect abundance in a soybean agroecosystem. *Arthropod-Plant Interactions*, 2(3): 125-135.

Diffenbaugh, N.S., C.H. Krupke, M.A. White, and C.E. Alexander, 2008. Global warming presents new challenges for maize pest management. *Environment Research Letters*, 3.

Dukes, J.S., 2000. Will the increasing atmospheric CO_2 concentration affect the success of invasive species?, *in* H.A. Mooney and R.J. Hobbs (eds.), *Invasive species in a changing world*. Island Press, Washington.

Dukes, J.S., 2002. Comparison of the effect of elevated CO_2 on an invasive species (*centaurea solstitialis*) in monoculture and community settings. *Plant Ecology*, 160(2): 225-234.

Dukes, J.S., N.R. Chiariello, S.R. Loarie, and C.B. Field, 2011. Strong response of an invasive plant species (*Centaurea solstitialis l.*) to global environmental changes. *Ecological Applications*, 21(6): 1887-1894.

Eastburn, D.M., M.M. Degennaro, E.H. Delucia, O. Dermody, and A.J. McElrone, 2010. Elevated atmospheric carbon dioxide and ozone alter soybean diseases at soyface. *Global Change Biology*, 16(1): 320-330.

Eastburn, D.M., A.J. McElrone, and D.D. Bilgin, 2011. Influence of atmospheric and climatic change on plant–pathogen interactions. *Plant Pathology*, 60(1): 54-69.

Erbs, M., R. Manderscheid, G., Jansen, S. Seddig, A. Pacholski, and H-J. Weigel. 2010. Effects of free-air CO_2 enrichment and nitrogen supply on grain quality parameters and elemental composition of wheat and barley grown in crop rotation. *Agriculture, Ecosystems and Environment*, 136:59-68.

Fargette, D., G. Konate, C. Fauquet, E. Muller, M. Peterschmitt, and J.M. Thresh, 2006. Molecular ecology and emergence of tropical plant viruses, *in Annual Review of Phytopathology*, pp. 235-260. Annual Reviews, Palo Alto.

Fiscus, E.L., F.L. Booker, and K.O. Burkey, 2005. Crop responses to ozone: Uptake, modes of action, carbon assimilation and partitioning. *Plant, Cell and Environment*, 28(8): 997-1011.

Fiscus, E.L., C.D. Reid, J.E. Miller, and A.S. Heagle, 1997. Elevated CO_2 reduces O_3 flux and O_3-induced yield losses in soybeans: Possible implications for elevated CO_2 studies. *Journal of Experimental Botany*, 48(2): 307-313.

Fuhrer, J., 2003. Agroecosystem responses to combinations of elevated CO_2, ozone, and global climate change. *Agriculture Ecosystems & Environment*, 97(1-3): 1-20.

Garrett, K.A., G.A. Forbes, S. Savary, P. Skelsey, A.H. Sparks, C. Valdivia, A.H.C. van Bruggen, L. Willocquet, A. Djurle, E. Duveiller, H. Eckersten, S. Pande, C.V. Cruz, and J. Yuen, 2011. Complexity in climate-change impacts: An analytical framework for effects mediated by plant disease. *Plant Pathology*, 60(1): 15-30.

Garrett, K.A., S.P. Dendy, E.E. Frank, M.N. Rouse, and S.E. Travers, 2006. Climate change effects on plant disease: Genomes to ecosystems, *in Annual review of phytopathology*, pp. 489-509.

Gaughan, J.B., N. Lacetera, S.E. Valtorta, H.H. Khalifa, L. Hahn, and T. Mader, 2009. Response of domestic animals to animal challenges, *in* K.L. Ebi, I. Burton and G.R. Mcgregor (eds.), *Biometeorology for adaptation to climate variability and change*. Springer, Dordrecht; London.

Gomi, T., M. Nagasaka, T. Fukuda, and H. Hagihara, 2007. Shifting of the life cycle and life-history traits of the fall webworm in relation to climate change. *Entomologia Experimentalis et Applicata*, 125(2): 179-184.

Gordo, O., and J.J. Sanz, 2006. Temporal trends in phenology of the honey bee *Apis mellifera* (l.) and the small white *pieris rapae* (l.) in the Iberian peninsula (1952–2004). *Ecological Entomology*, 31(3): 261-268.

Grantz, D.A., and H.-B. Vu, 2009. O_3 sensitivity in a potential C4 bioenergy crop: Sugarcane in California. *Crop Science*, 49(2): 643-650.

Gregory, P.J., S.N. Johnson, A.C. Newton, and J.S.I. Ingram, 2009. Integrating pests and pathogens into the climate change/food security debate. *Journal of Experimental Botany*, 60(10): 2827-2838.

Gutierrez, A., L. Ponti, and Q. Cossu, 2009. Effects of climate warming on olive and olive fly (*Bactrocera oleae* (Gmelin)) in California and Italy. *Climatic Change*, 95(1): 195-217.

Gutierrez, A., L. Ponti, T. d'Oultremont, and C. Ellis, 2008. Climate change effects on poikilotherm tritrophic interactions. *Climatic Change*, 87(0): 167-192.

Gutierrez, A.P., T. D'Oultremont, C.K. Ellis, and L. Ponti, 2006. Climatic limits of pink bollworm in Arizona and California: Effects of climate warming. *Acta Oecologica*, 30(3): 353-364.

Haag, C.R., M. Saastamoinen, J.H. Marden, and I. Hanski, 2005. A candidate locus for variation in dispersal rate in a butterfly metapopulation. *Proceedings of the Royal Society B-Biological Sciences*, 272(1580): 2449-2456.

Hance, T., J. van Baaren, P. Vernon, and G. Boivin, 2007. Impact of extreme temperatures on parasitoids in a climate change perspective, *in Annual review of entomology*, pp. 107-126. Annual Reviews, Palo Alto.

Harrington, R., S.J. Clark, S.J. Welham, P.J. Verrier, C.H. Denholm, M. HullÉ, D. Maurice, M.D. Rounsevell, N. Cocu, and European Union Examine Consortium, 2007. Environmental change and the phenology of European aphids. *Global Change Biology*, 13(8): 1550-1564.

Hatfield, J.L., K.J. Boote, B.A. Kimball, L.H. Ziska, R.C. Izaurralde, D. Ort, A.M. Thomson, and D. Wolfe, 2011. Climate impacts on agriculture: Implications for crop production. *Agronomy Journal*, 103(2): 351-370.

Hatfield, J.L., and J.H. Prueger, 2011. Spatial and temporal variation in evapotranspiration, *in* G. Gerosa (ed.) *Evapotranspiration - From measurements to agricultural and environmental applications.*

Hättenschwiler, S., and C. Körner, 2003. Does elevated CO_2 facilitate naturalization of the non-indigenous *Prunus laurocerasus* in Swiss temperate forests? *Functional Ecology*, 17(6): 778-785.

Hawkins, B.A., and M. Holyoak, 1998. Transcontinental crashes of insect populations? *American Naturalist*, 162(3): 480-484.

Heagle, A.S., 2003. Influence of elevated carbon dioxide on interactions between *Frankliniella occidentalis* and *Trifolium repens*. *Environmental Entomology*, 32(3): 421-424.

Heagle, A.S., R.L. Brandenburg, J.C. Burns, and J.E. Miller, 1994. Ozone and carbon dioxide effects on spider mites in white clover and peanut. *J. Environ. Qual.*, 23(6): 1168-1176.

Hegland, S.J., A. Nielsen, A. Lázaro, A.-L. Bjerknes, and Ø. Totland, 2009. How does climate warming affect plant-pollinator interactions? *Ecology Letters*, 12(2): 184-195.

Hill, J.K., H.M. Griffiths, and C.D. Thomas, 2011. Climate change and evolutionary adaptations at species' range margins. *Annual Review of Entomology*, 56(1): 143-159.

Himanen, S.J., A. Nissinen, W.-X. Dong, A.-M. Nerg, C.N. Stewart, G.M. Poppy, and J.K. Holopainen, 2008. Interactions of elevated carbon dioxide and temperature with aphid feeding on transgenic oilseed rape: Are bacillus thuringiensis (Bt) plants more susceptible to nontarget herbivores in future climate? *Global Change Biology*, 14(6): 1437-1454.

Holm, L.G., 1997. *World weeds : Natural histories and distribution.* Wiley, New York.

Holtum, J.A.M., and K. Winter, 2003. Photosynthetic CO_2 uptake in seedlings of two tropical tree species exposed to oscillating elevated concentrations of CO_2. *Planta*, 218(1): 152-158.

Hoover, J.K., and J.A. Newman, 2004. Tritrophic interactions in the context of climate change: A model of grasses, cereal aphids and their parasitoids. *Global Change Biology*, 10(7): 1197-1208.

Izaurralde, R.C., A.M. Thomson, J.A. Morgan, P.A. Fay, H.W. Polley, and J.L. Hatfield, 2011. Climate impacts on agriculture: Implications for forage and rangeland production. *Agronomy Journal*, 103(2): 371-381.

Jarvis, P.G., and K.G. McNaughton, 1986. Stomatal control of transpiration: Scaling up from leaf to region, *in* A. Macfadyen and E.D. Ford (eds.), *Advances in ecological research*, pp. 1-49. Academic Press.

Jeffree, C.E., and E.P. Jeffree, 1996. Redistribution of the potential geographical ranges of mistletoe and Colorado beetle in Europe in response to the temperature component of climate change. *Functional Ecology*, 10(5): 562-577.

Johnson, R., and D. Lincoln, 1990. Sagebrush and grasshopper responses to atmospheric carbon dioxide concentration. *Oecologia*, 84(1): 103-110.

ₗohnson, R.H., and D.E. Lincoln, 1991. Sagebrush carbon allocation patterns and grasshopper nutrition: The influence of CO_2 enrichment and soil mineral limitation. *Oecologia*, 87(1): 127-134.

Johnson, S., and J. McNicol, 2010. Elevated CO_2 and aboveground–belowground herbivory by the clover root weevil. *Oecologia*, 162(1): 209-216.

Jones, R.A.C., 2009. Plant virus emergence and evolution: Origins, new encounter scenarios, factors driving emergence, effects of changing world conditions, and prospects for control. *Virus Research*, 141(2): 113-130.

Jones, R.A.C., M.U. Salam, T.J. Maling, A.J. Diggle, and D.J. Thackray, 2010. Principles or predicting plant virus disease epidemics, *in* N.K. Vanalfen, G. Bruening and J.E. Leach (eds.), *Annual review of phytopathology, vol 48*, pp. 179-203. Annual Reviews, Palo Alto.

Joutei, B.A., J. Roy, G. Van Impe, and P. Lebrun, 2000. Effect of elevated CO_2 on the demography of a leaf-sucking mite feeding on bean. *Oecologia*, 123(1): 75-81.

Juroszek, P., and A. von Tiedemann, 2011. Potential strategies and future requirements for plant disease management under a changing climate. *Plant Pathology*, 60(1): 100-112.

Karban, R., and J.S. Thaler, 1999. Plant phase change and resistance to herbivory. *Ecology*, 80(2): 510-517.

Karl, T., J. Melillo, T. Peterson, and S.J. Hassol (Eds.), 2009. *Global climate change impacts in the United States*. Cambridge University Press.

Karowe, D., and A. Migliaccio, 2011. Performance of the legume-feeding herbivore, *Colias philodice* (lepidoptera: Pieridae) is not affected by elevated CO_2. *Arthropod-Plant Interactions*, 5(2): 107-114.

Klein, A.-M., B.E. Vaissière, J.H. Cane, I. Steffan-Dewenter, S.A. Cunningham, C. Kremen, and T. Tscharntke, 2007. Importance of pollinators in changing landscapes for world crops. *Proceedings of the Royal Society B: Biological Sciences*, 274(1608): 303-313.

Knowles, N., M.D. Dettinger, and D.R. Cayan, 2006. Trends in snowfall versus rainfall in the western United States. *Journal of Climate*, 19(18): 4545-4559.

Kocmánková, E., M. Trnka, J. Eitzinger, M. Dubrovský, P. Štěpánek, D. Semerádová, J. Balek, P. Skalák, A. Farda, J. Juroch, and Z. Žalud, 2011. Estimating the impact of climate change on the occurrence of selected pests at a high spatial resolution: A novel approach. *The Journal of Agricultural Science*, 149(02): 185-195.

Li, Z.-Y., T.-J. Liu, N.-W. Xiao, J.-S. Li, and F.-J. Chen, 2011. Effects of elevated CO_2 on the interspecific competition between two sympatric species of *Aphis gossypii* and *Bemisia tabaci* fed on transgenic Bt cotton. *Insect Science*, 18(4): 426-434.

Lincoln, D.E., and D. Couvet, 1989. The effect of carbon supply on allocation to allelochemicals and caterpillar consumption of peppermint. *Oecologia*, 78(1): 112-114.

Lincoln, D.E., N. Sionit, and B.R. Strain, 1984. Growth and feeding response of *Pseudoplusia includens* (lepidoptera: Noctuidae) to host plants grown in controlled carbon dioxide atmospheres. *Environmental Entomology*, 13(6): 1527-1530.

Lobell, D.B., M. Banziger, C. Magorokosho, and B. Vivek, 2011. Nonlinear heat effects on African maize as evidenced by historical yield trials. *Nature Clim. Change*, 1(1): 42-45.

Loladze, I., 2002. Rising atmospheric CO_2 and human nutrition: Toward globally imbalanced plant stoichiometry? *Trends in Ecology & Evolution*, 17(10): 457-461.

Long, S.P., E.A. Ainsworth, A.D.B. Leakey, and P.B. Morgan, 2005. Global food insecurity. Treatment of major food crops with elevated carbon dioxide or ozone under large-scale fully open-air conditions suggests recent models may have overestimated future yields. *Philosophical Transactions of the Royal Society B: Biological Sciences*, 360(1463): 2011-2020.

Long, S.P., E.A. Ainsworth, A.D.B. Leakey, J. Nösberger, and D.R. Ort, 2006. Food for thought: Lower-than-expected crop yield stimulation with rising CO_2 concentrations. *Science*, 312(5782): 1918-1921.

Malloch, G., F. Highet, L. Kasprowicz, J. Pickup, R. Neilson, and B. Fenton, 2006. Microsatellite marker analysis of peach-potato aphids (*Myzus persicae*, homoptera : Aphididae) from scottish suction traps. *Bulletin of Entomological Research*, 96(6): 573-582.

Margosian, M.L., K.A. Garrett, J.M.S. Hutchinson, and K.A. With, 2009. Connectivity of the American agricultural landscape: Assessing the national risk of crop pest and disease spread. *BioScience*, 59(2): 141-151.

Marks, S., and D.E. Lincoln, 1996. Antiherbivore defense mutualism under elevated carbon dioxide levels: A fungal endophyte and grass. *Environmental Entomology*, 25(3): 618-623.

Matros, A., S. Amme, B. Kettig, G.H. Buck-Sorlin, U.W.E. Sonnewald, and H.-P. Mock, 2006. Growth at elevated CO_2 concentrations leads to modified profiles of secondary metabolites in tobacco cv. Samsunnn and to increased resistance against infection with potato virus y. *Plant, Cell & Environment*, 29(1): 126-137.

May, R.M., and A.P. Dobson, 1986. Population dynamics and the rate of evolution of pesticide resistance, *in Pesticide resistance: Strategies and tactics for management*. National Research Council Committee on Strategies for the Management of Pesticide Resistant Pest Populations, National Academy Press, Washington, D.C.

McDonald, A., S. Riha, A. DiTommaso, and A. DeGaetano, 2009. Climate change and the geography of weed damage: Analysis of U.S. maize systems suggests the potential for significant range transformations. *Agriculture, Ecosystems, and Environment*, 130(3–4): 131-140.

Memmott, J., P.G. Craze, N.M. Waser, and M.V. Price, 2007. Global warming and the disruption of plant–pollinator interactions. *Ecology Letters*, 10(8): 710-717.

Mickley, L.J., D.J. Jacob, and B.D. Field, 2004. Effects of future climate change on regional air pollution episodes in the United States. *Geophysical Research Letters*, 31: L24103.

Mika, A.M., and J.A. Newman, 2010. Climate change scenarios and models yield conflicting predictions about the future risk of an invasive species in North America. *Agricultural and Forest Entomology*, 12(3): 213-221.

Mika, A.M., R.M. Weiss, O. Olfert, R.H. Hallett, and J.A. Newman, 2008. Will climate change be beneficial or detrimental to the invasive swede midge in North America? Contrasting predictions using climate projections from different general circulation models. *Global Change Biology*, 14(8): 1721-1733.

Miller, M., J. Belnap, S. Beatty, and R. Reynolds, 2006. Performance of *Bromus tectorum l.* in relation to soil properties, water additions, and chemical amendments in calcareous soils of southeastern Utah, USA. *Plant and Soil*, 288(1): 1-18.

Morales, F.J., and P.G. Jones, 2004. The ecology and epidemiology of whitefly-transmitted viruses in Latin America. *Virus Research*, 100(1): 57-65.

Muntifering, R.B., A.H. Chappelka, J.C. Lin, D.F. Karnosky, and G.L. Somers, 2006. Chemical composition and digestibility of *trifolium* exposed to elevated ozone and carbon dioxide in a free air (FACE) fumigation system. *Functional Ecology*, 20: 269-275.

Musolin, D.L., D. Tougou, and K. Fujisaki, 2010. Too hot to handle? Phenological and life-history responses to simulated climate change of the southern green stink bug *Nezara viridula* (heteroptera: Pentatomidae). *Global Change Biology*, 16(1): 73-87.

Navas-Castillo, J., E. Fiallo-Olivé, and S. Sánchez-Campos, 2011. Emerging virus diseases transmitted by whiteflies. *Annual Review of Phytopathology*, 49(1): 219-248.

Newman, J.A., 2005. Climate change and the fate of cereal aphids in southern Britain. *Global Change Biology*, 11(6): 940-944.

Newman, J.A., 2006. Using the output from global circulation models to predict changes in the distribution and abundance of cereal aphids in Canada: A mechanistic modeling approach. *Global Change Biology*, 12(9): 1634-1642.

Newton, P.C.D., H. Clark, C.C. Bell, and E.M. Glasgow, 1996. Interaction of soil moisture and elevated CO_2 on the above-ground growth rate, root length density and gas exchange of turves from temperate pasture. *Journal of Experimental Botany*, 47(6): 771-779.

O'Neill, B.F., A.R. Zangerl, E.H. DeLucia, C. Casteel, J.A. Zavala, and M.R. Berenbaum, 2011. Leaf temperature of soybean grown under elevated CO_2 increases *Aphis glycines* (hemiptera: Aphididae) population growth. *Insect Science*, 18(4): 419-425.

Oerke, E.C., 2006. Crop losses to pests. *Journal of Agricultural Science*, 144: 31-43.

Olfert, O., and R.M. Weiss, 2006a. Impact of climate change on potential distributions and relative abundances of *Oulema melanopus*, *Meligethes viridescens* and *Ceutorhynchus obstrictus* in Canada. *Agriculture, Ecosystems & Environment*, 113(1–4): 295-301.

Olfert, O., and R.M. Weiss, 2006b. Bioclimatic model of *Melanoplus sanguinipes* (fabricius) (orthoptera: Acrididae) populations in Canada and the potential impacts of climate change. *Journal of Orthoptera Research*, 15(1): 65-77.

Osbrink, W.L.A., J.T. Trumble, and R.E. Wagner, 1987. Host suitability of *Phaseolus lunata* for *Trichoplusia ni* (lepidoptera: Noctuidae) in controlled carbon dioxide atmospheres. *Environmental Entomology*, 16(3): 639-644.

Owensby, C.E., J.M. Ham, A.K. Knapp, and L.M. Auen, 1999. Biomass production and species composition change in a tallgrass prairie ecosystem after long-term exposure to elevated atmospheric CO_2. *Global Change Biology*, 5(5): 497-506.

Pangga, I.B., J. Hanan, and S. Chakraborty, 2011. Pathogen dynamics in a crop canopy and their evolution under changing climate. *Plant Pathology*, 60(1): 70-81.

Parmesan, C., 2006. Ecological and evolutionary responses to recent climate change. *Annual Review of Ecology, Evolution, and Systematics*, 37(1): 637-669.

Parmesan, C., and G. Yohe, 2003. A globally coherent fingerprint of climate change impacts across natural systems. *Nature*, 421(6918): 37-42.

Parton, W.J., J.A. Morgan, G. Wang, and S. Del Grosso, 2007. Projected ecosystem impact of the prairie heating and CO_2 enrichment experiment. *New Phytologist*, 174(4): 823-834.

Patterson, D.T., E.P. Flint, and J.L. Beyers, 1984. Effects of CO_2 enrichment on competition between a C4 weed and a C3 crop. *Weed Science*, 32(1): 101-105.

Patterson, D.T., J.K. Westbrook, R.J.V. Joyce, P.D. Lingren, and J. Rogasik, 1999. Weeds, insects, and diseases. *Climatic Change*, 43(4): 711-727.

Polley, H.W., H.B. Johnson, and J.D. Derner, 2003. Increasing CO_2 from subambient to superambient concentrations alters species composition and increases above-ground biomass in a C3/C4 grassland. *New Phytologist*, 160(2): 319-327.

Poorter, H., and M.-L. Navas, 2003. Plant growth and competition at elevated CO_2: On winners, losers and functional groups. *New Phytologist*, 157(2): 175-198.

153

Porter, J.H., M.L. Parry, and T.R. Carter, 1991. The potential effects of climatic change on agricultural insect pests. *Agricultural and Forest Meteorology*, 57(1–3): 221-240.

Potvin, C., and L. Vasseur, 1997. Long-term CO_2 enrichment of a pasture community: Species richness, dominance, and succession. *Ecology*, 78(3): 666-677.

Pujol Pereira, E.I., H. Chung, K. Scow, M.J. Sadowsky, C. van Kessel, and J. Six, 2011. Soil nitrogen transformations under elevated atmospheric CO_2 and O_3 during the soybean growing season. *Environmental Pollution*, 159(2): 401-407.

Rahman, A., and D. Wardle, 1990. Effects of climate change on cropping weeds in New Zealand, *in* R.A. Prestidge and R.P. Pottinger (eds.), *The impact of climate change on pests, diseases, weeds and beneficial organisms present in New Zealand agricultural and horticultural systems*. Ruakura Agricultural Centre, Hamilton.

Roos, J., R. Hopkins, A. Kvarnheden, and C. Dixelius, 2011. The impact of global warming on plant diseases and insect vectors in Sweden. *European Journal of Plant Pathology*, 129(1): 9-19.

Sacks, W.J., and C.J. Kucharik, 2011. Crop management and phenology trends in the U.S. corn belt: Impacts on yields, evapotranspiration and energy balance. *Agricultural and Forest Meteorology*, 151(7): 882-894.

Salt, D.T., G.L. Brooks, and J.B. Whittaker, 1995. Elevated carbon dioxide affects leaf-miner performance and plant growth in docks (rumex spp.). *Global Change Biology*, 1(2): 153-156.

Samuel, G., 1931. Some experiments on inoculating methods with plant viruses, and on local lesions. *Annals of Applied Biology*, 18(4): 494-507.

Savary, S., A. Nelson, A.H. Sparks, L. Willocquet, E. Duveiller, G. Mahuku, G. Forbes, K.A. Garrett, D. Hodson, J. Padgham, S. Pande, M. Sharma, J. Yuen, and A. Djurle, 2011. International agricultural research tackling the effects of global and climate changes on plant diseases in the developing world. *Plant Disease*, 95(10): 1204-1216.

Schlenker, W., and M.J. Roberts, 2009. Nonlinear temperature effects indicate severe damages to U.S. crop yields under climate change. *Proceedings of the National Academy of Sciences*, 106(37): 15594-15598.

Shaw, M.W., and T.M. Osborne, 2011. Geographic distribution of plant pathogens in response to climate change. *Plant Pathology*, 60(1): 31-43.

Smith, P.H.D., and T.H. Jones, 1998. Effects of elevated CO_2 on the chrysanthemum leaf-miner, *Chromatomyia syngenesiae*: A greenhouse study. *Global Change Biology*, 4(3): 287-291.

Smith, S.D., T.E. Huxman, S.F. Zitzer, T.N. Charlet, D.C. Housman, J.S. Coleman, L.K. Fenstermaker, J.R. Seemann, and R.S. Nowak, 2000. Elevated CO_2 increases productivity and invasive species success in an arid ecosystem. *Nature*, 408(6808): 79-82.

Song, L., J. Wu, C. Li, F. Li, S. Peng, and B. Chen, 2009. Different responses of invasive and native species to elevated CO_2 concentration. *Acta Oecologica*, 35(1): 128-135.

Srygley, R.B., R. Dudley, E.G. Oliveira, R. Aizprúa, N.Z. Pelaez, and A.J. Riveros, 2010. El Niño and dry season rainfall influence host-plant phenology and an annual butterfly migration from neotropical wet to dry forests. *Global Change Biology*, 16(3): 936-945.

Stephens, A.E.A., D.J. Kriticos, and A. Leriche, 2007. The current and future potential geographical distribution of the oriental fruit fly, *Bactrocera dorsalis* (diptera: Tephritidae). *Bulletin of Entomological Research*, 97(04): 369-378.

Stiling, P., and T. Cornelissen, 2007. How does elevated carbon dioxide (CO_2) affect plant–herbivore interactions? A field experiment and meta-analysis of CO_2-mediated changes on plant chemistry and herbivore performance. *Global Change Biology*, 13(9): 1823-1842.

Sun, Y.-C., J. Yin, F.-J. Chen, G. Wu, and F. Ge, 2011. How does atmospheric elevated CO_2 affect crop pests and their natural enemies? Case histories from China. *Insect Science*, 18(4): 393-400.

Thomsen, M.A., and C.M. D'Antonio, 2007. Mechanisms of resistance to invasion in a California grassland: The roles of competitor identity, resource availability, and environmental gradients. *Oikos*, 116(1): 17-30.

Thornton, P.K., J. van de Steeg, A. Notenbaert, and M. Herrero, 2009. The impacts of climate change on livestock and livestock systems in developing countries: A review of what we know and what we need to know. *Agricultural Systems*, 101(3): 113-127.

Thresh, J., 1983. Plant virus epidemiology and control: Current trends and future prospects, *in* R.T. Plumb and J.M. Thresh (eds.), *Plant virus epidemiology : The spread and control of insect-borne viruses*, pp. 349-360. Blackwell Scientific Publications ; Blackwell Mosby Book Distributors, Oxford; Boston; St. Louis, Mo.

Tobin, P.C., S. Nagarkatti, G. Loeb, and M.C. Saunders, 2008. Historical and projected interactions between climate change and insect voltinism in a multivoltine species. *Global Change Biology*, 14(5): 951-957.

Trnka, M., F. Muška, D. Semerádová, M. Dubrovský, E. Kocmánková, and Z. Žalud, 2007. European corn borer life stage model: Regional estimates of pest development and spatial distribution under present and future climate. *Ecological Modelling*, 207(2–4): 61-84.

Tu, C., F.L. Booker, K.O. Burkey, and S. Hu, 2009. Elevated atmospheric carbon dioxide and O_3 differentially alter nitrogen acquisition in peanut. *Crop Science*, 49(5): 1827-1836.

Tubiello, F.N., J.S. Amthor, K.J. Boote, M. Donatelli, W. Easterling, G. Fischer, R.M. Gifford, M. Howden, J. Reilly, and C. Rosenzweig, 2007. Crop response to elevated CO_2 and world food supply: A comment on "food for thought..." by Long et al. Science 312:1918–1921, 2006. *European Journal of Agronomy*, 26(3): 215-223.

Tylianakis, J.M., R.K. Didham, J. Bascompte, and D.A. Wardle, 2008. Global change and species interactions in terrestrial ecosystems. *Ecology Letters*, 11(12): 1351-1363.

Uauy, C., J. Brevis, X. Chen, I. Khan, L. Jackson, O. Chicaiza, A. Distelfeld, T. Fahima, and J. Dubcovsky, 2005. High-temperature adult-plant (htap) stripe rust resistance gene *Yr36* from *Triticum turgidum ssp. Dicoccoides* is closely linked to the grain protein content locus *Gpc-B1*. *TAG Theoretical and Applied Genetics*, 112(1): 97-105.

Van der Putten, W.H., M. Macel, and M.E. Visser, 2010. Predicting species distribution and abundance responses to climate change: Why it is essential to include biotic interactions across trophic levels. *Philosophical Transactions of the Royal Society B: Biological Sciences*, 365(1549): 2025-2034.

Walther, G.-R., 2010. Community and ecosystem responses to recent climate change. *Philosophical Transactions of the Royal Society B: Biological Sciences*, 365(1549): 2019-2024.

Walther, G.-R., E. Post, P. Convey, A. Menzel, C. Parmesan, T.J.C. Beebee, J.-M. Fromentin, O. Hoegh-Guldberg, and F. Bairlein, 2002. Ecological responses to recent climate change. *Nature*, 416(6879): 389-395.

Wand, S.J.E., G.F. Midgley, M.H. Jones, and P.S. Curtis, 1999. Responses of wild C_4 and C_3 grass (poaceae) species to elevated atmospheric CO_2 concentration: A meta-analytic test of current theories and perceptions. *Global Change Biology*, 5(6): 723-741.

Wang, Y., Z. Bao, Y. Zhu, and J. Hua, 2009. Analysis of temperature modulation of plant defense against biotrophic microbes. *Molecular Plant-Microbe Interactions*, 22(5): 498-506.

Webb, K.M., I. Oña, J. Bai, K.A. Garrett, T. Mew, C.M. Vera Cruz, and J.E. Leach, 2010. A benefit of high temperature: Increased effectiveness of a rice bacterial blight disease resistance gene. *New Phytologist*, 185(2): 568-576.

Welch, J.R., J.R. Vincent, M. Auffhammer, P.F. Moya, A. Dobermann, and D. Dawe, 2010. Rice yields in tropical/subtropical Asia exhibit large but opposing sensitivities to minimum and maximum temperatures. *Proceedings of the National Academy of Sciences*, 107(33): 14562-14567.

White, T.C.R., 1984. The abundance of invertebrate herbivores in relation to the availability of nitrogen in stressed food plants. *Oecologia*, 63(1): 90-105.

Whitham, S., S. McCormick, and B. Baker, 1996. The N gene of tobacco confers resistance to tobacco mosaic virus in transgenic tomato. *Proceedings of the National Academy of Sciences*, 93(16): 8776-8781.

Williams, A.L., K.E. Wills, J.K. Janes, J.K. Vander Schoor, P.C.D. Newton, and M.J. Hovenden, 2007. Warming and free-air CO$_2$ enrichment alter demographics in four co-occurring grassland species. *New Phytologist*, 176(2): 365-374.

Wilson, K.B., T.N. Carlson, and J.A. Bunce, 1999. Feedback significantly influences the simulated effect of CO$_2$ on seasonal evapotranspiration from two agricultural species. *Global Change Biology*, 5(8): 903-917.

Wittmann, E.J., and M. Baylis, 2000. Climate change: Effects on culicoides -transmitted viruses and implications for the UK. *The Veterinary Journal*, 160(2): 107-117.

Woodward, F.I., 1988. Temperature and the distribution of plant species, *in* S.P. Long and F.I. Woodward (eds.), *Plants and temperature*. Published for the Society for Experimental Biology by the Company of Biologists. University of Cambridge, Cambridge, UK.

Woodward, F.I., and B.G. Williams, 1987. Climate and plant distribution at global and local scales. *Plant Ecology*, 69(1): 189-197.

Wu, G., F.J. Chen, and F. Ge, 2006. Response of multiple generations of cotton bollworm *Helicoverpa armigera Hübner*, feeding on spring wheat, to elevated CO$_2$. *Journal of Applied Entomology*, 130(1): 2-9.

Zavala, J.A., C.L. Casteel, E.H. DeLucia, and M.R. Berenbaum, 2008. Anthropogenic increase in carbon dioxide compromises plant defense against invasive insects. *Proceedings of the National Academy of Sciences*, 105(13): 5129-5133.

Ziska, L.H., 2000. The impact of elevated CO$_2$ on yield loss from a C3 and C4 weed in field-grown soybean. *Global Change Biology*, 6(8): 899-905.

Ziska, L.H., 2001. Changes in competitive ability between a C4 crop and a C3 weed with elevated carbon dioxide. *Weed Science*, 49(5): 622-627.

Ziska, L.H., 2003. Evaluation of the growth response of six invasive species to past, present and future atmospheric carbon dioxide. *Journal of Experimental Botany*, 54(381): 395-404.

Ziska, L.H., 2003. Evaluation of yield loss in field sorghum from a C3 and C4 weed with increasing CO$_2$. *Weed Science*, 51(6): 914-918.

Ziska, L.H., and J.A. Bunce, 1998. The influence of increasing growth temperature and CO$_2$ concentration on the ratio of respiration to photosynthesis in soybean seedlings. *Global Change Biology*, 4(6): 637-643.

Ziska, L.H., and J.A. Bunce, 2007. Predicting the impact of changing CO$_2$ on crop yields: Some thoughts on food. *New Phytologist*, 175(4): 607-618.

Ziska, L.H., S. Faulkner, and J. Lydon, 2004. Changes in biomass and root:shoot ratio of field-grown Canada thistle (*Cirsium arvense*), a noxious, invasive weed, with elevated CO$_2$: Implications for control with glyphosate. *Weed Science*, 52(4): 584-588.

Ziska, L.H., J.B. Reeves, and B. Blank, 2005. The impact of recent increases in atmospheric CO$_2$ on biomass production and vegetative retention of cheatgrass (*Bromus tectorum*): Implications for fire disturbance. *Global Change Biology*, 11(8): 1325-1332.

Ziska, L.H., and J.R. Teasdale, 2000. Sustained growth and increased tolerance to glyphosate observed in a C3, quackgrass (*Elytrigia repens*), grown at elevated carbon dioxide. *Functional Plant Biology*, 27(2): 159-166.

Ziska, L.H., J.R. Teasdale, and J.A. Bunce, 1999. Future atmospheric carbon dioxide concentrations may increase tolerance to glyphosate. *Weed Science*, 47(608-615).

Chapter 5 References

Adam, N.R., G.W. Wall, B.A. Kimball, S.B. Idso, and A.N. Webber, 2004. Photosynthetic down-regulation over long-term CO$_2$ enrichment in leaves of sour orange (*Citrus aurantium*) trees. *New Phytologist*, 163(2): 341-347.

Adams, R.M., B.A. McCarl, K. Segerson, C. Rosenzweig, K.J. Bryant, B.L. Dixon, R. Connor, R.E. Evenson, and D. Ojima, 1999. The economic effects of climate change on U.S. Agriculture, *in* R.O. Mendelsohn and J.E. Neumann (eds.), *The impact of climate change on the United States economy*, pp. 19-54. Cambridge University Press, Cambridge; New York.

Adams, R.M., and D.E. Peck, 2008. Effects of climate change on drought frequency: Impacts and mitigation opportunities, *in* A. Garrido and A. Dinar (eds.), *Managing water resources in a time of global change: Mountains, valleys and flood plains*. Routledge, London; New York.

Ahmadi, E., H.R. Ghassemzadeh, M. Sadeghi, M. Moghaddam, and S.Z. Neshat, 2010. The effect of impact and fruit properties on the bruising of peach. *Journal of Food Engineering*, 97(1): 110-117.

Ainsworth, E.A., and S.P. Long, 2005. What have we learned from 15 years of free-air CO$_2$ enrichment (FACE)? A meta-analytic review of the responses of photosynthesis, canopy properties and plant production to rising CO$_2$. *New Phytologist*, 165(2): 351-372.

Ainsworth, E.A., A. Rogers, L.O. Vodkin, A. Walter, and U. Schurr, 2006. The effects of elevated CO$_2$ concentration on soybean gene expression. An analysis of growing and mature leaves. *Plant Physiology*, 142(1): 135-147.

Akin, D.E., S.L. Fales, L.L. Rigsby, and M.E. Snook, 1987. Temperature effects on leaf anatomy, phenolic acids, and tissue digestibility in tall fescue. *Agronomy Journal*, 79(2): 271-275.

Alagarswamy, G., and J.T. Ritchie, 1991. Phasic development in ceres-sorghum model, *in* T. Hodges (ed.) *Predicting crop phenology*, pp. 143-152. CRC Press, Boca Raton.

Alberto, A., L. Ziska, C. Cervancia, and P. Manalo, 1996. The influence of increasing carbon dioxide and temperature on competitive interactions between a C3 crop, rice *(Oryza sativa)* and a C4 weed *(Echinochloa glabrescens)*. *Functional Plant Biology*, 23(6): 795-802.

Allen, L.H., and J.C.V. Vu, 2009. Carbon dioxide and high temperature effects on growth of young orange trees in a humid, subtropical environment. *Agricultural and Forest Meteorology*, 149(5): 820-830.

Allen, R.D., and L. Aleman, 2011. Abiotic stress and cotton fiber development, *in* D.M. Oosterhuis (ed.) *Stress physiology in cotton*. The Cotton Foundation, Cordova, TN.

Alocilja, E.C., and J.T. Ritchie, 1991. A model for the phenology of rice, *in* T. Hodges (ed.) *Predicting crop phenology*, pp. 181-189. CRC Press, Boca Raton.

Amadi, C.O.E.-O., E. E.; Okonkwo, J. C.; Okocha, P. I., , 2009. Inter-relationships between yield and yield attributes of potato grown under supra-optimal ambient temperatures. *Global Journal of Pure and Applied Sciences*, 15: 5-14.

Amundson, J.L., T.L. Mader, R.J. Rasby, and Q.S. Hu, 2006. Environmental effects on pregnancy rate in beef cattle. *Journal of Animal Science*, 84(12): 3415-3420.

Anderson, L.J., H. Maherali, H.B. Johnson, H.W. Polley, and R.B. Jackson, 2001. Gas exchange and photosynthetic acclimation over subambient to elevated CO_2 in a C3–C4 grassland. *Global Change Biology*, 7(6): 693-707.

Arevalo, L.S., D.M. Oosterhuis, D. Coker, and R.S. Brown, 2008. Physiological response of cotton to high night temperature. *Amer. J. Plant Sci. And Biotechnol*, 2: 63-68.

Arias, R.A., T.L. Mader, and A.M. Parkhurst, 2011. Effects of diet type and metabolizable energy intake on tympanic temperature of steers fed during summer and winter seasons. *Journal of Animal Science*, 89(5): 1574-1580.

Arndt, C.H., 1945. Temperature-growth relations of the roots and hypocotyls of cotton seedlings. *Plant Physiology*, 20(2): 200-220.

Arnone, J.A., R.L. Jasoni, A.J. Lucchesi, J.D. Larsen, E.A. Leger, R.A. Sherry, Y. Luo, D.S. Schimel, and P.S.J. Verburg, 2011. A climatically extreme year has large impacts on C_4 species in tallgrass prairie ecosystems but only minor effects on species richness and other plant functional groups. *Journal of Ecology*, 99(3): 678-688.

ASWCC, 1997. *Ground water protection and management report for 1996.* Arkansas Soil and Water Conservation Commission.

Atkinson, C.J., J.M. Taylor, D. Wilkins, and R.T. Besford, 1997. Effects of elevated CO_2 on chloroplast components, gas exchange and growth of oak and cherry. *Tree Physiology*, 17(5): 319-325.

Badu-Apraku, B., R.B. Hunter, and M. Tollenaar, 1983. Effect of temperature during grain filling on whole plant and grain yield in maize (*Zea mays l.*). *Canadian Journal of Plant Science*, 63(2): 357-363.

Baker, J.T., and L.H. Allen Jr, 1993. Contrasting crop species responses to CO_2 and temperature: Rice, soybean and citrus. *Vegetatio*, 104-105: 239-260.

Baker, J.T., L.H. Allen Jr, and K.J. Boote, 1992. Response of rice to carbon dioxide and temperature. *Agricultural and Forest Meteorology*, 60(3–4): 153-166.

Baker, J.T., K.J. Boote, and J. L.H. Allen, 1995. Potential climate change effects on rice: Carbon dioxide and temperature, *in* C. Rosenzweig et al. (ed.) *Climate change and agriculture: Analysis of potential international impacts. ASA spec. Pub. No. 59.* ASA-CSSA-SSSA, Madison, WI.

Bale, J.S., G.J. Masters, I.D. Hodkinson, C. Awmack, T.M. Bezemer, V.K. Brown, J. Butterfield, A. Buse, J.C. Coulson, J. Farrar, J.E.G. Good, R. Harrington, S. Hartley, T.H. Jones, R.L. Lindroth, M.C. Press, I. Symrnioudis, A.D. Watt, and J.B. Whittaker, 2002. Herbivory in global climate change research: Direct effects of rising temperature on insect herbivores. *Global Change Biology*, 8(1): 1-16.

Ball, R.A., D.M. Oosterhuis, and A. Mauromoustakos, 1994. Growth dynamics of the cotton plant during water-deficit stress. *Agronomy Journal*, 86(5): 788-795.

Bange, M.P., S.P. Milroy, and P. Thongbai, 2004. Growth and yield of cotton in response to waterlogging. *Field Crops Research*, 88(2–3): 129-142.

Barber, H.N., and P.J.H. Sharpe, 1971. Genetics and physiology of sunscald of fruits. *Agricultural Meteorology*, 8(0): 175-191.

Barrow, J.R., 1983. Comparisons among pollen viability measurement methods in cotton. *Crop Science*, 23(4): 734-736.

Bartomeus, I., J.S. Ascher, D. Wagner, B.N. Danforth, S. Colla, S. Kornbluth, and R. Winfree, 2011. Climate-associated phenological advances in bee pollinators and bee-pollinated plants. *Proceedings of the National Academy of Sciences*, 108(51): 20645-20649.

Bates, J.D., T. Svejcar, R.F. Miller, and R.A. Angell, 2006. The effects of precipitation timing on sagebrush steppe vegetation. *Journal of Arid Environments*, 64(4): 670-697.

Bauer, G.A., G.M. Berntson, and F.A. Bazzaz, 2001. Regenerating temperate forests under elevated CO_2 and nitrogen deposition: Comparing biochemical and stomatal limitation of photosynthesis. *New Phytologist*, 152(2): 249-266.

Baylis, M., and A.K. Githeko, 2006. *T7.3: The effects of climate change on infectious diseases of animals.* Foresight http://www. bis.gov.uk/foresight/our-work/projects/published-projects/ infectious-diseases/reports-and-publications.

Beede, D.K., and R.J. Collier, 1986. Potential nutritional strategies for intensively managed cattle during thermal stress. *Journal of Animal Science*, 62(2): 543-554.

Bender, J., U. Hertstein, and C.R. Black, 1999. Growth and yield responses of spring wheat to increasing carbon dioxide, ozone and physiological stresses: A statistical analysis of 'ESPACE-wheat' results. *European Journal of Agronomy*, 10(3–4): 185-195.

Bennett, O.L., L.J. Erie, and A.J. MacKenzie, 1967. *Boll, fiber, and spinning properties of cotton as affected by management practices.* USDA Technical Bulletin No. 1372, Washington, DC.

Beppu, K., T. Ikeda, and I. Kataoka, 2001. Effect of high temperature exposure time during flower bud formation on the occurrence of double pistils in 'satohnishiki' sweet cherry. *Scientia Horticulturae*, 87(1-2): 77-84.

Beppu, K., S. Okamoto, A. Sugiyama, and I. Kataoka, 1997. Effects of temperature on flower development and fruit set of 'satohnishiki' sweet cherry. *Journal of the Japanese Society for Horticultural Science*, 65(4): 707-712.

Bernacchi, C.J., A.D.B. Leakey, L.E. Heady, P.B. Morgan, F.G. Dohleman, J.M. McGrath, K.M. Gillespie, V.E. Wittig, A. Rogers, S.P. Long, and D.R. Ort, 2006. Hourly and seasonal variation in photosynthesis and stomatal conductance of soybean grown at future CO_2 and ozone concentrations for 3 years under fully open-air field conditions. *Plant, Cell & Environment*, 29(11): 2077-2090.

Berry, J.K., J.A. Delgado, R. Khosla, and F.J. Pierce, 2003. Precision conservation for environmental sustainability. *Journal of Soil and Water Conservation*, 58(6): 332+.

Bibi, A.C., D.M. Oosterhuis, and E.D. Gonias, and J.M. Stewart, 2010. Comparison of a responses of a ruderal *Gossypium hirsutum l.* With commercial cotton genotypes under high temperature stress. *Amer. J. Plant Sci. Biotechnol*, 4: 87-92.

Bibi, A.C., D.M. Oosterhuis, and E.G. Gonias, 2008. Photosynthesis, quantum yield of photosystem II, and membrane leakage as affected by high temperatures in cotton genotypes. *Journal of Cotton Science*, 12: 150-159.

Bindi, M., L. Fibbi, and F. Miglietta, 2001. Free air CO_2 enrichment (FACE) of grapevine (*Vitis vinifera l.*): II. Growth and quality of grape and wine in response to elevated CO_2 concentrations. *European Journal of Agronomy*, 14(2): 145-155.

Bisognin, D.A., D.R. Muller, N.A. Streck, J.L. Andriolo, and D. Sausen, 2008. Development and yield of potato clones during spring and autumn. *Pesquisa Agropecuária Brasileira*, 43(6): 699-705.

Biswas, D.K., H. Xu, Y.G. Li, J.Z. Sun, X.Z. Wang, X.G. Han, and G.M. Jiang, 2008. Genotypic differences in leaf biochemical, physiological and growth responses to ozone in 20 winter wheat cultivars released over the past 60 years. *Global Change Biology*, 14(1): 46-59.

Bolhuis, C.G., and W. deGroot, 1959. Observations on the effect of varying temperature on the flowering and fruit set in three varieties of groundnut. *Neth. J. Agric. Sci*, 7: 317-326.

Boote, K.J., L.H. Allen, P.V.V. Prasad, J.T. Baker, R.W. Gesch, A.M. Snyder, D. Pan, and J.M.G. Thomas, 2005. Elevated temperature and CO2 impacts on pollination, reproductive growth, and yield of several globally important crops. *J. Agric. Meteorol*, 60: 469-474.

Boote, K.J., J.W. Jones, and G. Hoogenboom, 1998. Simulation of crop growth: CROPGRO model, *in* R. Peart and R.B. Curry (eds.), *Agricultural systems modeling and simulation*, pp. 651-692. Marcel Dekker, New York.

Boote, K.J., N.B. Pickering, and L.H.A. Jr., 1997. Plant modeling: Advances and gaps in our capability to project future crop growth and yield in response to global climate change, *in* L. H. Allen et al. (ed.) *Advances in carbon dioxide effects research. ASA Special Publication No. 61.* ASA-CSSA-SSSA, Madison, WI.

Bordonaba, G.J., and L.A. Terry, 2010. Manipulating the taste-related composition of strawberry fruits (*Fragaria ananassa*) from different cultivars using deficit irrigation. *Food Chemistry*, 122(4): 1020-1026.

Bordovsky, J.P., W.M. Lyle, R.J. Lascano, and D.R. Upchurch, 1992. Cotton irrigation management with LEPA systems. *Transactions of the American Society of Agricultural Engineering*, 35: 879-884.

Bouman, B.A.M., R.M. Lampayan, and T.P. Tuong, 2007. *Water management in irrigated rice : Coping with water scarcity.* International Rice Research Institute, Manila, The Philippines.

Bradow, J.M., and G.H. Davidonis, 2000. Quantitation of fiber quality and the cotton production-processing interface: A physiologist's perspective. *Journal of Cotton Science*, 4(34-64).

Brakke, M., and L.H. Allen, 1995. Gas-exchange of citrus seedlings at different temperatures, vapor-pressure deficits, and soil-water contents. *Journal of the American Society for Horticultural Science*, 120(3): 497-504.

Brown, P.W., and C.A. Zeiher, 1998. *A model to estimate cotton canopy temperature in the desert southwest*, paper presented at Proceedings of the Beltwide Cotton Conferences. National Cotton Council of America, Memphis, TN.

Brown, S., and D.M. Oosterhuis, 2010. High daytime temperature stress effects on the physiology of modern versus obsolete cultivars. *Amer. J. Plant Sci. Biotechnol*, 4: 93-96.

Bruce, R.R., and C.D. Shipp, 1962. Cotton fruiting as affected by soil moisture regime. *Agronomy Journal*, 54(1): 15-18.

Bruns, H.A., and H.K. Abbas, 2006. Planting date effects on Bt and non-Bt corn in the mid-south USA. *Agronomy Journal*, 98(1): 100-106.

Buban, T., 2000. The use of benzyladenine in orchard fruit growing: A mini review. *Plant Growth Regulation*, 32(2-3): 381-390.

Bunce, J.A., 1992. Stomatal conductance, photosynthesis and respiration of temperate deciduous tree seedlings grown outdoors at an elevated concentration of carbon-dioxide. *Plant Cell and Environment*, 15(5): 541-549.

Burke, J.J., J.R. Mahan, and J.L. Hatfield, 1988. Crop-specific thermal kinetic windows in relation to wheat and cotton biomass production. *Agronomy Journal*, 80(4): 553-556.

Burke, J.J., and D.F. Wanjura, 2009. Plant responses to temperature extremes, *in* J.M. Stewart, D.M. Oosterhuis, J.J. Heitholt and J.R. Mauney (eds.), *Physiology of cotton*, pp. 123-128. Springer, New York.

Bustan, A., E.E. Goldschmidt, and Y. Erner, 1996. *Integrating temperature effects on fruit growth into a citrus productivity model* paper presented at International Citrus Congress (8th : 1996: Sun City, South Africa), Sun City, South Africa, International Society of Citriculture.

Camejo, D., P. Rodríguez, M. Angeles Morales, J. Miguel Dell'Amico, A. Torrecillas, and J.J. Alarcón, 2005. High temperature effects on photosynthetic activity of two tomato cultivars with different heat susceptibility. *Journal of Plant Physiology*, 162(3): 281-289.

Caprio, J.M., and H.A. Quamme, 2002. Weather conditions associated with grape production in the Okanagan valley of British Columbia and potential impact of climate change. *Canadian Journal of Plant Science*, 82(4): 755-763.

Casierra-Posada, F.V., Y. A., 2007. Growth and yield of strawberry cultivars (*Fragaria* sp.) affected by flooding. *Revista Colombiana de Ciencias Horticolas*, 1: 21-32.

Centritto, M., 2002. The effects of elevated CO_2 and water availability on growth and physiology of peach (*Prunus persica*) plants. *Plant Biosystems*, 136(2): 177-188.

Centritto, M., 2005. Photosynthetic limitations and carbon partitioning in cherry in response to water deficit and elevated CO_2. *Agriculture Ecosystems & Environment*, 106(2-3): 233-242.

Centritto, M., H.S.J. Lee, and P.G. Jarvis, 1999. Increased growth in elevated CO_2: An early, short-term response? *Global Change Biology*, 5(6): 623-633.

Centritto, M., M.E. Lucas, and P.G. Jarvis, 2002. Gas exchange, biomass, whole-plant water-use efficiency and water uptake of peach (*Prunus persica*) seedlings in response to elevated carbon dioxide concentration and water availability. *Tree Physiology*, 22(10): 699-706.

Centritto, M., F. Magnani, H.S.J. Lee, and P.G. Jarvis, 1999. Interactive effects of elevated CO_2 and drought on cherry (*Prunus avium*) seedlings - II. Photosynthetic capacity and water relations. *New Phytologist*, 141(1): 141-153.

Charlier, T., 2002. Rice soaks up water along with tax dollars. *The Commercial Appeal, 06 October 2002.*

Chen, K., G.Q. Hu, and F. Lenz, 2001. Effects of doubled atmospheric CO_2 concentration on apple trees I. Growth analysis. *Gartenbauwissenschaft*, 66(6): 282-288.

Chen, K., G.Q. Hu, and F. Lenz, 2002. Effects of doubled atmospheric CO_2 concentration on apple trees II. Dry mass production. *Gartenbauwissenschaft*, 67(1): 28-33.

Chen, L.S., P. Li, and L. Cheng, 2008. Effects of high temperature coupled with high light on the balance between photooxidation and photoprotection in the sun-exposed peel of apple. *Planta*, 228(5): 745-756.

Cheng, L., F.L. Booker, K.O. Burkey, C. Tu, H.D. Shew, T.W. Rufty, E.L. Fiscus, J.L. Deforest, and S. Hu, 2011. Soil microbial responses to elevated CO_2 and O_3 in a nitrogen-aggrading agroecosystem. *PLoS ONE*, 6(6): e21377.

Chepil, W.S., and N.P. Woodruff, 1954. Estimations of wind erodibility of field surfaces. *J. Soil and Water Conserv*, 9: 257-265, 285.

Chowdhury, S., and I. Wardlaw, 1978. The effect of temperature on kernel development in cereals. *Australian Journal of Agricultural Research*, 29(2): 205-223.

Cleland, E.E., N.R. Chiariello, S.R. Loarie, H.A. Mooney, and C.B. Field, 2006. Diverse responses of phenology to global changes in a grassland ecosystem. *Proceedings of the National Academy of Sciences*, 103(37): 13740-13744.

Coakley, S.M., H. Scherm, and S. Chakraborty, 1999. Climate change and plant disease management. *Annual Review of Phytopathology*, 37: 399-426.

Cole, P., and P. McCloud, 1985. Salinity and climatic effects on the yields of citrus. *Australian Journal of Experimental Agriculture*, 25(3): 711-717.

Collaku, A., and S.A. Harrison, 2002. Losses in wheat due to waterlogging. *Crop Sci.*, 42(2): 444-450.

Conaty, W.C., D.K.Y. Tan, G.A. Constable, B.G. Sutton, D.J. Field, and E.A. Mamum, 2008. Genetic variation for waterlogging tolerance in cotton. *Journal of Cotton Science*, 12: 53-61.

Constable, G., and H. Rawson, 1980. Effect of leaf position, expansion and age on photosynthesis, transpiration and water use efficiency of cotton. *Functional Plant Biology*, 7(1): 89-100.

Cooper, N.T.W., T.J. Siebenmorgen, and P.A. Counce, 2008. Effects of nighttime temperature during kernel development on rice physicochemical properties. *Cereal Chemistry Journal*, 85(3): 276-282.

Cottee, N.S., D.K.Y. Tan, J.T. Cothren, M.P. Bange, and L.C. Campbell, 2007. *Screening cotton cultivars for thermotolerance under field conditions*, paper presented at Proceedings of the 4th World Cotton Research Conference. Lubbock, TX, USA, September, 2007.

Craine, J.M., A.J. Elmore, K.C. Olson, and D. Tolleson, 2010. Climate change and cattle nutritional stress. *Global Change Biology*, 16(10): 2901-2911.

Crisosto, C.H., D. Garner, J. Doyle, and K.R. Day, 1993. Relationship between fruit respiration, bruising susceptibility, and temperature in sweet cherries. *HortScience*, 28(2): 132-135.

Cruse, R.M., and C.G. Herndl, 2009. Balancing corn stover harvest for biofuels with soil and water conservation. *Journal of Soil and Water Conservation*, 64(4): 286-291.

Curtis, P.S., and X.Z. Wang, 1998. A meta-analysis of elevated CO_2 effects on woody plant mass, form, and physiology. *Oecologia*, 113(3): 299-313.

Dale, A. 2008. Raspberry production in greenhouses: physiological aspects. Acta Hort. 777:219-224.

Danka, R.G., and L.D. Beaman, 2007. Flight activity of USDA-ARS Russian honey bees (hymenoptera : Apidae) during pollination of lowbush blueberries in maine. *Journal of Economic Entomology*, 100(2): 267-272.

Darnell, R.L., and J.G. Williamson, 1997. Feasibility of blueberry production in warm climates. *Acta Hort. (ISHS)*, 446(251-256).

Davis, M.S., T.L. Mader, S.M. Holt, and A.M. Parkhurst, 2003. Strategies to reduce feedlot cattle heat stress: Effects on tympanic temperature. *Journal of Animal Science*, 81(3): 649-661.

De Boeck, H.J., C.M.H.M. Lemmens, C. Zavalloni, B. Gielen, S. Malchair, M. Carnol, R. Merckx, J. Van den Berge, R. Ceulemans, and I. Nijs, 2008. Biomass production in experimental grasslands of different species richness during three years of climate warming. *Biogeosciences*, 5(2): 585-594.

de Orduna, R.M., 2010. Climate change associated effects on grape and wine quality and production. *Food Research International*, 43(7): 1844-1855.

de Steiguer, J.E., 2008. Semi-arid rangelands and carbon offset markets: A look at the economic prospects. *Rangelands*, 30(2): 27-32.

DeBach, P., and S. R.A., 1963. Competitive displacement between ecological homologues. *Hilgardia*, 34: 105-166.

Debinski, D.M., H. Wickham, K. Kindscher, J.C. Caruthers, and M. Germino, 2010. Montane meadow change during drought varies with background hydrologic regime and plant functional group. *Ecology*, 91(6): 1672-1681.

Delgado, J.A., P.M. Groffman, M.A. Nearing, T. Goddard, D. Reicosky, R. Lal, N.R. Kitchen, C.W. Rice, D. Towery, and P. Salon, 2011. Conservation practices to mitigate and adapt to climate change. *Journal of Soil and Water Conservation*, 66(4): 118A-129A.

Delgado, J.A., and A.R. Mosier, 1996. Mitigation alternatives to decrease nitrous oxides emissions and urea-nitrogen loss and their effect on methane flux. *Journal of Environmental Quality*, 25(5): 1105-1111.

DeLucia, E.H., and R.B. Thomas, 2000. Photosynthetic responses to CO2 enrichment of four hardwood species in a forest understory. *Oecologia*, 122(1): 11-19.

Derner, J.D., K.R. Hickman, and H.W. Polley, 2011. Decreasing precipitation variability does not elicit major aboveground biomass or plant diversity responses in a mesic rangeland. *Rangeland Ecology & Management*, 64(4): 352-357.

Derner, J.D., H.B. Johnson, B.A. Kimball, P.J. Pinter, H.W. Polley, C.R. Tischler, T.W. Boutton, R.L. Lamorte, G.W. Wall, N.R. Adam, S.W. Leavitt, M.J. Ottman, A.D. Matthias, and T.J. Brooks, 2003. Above- and below-ground responses of C3 –C4 species mixtures to elevated CO_2 and soil water availability. *Global Change Biology*, 9(3): 452-460.

Dettinger, M.D., and S. Earman, 2007. Western ground water and climate change - pivotal to supply sustainability or vulnerable in its own right? *Ground Water News and Views, Association of Ground Water Scientists and Engineers Newsletter*, 4(1): 4-5.

Dhillon, T., S.P. Pearce, E.J. Stockinger, A. Distelfeld, C. Li, A.K. Knox, I. Vashegyi, A. Vágújfalvi, G. Galiba, and J. Dubcovsky, 2010. Regulation of freezing tolerance and flowering in temperate cereals: The VRN-1 connection. *Plant Physiology*, 153(4): 1846-1858.

Diffenbaugh, N.S., M.A. White, G.V. Jones, and M. Ashfaq, 2011. Climate adaptation wedges: A case study of premium wine in the western United States. *Environmental Research Letters*, 6(2).

Dijkstra, F.A., D. Blumenthal, J.A. Morgan, E. Pendall, Y. Carrillo, and R.F. Follett, 2010. Contrasting effects of elevated CO_2 and warming on nitrogen cycling in a semiarid grassland. *New Phytologist*, 187(2): 426-437.

Dixon, J., and E.W. Hewett, 2000. Factors affecting apple aroma/flavour volatile concentration: A review. *New Zealand Journal of Crop and Horticultural Science*, 28(3): 155-173.

DOI, 2011. *Reclamation: Managing water in the west. Literature synthesis on climate change implications for water and environmental resources. Technical memorandum 86-68210-2010-03*. Department of Interior, Bureau of Reclamation, Denver, CO.

Dokoozlian, N.K., and W.M. Kliewer, 1996. Influence of light on grape berry growth and composition varies during fruit development. *Journal of the American Society for Horticultural Science*, 121(5): 869-874.

Dominati, E., M. Patterson, and A. Mackay, 2010. A framework for classifying and quantifying the natural capital and ecosystem services of soils. *Ecological Economics*, 69(9): 1858-1868.

Downes, R., 1972. Effect of temperature on the phenology and grain yield of *Sorghum bicolor*. *Australian Journal of Agricultural Research*, 23(4): 585-594.

Druta, A., 2001. Effect of long term exposure to high CO_2 concentrations on photosynthetic characteristics of *Prunus avium l.* plants. *Photosynthetica*, 39(2): 289-297.

Dufault, R.J., B. Ward, and R.L. Hassell, 2009. Dynamic relationships between field temperatures and romaine lettuce yield and head quality. *Scientia Horticulturae*, 120(4): 452-459.

Dukes, J.S., N.R. Chiariello, E.E. Cleland, L.A. Moore, M.R. Shaw, S. Thayer, T. Tobeck, H.A. Mooney, and C.B. Field, 2005. Responses of grassland production to single and multiple global environmental changes. *PLoS Biol*, 3(10): e319.

Duvick, D.N., and K.G. Cassman, 1999. Post–green revolution trends in yield potential of temperate maize in the north-central United States. *Crop Science*, 39(6): 1622-1630.

Eagle, A.J., L.R. Henry, L.P. Olander, K.Haugen-Kozyra, N.Millar, and G.P. Robertson, 2010. *Greenhouse gas mitigation potential of agricultural land management in the United States: A synthesis of the literature. Technical working group on agricultural greenhouse gases (T-AGG)* Nicholas Institute for Environmental Policy Solutions, Duke University, Durham, NC http://nicholasinstitute.duke.edu/ecosystem/land/TAGGDLitRev.

Eastburn, D.M., M.M. Degennaro, E.H. Delucia, O. Dermody, and A.J. McElrone, 2010. Elevated atmospheric carbon dioxide and ozone alter soybean diseases at soyface. *Global Change Biology*, 16(1): 320-330.

Eaton, F.M., and D.R. Ergle, 1952. Fiber properties and carbohydrate and nitrogen levels of cotton plants as influenced by moisture supply and fruitfulness. *Plant Physiology*, 27(3): 541-562.

Edwards, N.T., and R.J. Norby, 1999. Below-ground respiratory responses of sugar maple and red maple saplings to atmospheric CO_2 enrichment and elevated air temperature. *Plant and Soil*, 206(1): 85-97.

Eigenberg, R.A., T.M. Brown-Brandl, J.A. Nienaber, and G.L. Hahn, 2005. Dynamic response indicators of heat stress in shaded and non-shaded feedlot cattle, part 2: Predictive relationships. *Biosystems Engineering*, 91(1): 111-118.

Else, M., and C. Atkinson, 2010. Climate change impacts on UK top and soft fruit production. *Outlook on Agriculture*, 39(4): 257-262.

Elsner, M., L. Cuo, N. Voisin, J. Deems, A. Hamlet, J. Vano, K. Mickelson, S.-Y. Lee, and D. Lettenmaier, 2010. Implications of 21st century climate change for the hydrology of Washington state. *Climatic Change*, 102(1): 225-260.

ERS, 2012. *Wheat data: Yearbook tables* http://www.ers.usda.gov/data/wheat/YBtable21.asp.

Estiarte, M., J. Penuelas, B.A. Kimball, S.B. Idso, R.L. Lamorte, P.J. Pinter, G.W. Wall, and R.L. Garcia, 1994. Elevated CO_2 effects on stomatal density of wheat and sour orange trees. *Journal of Experimental Botany*, 45(280): 1665-1668.

Evans, L.T., 1993. *Crop evolution, adaptation, and yield.* Cambridge University Press, Cambridge; New York.

Ezin, V., R.D.L. Pena, and A. Ahanchede, 2010. Flooding tolerance of tomato genotypes during vegetative and reproductive stages. *Brazilian Journal of Plant Physiology*, 22: 131-142.

Farquhar, G.D., S. von Caemmerer, and J.A. Berry, 1980. A biochemical model of photosynthetic CO_2 assimilation in leaves of C3 species. *Planta (Berlin)*, 149: 78-90.

Faver, K.L., T.J. Gerik, P.M. Thaxton, and K.M. El-Zik, 1996. Late season water stress in cotton: II. Leaf gas exchange and assimilation capacity. *Crop Science*, 36(4): 922-928.

Favis-Mortlock, D.T., and A.J.T. Guerra, 1999. The implications of general circulation model estimates of rainfall for future erosion: A case study from Brazil. *CATENA*, 37(3–4): 329-354.

Favis-Mortlock, D.T., and M.R. Savabi, 1996. Shifts in rates and spatial distributions of soil erosion and deposition under climate change., *in* M.G. Anderson and S.M. Brooks (eds.), *Advances in hillslope processes. Vol. 1.* John Wiley & Sons, Chichester.

Fay, P.A., J.M. Blair, M.D. Smith, J.B. Nippert, J.D. Carlisle, and A.K. Knapp, 2011. Relative effects of precipitation variability and warming on grassland ecosystem function. *Biogeosciences Discuss.*, 8(4): 6859-6900.

Fay, P.A., D.M. Kaufman, J.B. Nippert, J.D. Carlisle, and C.W. Harper, 2008. Changes in grassland ecosystem function due to extreme rainfall events: Implications for responses to climate change. *Global Change Biology*, 14(7): 1600-1608.

Felicetti, D.A., and L.E. Schrader, 2008. Changes in pigment concentrations associated with the degree of sunburn browning of 'fuji' apple. *Journal of the American Society for Horticultural Science*, 133(1): 27-34.

Feng, Z., K. Kobayashi, and E.A. Ainsworth, 2008. Impact of elevated ozone concentration on growth, physiology, and yield of wheat (*Triticum aestivum l.*): A meta-analysis. *Global Change Biology*, 14(11): 2696-2708.

Fernandez, G.E., F.J. Louws, J.R. Ballington, and E.B. Poling, 1998. Growing raspberries in North Carolina, North Carolina Cooperative Extension Service AG-569.

Ferris, R., R.H. Ellis, T.R. Wheeler, and P. Hadley, 1998. Effect of high temperature stress at anthesis on grain yield and biomass of field-grown crops of wheat. *Annals of Botany*, 82(5): 631-639.

Fischlin, A., G.F. Midgley, J.T. Price, R. Leemans, B. Gopal, C. Turley, M.D.A. Rounsevell, O.P. Dube, J. Tarazona, and A.A. Velichko, 2007. Ecosystems, their properties, goods, and services, *in* M.L. Parry, O.F. Canziani, J.P. Palutikof, P.J.V.D. Linden and C.E. Hanson (eds.), *Climate change 2007: Impacts, adaptation and vulnerability. Contribution of Working Group II to the Fourth Assessment Report of the Intergovernmental Panel on Climate Change*, pp. 211-272. Cambridge University Press, Cambridge.

Fitzsimons, T., and D.M. Oosterhuis, 2011. *Relationships of cotton productions and high temperatures for Lee County, Arkansas*, paper presented at ASA/CSSA/SSSA Annual Meetings. San Antonio, TX, Oct 17-20, 2011.

Flore, J.A., and D.R. Layne, 1999. Photoassimilate production and distribution in cherry. *HortScience*, 34: 1015-1019.

Forshey, C.G., 1976. *Factors affecting chemical thinning of apples.* New York's Food and Life Sci. Bul. 64.

Frank, K.L., T.L. Mader, J.A. Harrington, G.L. Hahn, and M.S. Davis, 2001. *Climate change effects on livestock production in the Great Plains*, paper presented at 6th International Livestock Environment Symposium, American Society of Agricultural Engineers, St. Joseph, MI.

Frantz, J.M., G. Ritchie, N.N. Cometti, J. Robinson, and B. Bugbee, 2004. Exploring the limits of crop productivity: Beyond the limits of tipburn in lettuce. *Journal of the American Society for Horticultural Science*, 129(3): 331-338.

Friedel, M.H., 1991. Range condition assessment and the concept of thresholds: A viewpoint. *Journal of Range Management*, 44(5): 422-426.

Frumhoff, P., J. McCarthy, J. Mellilo, S. Moser, and D. Wuebbles, 2006. *Climate change in the U.S. Northeast.* Union of Concerned Scientists Publications, Cambridge (also available at http://www.climatechoices.org).

Fuhrer, J., 2009. Ozone risk for crops and pastures in present and future climates. *Naturwissenschaften*, 96(2): 173-194.

Gadgil, S., P.R. Seshagiri Rao, and K. Narahari Rao, 2002. Use of climate information for farm-level decision making: Rainfed groundnut in southern India. *Agricultural Systems*, 74(3): 431-457.

Garcia-Sanchez, F., and J.P. Syvertsen, 2006. Salinity tolerance of cleopatra mandarin and carrizo citrange rootstock seedlings is affected by CO_2 enrichment during growth. *Journal of the American Society for Horticultural Science*, 131(1): 24-31.

Garrett, K.A., S.P. Dendy, E.E. Frank, M.N. Rouse, and S.E. Travers, 2006. Climate change effects on plant disease: Genomes to ecosystems, *in Annual review of phytopathology*, pp. 489-509.

Gates, R.N., C.L. Quarin, and C.G.S. Pedreira, 2004. Bahiagrass, *in* L.E. Moser, B.L. Burson and L.E. Sollenberger (eds.), *Warm-season C4 grasses*, pp. 651-680. American Society of Agronomy : Crop Science Society of America : Soil Science Society of America, Madison, Wis.

Gaughan, J.B., J. Goopy, and J. Spark, 2002a. *Excessive heat load index for feedlot cattle.* Meat and Livestock-Australia Project Report, FLOT.316.

Gaughan, J.B., N. Lacetera, S.E. Valtorta, H.H. Khalifa, L. Hahn, and T. Mader, 2009. Response of domestic animals to animal challenges, *in* K.L. Ebi, I. Burton and G.R. Mcgregor (eds.), *Biometeorology for adaptation to climate variability and change.* Springer, Dordrecht; London.

Gaughan, J.B., and T.L. Mader, 2007. Managing heat stress of feedlot cattle through nutrition, *in Recent advances in animal nutrition in Australia, July 2007*, pp. 209-219. Univ. New England, Armidale, NSW Australia.

Gaughan, J.B., T.L. Mader, S.M. Holt, G.L. Hahn, and B.A. Young., 2002b. Review of current assessment of cattle and microclimate during periods of high heat load. *Anim. Prod. in Austr* 24: 77-80.

Gaughan, J.B., T.L. Mader, S.M. Holt, M.J. Josey, and K.J. Rowan, 1999. Heat tolerance of boran and tuli crossbred steers. *Journal of Animal Science*, 77(9): 2398-2405.

Gaughan, J.B., T.L. Mader, S.M. Holt, and A. Lisle, 2008. A new heat load index for feedlot cattle. *Journal of Animal Science*, 86(1): 226-234.

Gent, M.P.N., 2007. Effect of degree and duration of shade on quality of greenhouse tomato. *HortScience*, 42(3): 514-520.

Gentile, R., M. Dodd, M. Lieffering, S. Brock, P. Theobald, and P. Newton, 2011. Effects of long-term exposure to enriched CO_2 on the nutrient-supplying capacity of a grassland soil. *Biology and Fertility of Soils*: 1-6.

Gerik, T.J., K.L. Faver, P.M. Thaxton, and K.M. El-Zik, 1996. Late season water stress in cotton: I. Plant growth, water use, and yield. *Crop Science*, 36(4): 914-921.

Germanà, C., A. Continella, and E. Tribulato, 2003. Net shading influence on floral induction on citrus trees. *Acta Hort. (ISHS)*, 614: 527-533.

Ghosh, S.C.A., Koh-ichiro; Kusutani, Akihito; Toyota, Masanori, 2000. Effects of temperature at different growth stages on nonstructural carbohydrate, nitrate reductase activity and yield of potato. *Environment Control in Biology*, 38: 197-206.

Gipson, J.R., and H.E. Joham, 1968. Influence of night temperature on growth and development of cotton (*Gossypium birsutum l.*). I. Fruiting and boll development. *Agronomy Journal*, 60(3): 292-295.

Glenn, D.M., 2009. Particle film mechanisms of action that reduce the effect of environmental stress in 'empire' apple. *Journal of the American Society for Horticultural Science*, 134(3): 314-321.

Goncalves, B., V. Falco, J. Moutinho-Pereira, E. Bacelar, F. Peixoto, and C. Correia, 2009. Effects of elevated CO_2 on grapevine (*Vitis vinifera l.*): Volatile composition, phenolic content, and in vitro antioxidant activity of red wine. *Journal of Agricultural and Food Chemistry*, 57(1): 265-273.

Gote, G.N., and P.R. Padghan, 2009. Studies on different thermal regimes and thermal sensitivity analysis of tomato genotypes. *Asian Journal of Environmental Science*, 3: 158-161.

Grauslund, J., 1978. Effects of temperature, shoot-tipping, and carbaryl on fruit set of apple trees. *Acta Hort. (ISHS)*, 80: 207-212.

Grazia, J.d., P.A. Tittonell, and Á. Chiesa, 2001. Effects of sowing date, radiation and nitrogen nutrition on growth pattern and yield of lettuce (*Lactuca sativa l.*) crop. *Investigación Agraria, Producción y Protección Vegetales*, 16: 355-365.

Greenland, R.G., 2000. Optimum height at which to kill barley used as a living mulch in onions. *HortScience*, 35(5): 853-855.

Grimes, D.W., W.L. Dickens, and W.D. Anderson, 1969. Functions for cotton (*Gossypium hirsutum l.*) production from irrigation and nitrogen fertilization variables. II. Yield components and quality characteristics. *Agronomy Journal*, 61: 773-776.

Grimes, D.W., and H. Yamada, 1982. Relation of cotton growth and yield to minimum leaf water potential. *Crop Science*, 22(1): 134-139.

Groninger, J.W., J.R. Seiler, S.M. Zedaker, and P.C. Berrang, 1996. Effects of CO_2 concentration and water availability on growth and gas exchange in greenhouse-grown miniature stands of Loblolly Pine and Red Maple. *Functional Ecology*, 10(6): 708-716.

Groves, F.E. (2009). *Improvement of cotton through selective use of lint and seed parameters*. University of Arkansas, Fayetteville: ProQuest, LLC, 2010.

Guardiola, J.L., and A. Garcia-Luis, 2000. Increasing fruit size in citrus. Thinning and stimulation of fruit growth. *Plant Growth Regulation*, 31(1-2): 121-132.

Gutierrez, A.P., 2005. *UC/IPM progress report.*

Gutierrez, A.P., L. Ponti, C.K. Ellis, and T. d'Oultremont, 2006. *Analysis of climate effects on agricultural systems.* . A report from the California Climate Change Center. White Paper CEC-500-2005-188-SF, University of California, Berkeley.

Hahn, G.L., 1995. Environmental management for improved livestock performance, health and well-being. *Japanese journal of livestock management*, 30(3): 113-127 %U http://ci.nii.ac.jp/naid/110003852999/en/.

Hahn, G.L., T. Brown-Brandl, R.A. Eigenberg, J.B. Gaughan, T.L. Mader, and J.A. Nienaber, 2005. *Climate change and livestock: Challenges and adaptive responses of animals and production systems*, paper presented at International Conference on Biometeorology. September 2005. Bavaria, Germany.

Hahn, G.L., T.L. Mader, and R.A. Eigenberg, 2003. Perspective on development of thermal indices for animal studies and management, *in* N. Lacetera (ed.) *Interactions between climate and animal production (EAAP Technical Series 7)*. Wageningen Academic Publishers, Wageningen, the Netherlands.

Hahn, G.L., T.L. Mader, J.B. Gaughan, Q. Hu, and J.A. Nienaber., 1999. *Waves and their impacts on feedlot cattle*, paper presented at Proceedings of the 15th International Congress of Biometeorology and International Congress on Urban Climatology. Sydney, Australia.

Hahn, G.L., and T.L. Mader., 1997. *Heat waves in relation to thermoregulation, feeding behavior and mortality of feedlot cattle*, paper presented at Proceedings of the 5th International Livestock Environ Symp. American Society of Agricultural Engineers. Joseph, Mich.

Hall, A.E., 2001. *Crop responses to environment*. CRC Press, Boca Raton.

Hance, T., J. van Baaren, P. Vernon, and G. Boivin, 2007. Impact of extreme temperatures on parasitoids in a climate change perspective, *in Annual review of entomology*, pp. 107-126. Annual Reviews, Palo Alto.

Hancock, A.M., B. Brachi, N. Faure, M.W. Horton, L.B. Jarymowycz, F.G. Sperone, C. Toomajian, F. Roux, and J. Bergelson, 2011. Adaptation to climate across the *Arabidopsis thaliana* genome. *Science*, 333(6052): 83-86.

Hao, X., Wang, Q. and Khosla, S. , 2008. Responses of greenhouse tomatoes to summer CO_2 enrichment. *Acta Hort. (ISHS)*, 797: 241-246.

Harbut, R.M., J.A. Sullivan, and J.T.A. Proctor, 2010. Temperature affects dry matter production and net carbon exchange rate of lower-ploidy *Fragaria* species and species hybrids. *Canadian Journal of Plant Science*, 90(6): 885-892.

Harris, M.O., J.J. Stuart, M. Mohan, S. Nair, R.J. Lamb, and O. Rohfritsch, 2003. Grasses and gall midges: Plant defense and insect adaptation. *Annual Review of Entomology*, 48(1): 549-577.

Hatchett, J.H., K.J. Starks, and J.A. Webster, 1987. Insect and mite pests of wheat, *in* E.G. Heyne (ed.) *Wheat and wheat improvement*, pp. 625-675. American Society of Agronomy, Crop Science Society of America, Soil Science Society of America, Madison, Wis., USA.

Hatfield, J.L., 2006. Multifunctionality of agriculture and farming system design: Perspectives from the United States. *Bibliotecha Fragmenta Agronomica*, 11: 43-52.

Hatfield, J.L., K.J. Boote, B.A. Kimball, L.H. Ziska, R.C. Izaurralde, D. Ort, A.M. Thomson, and D. Wolfe, 2011. Climate impacts on agriculture: Implications for crop production. *Agronomy Journal*, 103(2): 351-370.

Hatfield, J.L., T.J. Sauer, and J.H. Prueger, 2001. Managing soils to achieve greater water use efficiency. *Agron. J.*, 93(2): 271-280.

Hauagge, R., 'IPR julieta', a new early low chill requirement apple cultivar. *Acta Hort. (ISHS)*, 872: 193-196.

Havstad, K.M., D.C. Peters, B. Allen-Diaz, J. Bartolome, B.T. Bestel-meyer, D. Briske, J. Brown, M. Brunson, J.E. Herrick, L. Hunt-singer, P. Johnson, L. Joyce, R. Pieper, A.J. Svejcar, and J. Yao, 2009. The western United States rangelands, a major resource, *in* W.F. Wedin and S.L. Fales (eds.), *Grassland : Quietness and strength for a new American agriculture*. American Society of Agronomy, Crop Science Society of America, Soil Society of America, Madison, WI.

Hayhoe, K., C.P. Wake, T.G. Huntington, L.F. Luo, M.D. Schwartz, J. Sheffield, E. Wood, B. Anderson, J. Bradbury, A. DeGaetano, T.J. Troy, and D. Wolfe, 2007. Past and future changes in climate and hydrological indicators in the U.S. Northeast. *Climate Dynamics*, 28(4): 381-407.

Heagle, A.S., 1989. Ozone and crop yield. *Annual Review of Phytopa-thology*, 27: 397-423.

Heal, G.M., 2000. *Nature and the marketplace : Capturing the value of ecosystem services*. Island Press, Washington, D.C.

Hedhly, A., J.I. Hormaza, and M. Herrera, 2004. Effect of temperature on pollen tube kinetics and dynamics in sweet cherry, *Prunus avium* (rosaceae). *American Journal of Botany*, 91(4): 558-564.

Hegland, S.J., A. Nielsen, A. Lázaro, A.-L. Bjerknes, and Ø. Totland, 2009. How does climate warming affect plant-pollinator interac-tions? *Ecology Letters*, 12(2): 184-195.

Heide, O.M., and A.K. Prestrud, 2005. Low temperature, but not pho-toperiod, controls growth cessation and dormancy induction and release in apple and pear. *Tree Physiology*, 25(1): 109-114.

Heisler-White, J.L., J.M. Blair, E.F. Kelly, K. Harmoney, and A.K. Knapp, 2009. Contingent productivity responses to more extreme rainfall regimes across a grassland biome. *Global Change Biology*, 15(12): 2894-2904.

Henderson, M.S., and D.L. Robinson, 1982. Environmental influences on yield and in vitro true digestibility of warm-season perennial grasses and the relationships to fiber components. *Agronomy Journal*, 74(6): 943-946.

Heong, K.L., P.S. Teng, and K. Moody, 1995. Managing rice pests with less chemicals. *GeoJournal*, 35(3): 337-349.

Herrero, M.P., and R.R. Johnson, 1980. High temperature stress and pollen viability of maize. *Crop Sci.*, 20(6): 796-800.

Hesketh, J.D., D.L. Myhre, and C.R. Willey, 1973. Temperature control of time intervals between vegetative and reproductive events in soybeans. *Crop Sci.*, 13(2): 250-254.

Heuvelink, E., M. Bakker, L.F.M. Marcelis, M. Raaphorst, S. Pascale, G. de Scarascia Mugnozza, A. Maggio, and E. Schettini, 2008. Climate and yield in a closed greenhouse. *Acta Horticulturae*, 801: 1083-1092.

Hileman, D.R., G. Huluka, P.K. Kenjige, N. Sinha, N.C. Bhattacha-rya, P.K. Biswas, K.F. Lewin, J. Nagy, and G.R. Hendrey, 1994. Canopy photosynthesis and transpiration of field-grown cotton exposed to free-air CO_2 enrichment (FACE) and differential irriga-tion. *Agricultural and Forest Meteorology*, 70(1–4): 189-207.

Hill, A.R., 1996. Nitrate removal in stream riparian zones. *Journal of Environmental Quality.*, 25(4): 743-755.

Hodges, H.F., K.R. Reddy, J.M. McKinnon, and V.R. Reddy, 1993. *Temperature effects on cotton*. Mississippi State University, MS, Mississippi Agr. & Forestry Exp. Sta.

Hodges, T., and J.T. Ritchie, 1991. The CERES-Wheat phenology model, *in* T. Hodges (ed.) *Predicting crop phenology*, pp. 115-131. CRC Press, Boca Raton.

Hopkins, A., and A. Del Prado, 2007. Implications of climate change for grassland in Europe: Impacts, adaptations and mitigation options: A review. *Grass & Forage Science*, 62(2): 118-126.

Hovenden, M.J., K.E. Wills, J.K. Vander Schoor, A.L. Williams, and P.C.D. Newton, 2008. Flowering phenology in a species-rich tem-perate grassland is sensitive to warming but not elevated CO_2. *New Phytologist*, 178(4): 815-822.

Howitt, R., J. Medellín-Azuara, and D. MacEwan, 2010. Climate change, markets, and technology. *Choices*, 25(3), http://www.choicesmagazine.org/magazine/article.php?article=148.

Huang, H., and M. Khanna, 2010. *An econometric analysis of U.S. crop yields and cropland acreages: Implications for the impact of climate change*, paper presented at AAEA annual meeting. Denver, Colorado, 25-27July

Hubbard, K.G., and F.J. Flores-Mendoza, 1995. Relating United States crop land use to natural resources and climate change. *Journal of Climate*, 8(2): 329-335.

Hutton, R.J., and J.J. Landsberg, 2000. Temperature sums experienced before harvest partially determine the post-maturation juicing quality of oranges grown in the Murrumbidgee Irrigation Areas (MIA) of New South Wales. *Journal of the Science of Food and Agriculture*, 80(2): 275-283.

Idso, S.B., and B.A. Kimball, 1997. Effects of long-term atmospheric CO_2 enrichment on the growth and fruit production of sour orange trees. *Global Change Biology*, 3(2): 89-96.

Idso, S.B., B.A. Kimball, G.W. Wall, R.L. Garcia, R. LaMorte, P.J. Pinter Jr, J.R. Mauney, G.R. Hendrey, K. Lewin, and J. Nagy, 1994. Effects of free-air CO_2 enrichment on the light response curve of net photosynthesis in cotton leaves. *Agricultural and Forest Meteorology*, 70(1–4): 183-188.

Iglesias, A., K. Avis, M.Benzie, P. Fisher, M. Harley, N. Hodgson, L. Horrocks, M. Moneo, and J. Webb, 2007. *Adaptation to climate change in the agricultural sector. Report to European commis-sion directorate - general for agriculture and rural development. ED05334* http://ec.europa.eu/agriculture/analysis/external/climate/.

Iglesias, D.J., M. Cercós, J.M. Colmenero-Flores, M.A. Naranjo, G. Ríos, E. Carrera, O. Ruiz-Rivero, I. Lliso, R. Morillon, F.R. Tadeo, and M. Talon, 2007. Physiology of citrus fruiting. *Brazilian Jour-nal of Plant Physiology*, 19: 333-362.

Immerzeel, W.W., L.P.H. van Beek, and M.F.P. Bierkens, 2010. Climate change will affect the Asian water towers. *Science*, 328(5984): 1382-1385.

IPCC, 2007b. *Climate change 2007 : Impacts, adaptation and vulner-ability: Contribution of Working Group II to the Fourth Assessment Report of the Intergovernmental Panel on Climate Change*. Cam-bridge University Press, Cambridge, U.K.; New York.

Ito, J., S. Hasegawa, K. Fujita, S. Ogasawara, and T. Fujiwara, 2002. Changes in water relations induced by CO_2 enrichment govern diurnal stem and fruit diameters of Japanese pear. *Plant Science*, 163(6): 1169-1176.

Jackson, J.E., 1980. Light interception and utilization by orchard systems, *in Horticultural reviews*, pp. 208-267. John Wiley & Sons, Inc.

Jackson, J.E., P.J.C. Hamer, and M.F. Wickenden, 1983. Effects of early spring temperatures on the set of fruits of Cox's orange pippin apple and year-to-year variation in its yields. *Acta Hort. (ISHS)*, 139: 75-82.

Jentsch, A., J. Kreyling, M. Elmer, E. Gellesch, B. Glaser, K. Grant, R. Hein, M. Lara, H. Mirzae, S.E. Nadler, L. Nagy, D. Otieno, K. Pritsch, U. Rascher, M. Schädler, M. Schloter, B.K. Singh, J. Stadler, J. Walter, and C. Wellstein, 2011. Climate extremes initiate ecosystem-regulating functions while maintaining productivity. *Journal of Ecology*, 99(3): 689-702.

Jones, G., M. White, O. Cooper, and K. Storchmann, 2005. Climate change and global wine quality. *Climatic Change*, 73(3): 319-343.

Jones, G.V., 2005. Climate change in the western United States grape growing regions. *Acta Hort. (ISHS)*, 689: 41-60.

Jones, G.V., and G.B. Goodrich, 2007. Influence of climate variability on wine regions in the western USA and on wine quality in the Napa Valley. *Climate Research*, 35(3): 241-154.

Kadir, S., G. Sidhu, and K. Al-Khatib, 2006. Strawberry (*Fragaria ×ananassa duch.*) growth and productivity as affected by temperature. *HortScience*, 41(6): 1423-1430.

Kakani, V.G., K.R. Reddy, S. Koti, T.P. Wallace, P.V.V. Prasad, V.R. Reddy, and D. Zhao, 2005. Differences in in vitro pollen germination and pollen tube growth of cotton cultivars in response to high temperature. *Annals of Botany*, 96(1): 59-67.

Karlen, D.L., N.C. Wollenhaupt, D.C. Erbach, E.C. Berry, J.B. Swan, N.S. Eash, and J.L. Jordahl, 1994a. Crop residue effects on soil quality following 10-years of no-till corn. *Soil and Tillage Research*, 31(2–3): 149-167.

Karlen, D.L., N.C. Wollenhaupt, D.C. Erbach, E.C. Berry, J.B. Swan, N.S. Eash, and J.L. Jordahl, 1994b. Long-term tillage effects on soil quality. *Soil and Tillage Research*, 32(4): 313-327.

Kean, S., 2010. Besting Johnny Appleseed. *Science*, 328(5976): 301-303.

Kelm, M.A., J.A. Flore, and C.W. Beninger, 2005. Effect of elevated CO_2 levels and leaf area removal on sorbitol, sucrose, and phloridzin content in 'gala'/malling 9 apple leaves. *Journal of the American Society for Horticultural Science*, 130(3): 326-330.

Keutgen, N., and K. Chen, 2001. Responses of citrus leaf photosynthesis, chlorophyll fluorescence, macronutrient and carbohydrate contents to elevated CO_2. *Journal of Plant Physiology*, 158(10): 1307-1316.

Kim, H.Y., T. Horie, H. Nakagawa, and K. Wada, 1996. Effects of elevated CO_2 concentration and high temperature on growth and yield of rice 2: The effect on yield and its components of ahihikari rice. *Japanese Journal of Crop Science*, 65(4): 644-651.

Kim, H.-Y., M. Lieffering, K. Kobayashi, M. Okada, and S.H.U. Miura, 2003. Seasonal changes in the effects of elevated CO_2 on rice at three levels of nitrogen supply: A free air CO_2 enrichment (FACE) experiment. *Global Change Biology*, 9(6): 826-837.

Kimball, B.A., 2011. Effects and interactions with water, nitrogen and temperature, *in* D. Hillel and C. Rosenzweig (eds.), *Handbook of climate change and agroecosystems : Impacts, adaptation, and mitigation*, pp. 87-107. Imperial College Press ; Distributed by World Scientific Publishing Co., London; Singapore; Hackensack, NJ.

Kimball, B.A., S.B. Idso, S. Johnson, and M.C. Rillig, 2007. Seventeen years of carbon dioxide enrichment of sour orange trees: Final results. *Global Change Biology*, 13(10): 2171-2183.

Kimball, B.A., K. Kobayashi, and M. Bindi, 2002. Responses of agricultural crops to free-air CO_2 enrichment. *Advances in Agronomy, Vol 77*, 77: 293-368.

Kiniry, J.R., and R. Bonhomme, 1991. Predicting maize phenology, *in* T. Hodges (ed.) *Predicting crop phenology*, pp. 115-131. CRC Press, Boca Raton.

Kittock, D.L., W.C. Hofmann, and E.L. Turcotte, 1988. Estimation of heat tolerance improvement in recent American pima cotton cultivars. *Journal of Agronomy & Crop Science*, 161(5): 305-309.

Klamkowski, K., and W. Treder, 2008. Response to drought stress of three strawberry cultivars grown under greenhouse conditions. *Journal of Fruit and Ornamental Plant Research*, 16: 179-188.

Klein, J.A., J. Harte, and X.-Q. Zhao, 2007. Experimental warming, not grazing, decreases rangeland quality on the Tibetan plateau. *Ecological Applications*, 17(2): 541-557.

Knapp, A.K., C. Beier, D.D. Briske, A.T. Classen, Y. Luo, M. Reichstein, M.D. Smith, S.D. Smith, J.E. Bell, P.A. Fay, J.L. Heisler, S.W. Leavitt, R. Sherry, B. Smith, and E. Weng, 2008. Consequences of more extreme precipitation regimes for terrestrial ecosystems. *BioScience*, 58(9): 811-821.

Knowles, N., M.D. Dettinger, and D.R. Cayan, 2006. Trends in snowfall versus rainfall in the western United States. *Journal of Climate*, 19(18): 4545-4559.

Kobayashi, T., K. Ishiguro, T. Nakajima, H.Y. Kim, M. Okada, and K. Kobayashi, 2006. Effects of elevated atmospheric CO_2 concentration on the infection of rice blast and sheath blight. *Phytopathology*, 96(4): 425-431.

Kobza, J., and G.E. Edwards, 1987. Influences of leaf temperature on photosynthetic carbon metabolism in wheat. *Plant Physiology*, 83: 69-74.

Körner, C., 2006. Plant CO_2 responses: An issue of definition, time and resource supply. *New Phytologist*, 172(3): 393-411.

Kreikemeier, W.M., and T.L. Mader, 2004. Effects of growth-promoting agents and season on yearling feedlot heifer performance. *Journal of Animal Science*, 82(8): 2481-2488.

Kubiske, M.E., and K.S. Pregitzer, 1996. Effects of elevated CO_2 and light availability on the photosynthetic light response of trees of contrasting shade tolerance. *Tree Physiology*, 16(3): 351-358.

Kucharik, C.J., 2006. A multidecadal trend of earlier corn planting in the central USA. *Agronomy Journal*, 98(6): 1544-1550.

Kucharik, C.J., and S.P. Serbin, 2008. Impacts of recent climate change on Wisconsin corn and soybean yield trends. *Environmental Research Letters*, 3(3): 034003.

Kunkel, K.E., K. Andsager, and D.R. Easterling, 1999. Long-term trends in extreme precipitation events over the conterminous United States and Canada. *Journal of Climate*, 12(8): 2515-2527.

Laing, D.R., P.G. Jones, and J.H. Davis, 1984. Common bean (*Phaseolus vulgaris l.*), *in* P.R. Goldsworthy and N.M. Fisher (eds.), *The physiology of tropical field crops*, pp. 305-351. Wiley, Chichester; New York.

Lal, R., 1997. *Soil processes and the carbon cycle*. CRC, Boca Raton.

Lal, R., J.A. Delgado, P.M. Groffman, N. Millar, C. Dell, and A. Rotz, 2011. Management to mitigate and adapt to climate change. *Journal of Soil and Water Conservation*, 66(4): 276-285.

Langley, J.A., and J.P. Megonigal, 2010. Ecosystem response to elevated CO_2 levels limited by nitrogen-induced plant species shift. *Nature*, 466(7302): 96-99.

Larson, K.D., S.T. Koike, and F.G. Zalom, 2005. Bed mulch treatment affects strawberry fruit bronzing and yield performance. *HortScience*, 40(1): 72-75.

Leakey, A.D.B., E.A. Ainsworth, C.J. Bernacchi, A. Rogers, S.P. Long, and D.R. Ort, 2009a. Elevated CO_2 effects on plant carbon, nitrogen, and water relations: Six important lessons from FACE. *Journal of Experimental Botany*, 60(10): 2859-2876.

Leakey, A.D.B., C.J. Bernacchi, F.G. Dohleman, D.R. Ort, and S.P. Long, 2004. Will photosynthesis of maize (*Zea mays*) in the U.S. corn belt increase in future CO_2 rich atmospheres? An analysis of diurnal courses of CO_2 uptake under free-air concentration enrichment (FACE). *Global Change Biology*, 10(6): 951-962.

Leakey, A.D.B., M. Uribelarrea, E.A. Ainsworth, S.L. Naidu, A. Rogers, D.R. Ort, and S.P. Long, 2006. Photosynthesis, productivity, and yield of maize are not affected by open-air elevation of CO_2 concentration in the absence of drought. *Plant Physiology*, 140(2): 779-790.

Leakey, A.D.B., F. Xu, K.M. Gillespie, J.M. McGrath, E.A. Ainsworth, and D.R. Ort, 2009b. Genomic basis for stimulated respiration by plants growing under elevated carbon dioxide. *Proceedings of the National Academy of Sciences*, 106(9): 3597-3602.

Leavitt, S.W., S.B. Idso, B.A. Kimball, J.M. Burns, A. Sinha, and L. Stott, 2003. The effect of long-term atmospheric CO_2 enrichment on the intrinsic water-use efficiency of sour orange trees. *Chemosphere*, 50(2): 217-222.

Ledesma, N., and N. Sugiyama, 2005. Pollen quality and performance in strawberry plants exposed to high-temperature stress. *Journal of the American Society for Horticultural Science*, 130(3): 341-347.

Ledesma, N.A., M. Nakata, and N. Sugiyama, 2008. Effect of high temperature stress on the reproductive growth of strawberry cvs. 'nyoho' and 'toyonoka'. *Scientia Horticulturae*, 116(2): 186-193.

Leuzinger, S., Y. Luo, C. Beier, W. Dieleman, S. Vicca, and C. Körner, 2011. Do global change experiments overestimate impacts on terrestrial ecosystems? *Trends in Ecology and Evolution*, 26(5): 236-241.

Lewis, H., 2000. Environmental regulation of yield and quality components in American upland cotton. *in Proceedings of the conference on Genetic Control of Fiber and Seed Quality*. Cotton Incorporated, Cary, NC.

Li, H., T. Li, R.J. Gordon, S.K. Asiedu, and K. Hu, 2010. Strawberry plant fruiting efficiency and its correlation with solar irradiance, temperature and reflectance water index variation. *Environmental and Experimental Botany*, 68(2): 165-174.

Lin-Wang, K.U.I., D. Micheletti, J. Palmer, R. Volz, L. Lozano, R. Espley, R.P. Hellens, D. ChagnÈ, D.D. Rowan, M. Troggio, I. Iglesias, and A.C. Allan, 2011. High temperature reduces apple fruit colour via modulation of the anthocyanin regulatory complex. *Plant, Cell & Environment*, 34(7): 1176-1190.

Lobell, D.B., and G.P. Asner, 2003. Climate and management contributions to recent trends in U.S. agricultural yields. *Science*, 299(5609): 1032.

Lobell, D.B., M. Banziger, C. Magorokosho, and B. Vivek, 2011. Nonlinear heat effects on African maize as evidenced by historical yield trials. *Nature Clim. Change*, 1(1): 42-45.

Lobell, D.B., K.G. Cassman, and C.B. Field, 2009. Crop yield gaps: Their importance, magnitudes, and causes. *Annual Review of Environment and Resources*, 34(1): 179-204.

Lobell, D.B., and C.B. Field, 2007. Global scale climate–crop yield relationships and the impacts of recent warming. *Environmental Research Letters*, 2(1): 014002.

Lobell, D.B., and C.B. Field, 2008. Estimation of the carbon dioxide (CO_2) fertilization effect using growth rate anomalies of CO_2 and crop yields since 1961. *Global Change Biology*, 14(1): 39-45.

Lobell, D.B., C.B. Field, K.N. Cahill, and C. Bonfils, 2006. Impacts of future climate change on California perennial crop yields: Model projections with climate and crop uncertainties. *Agricultural and Forest Meteorology*, 141(2–4): 208-218.

Lobell, D.B., W. Schlenker, and J. Costa-Roberts, 2011. Climate trends and global crop production since 1980. *Science*, 333: 208-218.

Loka, D.A., and D.M. Oosterhuis, 2010. Effect of high night temperatures on cotton respiration, ATP levels and carbohydrate content. *Environmental and Experimental Botany*, 68(3): 258-263.

Loka, D.A., D.M. Oosterhuis, and G.L. Ritchie, 2011. Water-deficit stress in cotton, *in* D.M. Oosterhuis (ed.) *Stress physiology in cotton*. The Cotton Foundation, Cordova, TN.

Luck, J., M. Spackman, A. Freeman, P. Trebicki, W. Griffiths, K. Finlay, and S. Chakraborty, 2011. Climate change and diseases of food crops. *Plant Pathology*, 60(1): 113-121.

Luck, J.D., T. Mueller, A.C. Pike, and S.A. Shearer, 2010. Grassed waterway planning model evaluated for agricultural fields in the western coal field physiographic region of Kentucky. *Journal of Soil and Water Conservation*, 65(5): 280-288.

Luo, Q., W. Bellotti, M. Williams, and E. Wang, 2009. Adaptation to climate change of wheat growing in South Australia: Analysis of management and breeding strategies. *Agriculture, Ecosystems and Environment*, 129(1–3): 261-267.

Luo, Y., R. Sherry, X. Zhou, and S. Wan, 2009. Terrestrial carbon-cycle feedback to climate warming: Experimental evidence on plant regulation and impacts of biofuel feedstock harvest. *GCB Bioenergy*, 1(1): 62-74.

Luo, Y., B.O. Su, W.S. Currie, J.S. Dukes, A. Finzi, U. Hartwig, B. Hungate, R.E. Mc Murtrie, R.A.M. Oren, W.J. Parton, D.E. Pataki, M.R. Shaw, D.R. Zak, and C.B. Field, 2004. Progressive nitrogen limitation of ecosystem responses to rising atmospheric carbon dioxide. *BioScience*, 54(8): 731-739.

Mader, T., M. Davis, and J. Gaughan, 2007. Effect of sprinkling on feedlot microclimate and cattle behavior. *International Journal of Biometeorology*, 51(6): 541-551.

Mader, T., K. Frank, J. Harrington, G. Hahn, and J. Nienaber, 2009. Potential climate change effects on warm-season livestock production in the Great Plains. *Climatic Change*, 97(3): 529-541.

Mader, T.L., 2003. Environmental stress in confined beef cattle. *Journal of Animal Science*, 81(14 suppl 2): E110-E119.

Mader, T.L., L. J. Johnson, T. M. Brown-Brandl, and J. B. Gaughan, 2008. *Climate conditions in bedded confinement buildings'* paper presented at the ASABE International Livestock Environment Symposium. Aug. 31-Sept 04, 2008. Iguassu Falls City, Brazil.

Mader, T.L., J.M. Dahlquist, and J.B. Gaughan, 1997. Wind protection effects and airflow patterns in outside feedlots. *Journal of Animal Science*, 75(1): 26-36.

Mader, T.L., J.M. Dahlquist, G.L. Hahn, and J.B. Gaughan, 1999. Shade and wind barrier effects on summertime feedlot cattle performance. *Journal of Animal Science*, 77(8): 2065-2072.

Mader, T.L., and M.S. Davis, 2004. Effect of management strategies on reducing heat stress of feedlot cattle: Feed and water intake. *Journal of Animal Science*, 82(10): 3077-3087.

Mader, T.L., M.S. Davis, and T. Brown-Brandl, 2006. Environmental factors influencing heat stress in feedlot cattle. *Journal of Animal Science*, 84(3): 712-719.

Mader, T.L., L.J. Johnson, and J.B. Gaughan, 2010. A comprehensive index for assessing environmental stress in animals. *Journal of Animal Science*, 88(6): 2153-2165.

Mader, T.L., L.J. Johnson, and J.B. Gaughan, 2011. Erratum to "A comprehensive index for assessing environmental stress in animals" (J. Anim. Sci. 88:2153–2165). *Journal of Animal Science*, 89(9): 2955.

Mader, T.L., and W.M. Kreikemeier, 2006. Effects of growth-promoting agents and season on blood metabolites and body temperature in heifers. *Journal of Animal Science*, 84(4): 1030-1037.

Maiti, R.K., 1996. *Sorghum science*. Science Pub., Lebanon, NH.

Manderscheid, R., and H.J. Weigel, 1997. Photosynthetic and growth responses of old and modern spring wheat cultivars to atmospheric CO_2 enrichment. *Agriculture, Ecosystems, and Environment*, 64(1): 65-73.

Marani, A., and A. Amirav, 1971. Effects of soil moisture stress on two varieties of upland cotton in Israel I. The coastal plain region. *Experimental Agriculture*, 7(03): 213-224.

Marani, A., D.N. Baker, V.Ř. Reddy, and J.M. McKinion, 1985. Effect of water stress on canopy senescence and carbon exchange rates in cotton. *Crop Science*, 25(5): 798-802.

Marini, R.P., D. Sowers, and M.C. Marini, 1991. Peach fruit quality is affected by shade during final swell of fruit growth *J. Am. Soc. Hort. Sci.*, 116: 383-389.

Marcellos, H., 1977. Wheat frost injury — freezing stress and photosynthesis. *Australian Journal of Agricultural Research*, 28(4): 557-564.

Markelz, R.J.C., R.S. Strellner, and A.D.B. Leakey, 2011. Impairment of C4 photosynthesis by drought is exacerbated by limiting nitrogen and ameliorated by elevated CO_2 in maize. *Journal of Experimental Botany*, 62(9): 3235-3246.

Martin, P., and S.N. Johnson, 2011. Evidence that elevated CO_2 reduces resistance to the European large raspberry aphid in some raspberry cultivars. *Journal of Applied Entomology*, 135(3): 237-240.

Massey, J.H., E.F. Scherder, R.E. Talbert, R.M. Zablotowicz, M.A. Locke, M.A. Weaver, M.C.Smith, and R.W. Steinriede, 2003. *Reduced water use and methane emissions from rice grown using intermittent irrigation*, paper presented at Proceedings of the 33rd Annual Mississippi Water Resources Conference. Mississippi Water Resources Research Institute.

Matsushima, S., T. Tanaka, and T. Hoshino, 1964. Analysis of yield determining process and its application to yield-prediction and culture improvement of lowland rice. LXX. Combined effect of air temperature and water temperature at different stages of growth on the grain yield and its components of lowland rice. *Proc. Crop Sci. Soc. Jpn*, 33: 53-58.

Mauney, J.R., B.A. Kimball, P.J. Pinter Jr, R.L. LaMorte, K.F. Lewin, J. Nagy, and G.R. Hendrey, 1994. Growth and yield of cotton in response to a free-air carbon dioxide enrichment (FACE) environment. *Agricultural and Forest Meteorology*, 70(1–4): 49-67.

Mayer, P.M., S.K. Reynolds, M.D. McCutchen, and T.J. Canfield, 2007. Meta-analysis of nitrogen removal in riparian buffers. *Journal of Environmental Quality*, 36(4): 1172-1180.

McArtney, S., M. Parker, J. Obermiller, and T. Hoyt, 2011. Effects of 1-methylcyclopropene on firmness loss and the development of rots in apple fruit kept in farm markets or at elevated temperatures. *HortTechnology*, 21(4): 494-499.

McCarthy, J., O. Canziani, N. Leary, D. Kokken, and K. White, 2001. Climate change 2001: Impacts, adaptation and vulnerability. Annex B: Glossary of terms, *in Contribution of Working Group II to the Third Assessment Report of the Intergovernmental Panel on Climate Change*. Cambridge University Press Cambridge and New York.

McElrone, A.J., C.D. Reid, K.A. Hoye, E. Hart, and R.B. Jackson, 2005. Elevated CO_2 reduces disease incidence and severity of a red maple fungal pathogen via changes in host physiology and leaf chemistry. *Global Change Biology*, 11(10): 1828-1836.

McGrath, J.M., and D.B. Lobell, 2011. An independent method of deriving the carbon dioxide fertilization effect in dry conditions using historical yield data from wet and dry years. *Global Change Biology*, 17(8): 2689-2696.

McKeown, A., J. Warland, and M.R. McDonald, 2005. Long-term marketable yields of horticultural crops in southern Ontario in relation to seasonal climate. *Canadian Journal of Plant Science*, 85(2): 431-438.

McKeown, A.W., J. Warland, M.R. McDonald, and C.M. Hutchinson, 2004. Cool season crop production trends: A possible signal for global warming. *Acta Hort. (ISHS)*, 638: 241-248.

McMichael, B.L., and J.J. Burke, 1994. Metabolic activity of cotton roots in response to temperature. *Environmental and Experimental Botany*, 34(2): 201-206.

McMichael, B.L., and J.D. Hesketh, 1982. Field investigations of the response of cotton to water deficits. *Field Crops Research*, 5(0): 319-333.

McWilliams, D., 2003. *Drought strategies for cotton. Cooperative extension service circular 582*. New Mexico State Univ. College of Agriculture and Home Economics, Las Cruces, NM.

Medlyn, B.E., C.V.M. Barton, M.S.J. Broadmeadow, R. Ceulemans, P. De Angelis, M. Forstreuter, M. Freeman, S.B. Jackson, S. Kellomäki, E. Laitat, A. Rey, P. Roberntz, B.D. Sigurdsson, J. Strassemeyer, K. Wang, P.S. Curtis, and P.G. Jarvis, 2001. Stomatal conductance of forest species after long-term exposure to elevated CO_2 concentration: A synthesis. *New Phytologist*, 149(2): 247-264.

Meyer, V.G., 1966. Environmental effects on the differentiation of abnormal cotton flowers. *American Journal of Botany*, 53(10): 976-980.

Miller, N.L., K.E. Bashford, and E. Strem, 2003. Potential impacts of climate change on California hydrology 1. *JAWRA Journal of the American Water Resources Association*, 39(4): 771-784.

Mishra, V., and K.A. Cherkauer, 2010. Retrospective droughts in the crop growing season: Implications to corn and soybean yield in the midwestern United States. *Agricultural and Forest Meteorology*, 150(7–8): 1030-1045.

Mohammed, A.R., and L. Tarpley, 2009. High nighttime temperatures affect rice productivity through altered pollen germination and spikelet fertility. *Agricultural and Forest Meteorology*, 149(6–7): 999-1008.

Montgomery, D.R., 2007. Soil erosion and agricultural sustainability. *Proceedings of the National Academy of Sciences*, 104(33): 13268-13272.

Moran, K.K., and J.D. Jastrow, 2010. Elevated carbon dioxide does not offset loss of soil carbon from a corn–soybean agroecosystem. *Environmental Pollution*, 158(4): 1088-1094.

Morgan, J.A., 2005. Rising atmospheric CO_2 and global climate change: Responses and management implications for grazing lands, *in* S.G. Reynolds and J. Frame (eds.), *Grasslands : Developments, opportunities, perspectives*. Food and Agricultural Organization of the United Nations; Science Publishers, Inc., Rome; Enfield, NH.

Morgan, J.A., J.D. Derner, D.G. Milchunas, and E. Pendall, 2008. Management implications of global change for Great Plains rangelands. *Rangelands*, 30(3): 18-22.

Morgan, J.A., D.R. LeCain, E. Pendall, D.M. Blumenthal, B.A. Kimball, Y. Carrillo, D.G. Williams, J. Heisler-White, F.A. Dijkstra, and M. West, 2011. C4 grasses prosper as carbon dioxide eliminates desiccation in warmed semi-arid grassland. *Nature*, 476(7359): 202-205.

Morgan, J.A., D.G. Milchunas, D.R. LeCain, M. West, and A.R. Mosier, 2007. Carbon dioxide enrichment alters plant community structure and accelerates shrub growth in the shortgrass steppe. *Proceedings of the National Academy of Sciences*, 104(37): 14724-14729.

Morgan, J.A., D.E. Pataki, C. Körner, H. Clark, S.J. Grosso, J.M. Grünzweig, A.K. Knapp, A.R. Mosier, P.C.D. Newton, P.A. Niklaus, J.B. Nippert, R.S. Nowak, W.J. Parton, H.W. Polley, and M.R. Shaw, 2004. Water relations in grassland and desert ecosystems exposed to elevated atmospheric CO_2. *Oecologia*, 140(1): 11-25.

Mori, K., N. Goto-Yamamoto, M. Kitayama, and K. Hashizume, 2007. Loss of anthocyanins in red-wine grape under high temperature. *Journal of Experimental Botany*, 58(8): 1935-1945.

Moss, G.I., 1976. Temperature effects on flower initiation in sweet orange (citrus-sinensis). *Australian Journal of Agricultural Research*, 27(3): 399-407.

Moutinho-Pereira, J., B. Goncalves, E. Bacelar, J.B. Cunha, J. Coutinho, and C.M. Correia, 2009. Effects of elevated CO_2 on grapevine (*Vitis vinifera l.*): Physiological and yield attributes. *Vitis*, 48(4): 159-165.

Moya, T.B., L.H.H. Ziska, O.S. Namuco, and D. Olszyk, 1998. Growth dynamics and genotypic variation in tropical, field-grown paddy rice (*Oryza sativa l.*) in response to increasing carbon dioxide and temperature. *Global Change Biology*, 4(6): 645-656.

Muchow, R.C., T.R. Sinclair, and J.M. Bennett, Temperature and solar radiation effects on potential maize yield across locations. *Agron. J.*, 82(2): 338-343.

Mueller, T.G., H. Cetin, C.R. Dillon, R.A. Fleming, A.D. Karathanasis, and S.A. Shearer, 2005. Erosion probability maps: Calibrating precision agriculture data with soil surveys using logistic regression. *Journal of Soil and Water Conservation*, 60(6): 462+.

Munson, S.M., J. Belnap, and G.S. Okin, 2011. Responses of wind erosion to climate-induced vegetation changes on the Colorado Plateau. *Proceedings of the National Academy of Sciences*, 108(10): 3854-3859.

Musick, J.T., O.R. Jones, B.A. Stewart, and D.A. Dusek, 1994. Water-yield relationships for irrigated and dryland wheat in the U.S. Southern Plains. *Agron. J.*, 86(6): 980-986.

Musser, F.R., and A.M. Shelton, 2005. The influence of post-exposure temperature on the toxicity of insecticides to *Ostrinia nubilalis* (lepidoptera: Crambidae). *Pest Management Science*, 61(5): 508-510.

NAS, 2007. *Status of pollinators in North America.* National Academy of Sciences, National Academies Press, Washington, D.C.

NASS, 2010. National Agricultural Statistical Service Web Page. http://www.nass.usda.gov/Data_and_Statistics/Quick_Stats/index.asp

NASS, 2011. *Small grains 2011 summary* www.nass.usda.gov.

Nayak, A., D. Marks, D.G. Chandler, and M. Seyfried, 2010. Long-term snow, climate, and streamflow trends at the Reynolds Creek Experimental Watershed, Owyhee Mountains, Idaho, United States. *Water Resour. Res.*, 46(6): W06519.

Nearing, M.A., 2001. Potential changes in rainfall erosivity in the U.S., with climate change during the 21st century. *Journal of Soil and Water Conservation*, 56(3): 229+.

Nearing, M.A., L.D. Ascough, and J.M. Laflen, 1990. Sensitivity analysis of the WEPP hillslope profile erosion model. *Transactions of the American Society of Agricultural Engineering*, 33: 839-849.

Nearing, M.A., V. Jetten, C. Baffaut, O. Cerdan, A. Couturier, M. Hernandez, Y. Le Bissonnais, M.H. Nichols, J.P. Nunes, C.S. Renschler, V. Souchère, and K. van Oost, 2005. Modeling response of soil erosion and runoff to changes in precipitation and cover. *CATENA*, 61(2–3): 131-154.

Newman, J.A., 2006. Using the output from global circulation models to predict changes in the distribution and abundance of cereal aphids in Canada: A mechanistic modeling approach. *Global Change Biology*, 12(9): 1634-1642.

Newman, Y.C., L.E. Sollenberger, K.J. Boote, L.H. Allen, J.C.V. Vu, and M.B. Hall, 2005. Temperature and carbon dioxide effects on nutritive value of rhizoma peanut herbage. *Crop Sci.*, 45(1): 316-321.

Niu, S., R.A. Sherry, X. Zhou, S. Wan, and Y. Luo, 2010. Nitrogen regulation of the climate–carbon feedback: Evidence from a long-term global change experiment. *Ecology*, 91(11): 3261-3273.

Nolan, T., J. Connolly, and M. Wachendorf, 2001. Mixed grazing and climatic determinants of white clover (*Trifolium repens l.*) content in a permanent pasture. *Annals of Botany*, 88(suppl 1): 713-724.

Norby, R., T. Long, J. Hartz-Rubin, and E. O'Neill, 2000. Nitrogen resorption in senescing tree leaves in a warmer, CO_2-enriched atmosphere. *Plant and Soil*, 224(1): 15-29.

Norby, R.J., S.D. Wullschleger, C.A. Gunderson, D.W. Johnson, and R. Ceulemans, 1999. Tree responses to rising CO_2 in field experiments: Implications for the future forest. *Plant, Cell & Environment*, 22(6): 683-714.

Noy-Meir, I., 1973. Desert ecosystems: Environment and producers. *Annual Review of Ecology and Systematics*, 4(1): 25-51.

NRC, 1981. *Effect of environment on nutrient requirements of domestic animals.* National Research Council, National Academy Press, Washington, D.C.

NRC, 1987. *Predicting feed intake of food-producing animals.* National Research Council, National Academy Press.

NRC, 1996. *Nutrient requirements of beef cattle. 7th edition.* National Research Council, National Academy Press, Washington, D.C.

Oerke, E.C., 2006. Crop losses to pests. *Journal of Agricultural Science*, 144: 31-43.

Ojima, D., L. Garcia, E. Elgaali, K. Miller, T.G.F. Kittel, and J. Lackett, 1999. Potential climate change impacts on water resources in the Great Plains 1. *JAWRA Journal of the American Water Resources Association*, 35(6): 1443-1454.

Olien, W.C., and M.J. Bukovac, 1978. Effect of temperature on rate of ethylene evolution from ethephon and from ethephon-treated leaves of sour cherry. *Journal of the American Society for Horticultural Science*, 103(2): 199-202.

Oliveira, S.K.L., L.C. Grangeiro, M.Z. Negreiros, B.S. de Souza, and S.R.R. de Souza, 2006. Lettuce cultivation with agrotextile protection under conditions of high temperatures and intensity of radiation. *Caatinga*, 19: 112-116.

Olmstead, A.L., and P.W. Rhode, 2011. Adapting North American wheat production to climatic challenges, 1839–2009. *Proceedings of the National Academy of Sciences*, 108(2): 480-485.

O'Neal, M.R., M.A. Nearing, R.C. Vining, J. Southworth, and R.A. Pfeifer, 2005. Climate change impacts on soil erosion in midwest United States with changes in crop management. *CATENA*, 61(2–3): 165-184.

Ong, C.K., 1986. *Proceedings of an international symposium, Agro-climatological Factors Affecting Phenology of Groundnut. 21-26 Aug. 1985.* ICRISAT Sahelian Center, Niamey, Niger. ICRISAT, Patancheru, A.P. 502 324, India.

Onoda, Y., T. Hirose, and K. Hikosaka, 2009. Does leaf photosynthesis adapt to CO_2-enriched environments? An experiment on plants originating from three natural CO_2 springs. *New Phytologist*, 182(3): 698-709.

Oosterhuis, D.M., 1999. *Yield response to environmental extremes in cotton*, paper presented at 1999 Cotton Research Meeting, Report 193. Arkansas Agric. Exp. Stn., Fayetteville, AR.

Oosterhuis, D.M., 2002. Day or night high temperature: A major cause of yield variability. *Cotton Grower*, 46(8–9).

Oosterhuis, D.M., F.M. Bourland, A.C. Bibi, E.D. Gonias, D. Loka, and D.K. Storch, 2009. Screening for temperature tolerance in cotton, *in Summaries of cotton research in 2008*, pp. 37-41. Univ. Arkansas Agric. Exp. Sta., Research Series.

Oosterhuis, D.M., and J.L. Snider, 2011. High temperature stress on floral development and yield of cotton *in* D.M. Oosterhuis (ed.) *Stress physiology in cotton.* Cotton Foundation, Cordova, TN.

Ortiz, R., K.D. Sayre, B. Govaerts, R. Gupta, G.V. Subbarao, T. Ban, D. Hodson, J.M. Dixon, J. Iván Ortiz-Monasterio, and M. Reynolds, 2008. Climate change: Can wheat beat the heat? *Agriculture, Ecosystems & Environment*, 126(1–2): 46-58.

Osborne, C.P., J.L. Roche, R.L. Garcia, B.A. Kimball, G.W. Wall, P.J. Pinter, R.L.L. Morte, G.R. Hendrey, and S.P. Long, 1998. Does leaf position within a canopy affect acclimation of photosynthesis to elevated CO_2? *Plant Physiology*, 117(3): 1037-1045.

Ottman, M.J., B.A. Kimball, J.W. White, and G.W. Wall, 2012. Wheat growth response to increased temperature from varied planting dates and supplemental infrared heating. *Agron. J.*, 104(1): 7-16.

Pan, Q., Z. Wang, and B. Quebedeaux, 1998. Responses of the apple plant to CO_2 enrichment: Changes in photosynthesis, sorbitol, other soluble sugars, and starch. *Australian Journal of Plant Physiology*, 25(3): 293-297.

Paterson, A., Y.S. Saranga, M.M. Menz, C.X.J. Jiang, and R.W. Wright, 2003. QTL analysis of genotype × environment interactions affecting cotton fiber quality. *TAG Theoretical and Applied Genetics*, 106(3): 384-396.

Pearson, R.W., L.F. Ratliff, and H.M. Taylor, 1970. Effect of soil temperature, strength, and pH on cotton seedling root elongation. *Agronomy Journal*, 62(2): 243-246.

Peet, M., S. Sato, and C.P. Clément, E, 2003. Heat stress increases sensitivity of pollen, fruit and seed production in tomatoes (*Lycopersicon Esculentum Mill.*) to non-optimal vapor pressure deficits. *Acta Hort. (ISHS)*, 618: 209-215.

Peet, M.M., and D.W. Wolfe., 2000. Crop ecosystem responses to climate change - vegetable crops, *in* K.R. Reddy and H.F. Hodges (eds.), *Climate change and global crop productivity*. CABI Pub., Wallingford, Oxon, UK; New York, NY.

Pendall, E., Y.U.I. Osanai, A.L. Williams, and M.J. Hovenden, 2011. Soil carbon storage under simulated climate change is mediated by plant functional type. *Global Change Biology*, 17(1): 505-514.

Peng, S., J. Huang, J.E. Sheehy, R.C. Laza, R.M. Visperas, X. Zhong, G.S. Centeno, G.S. Khush, and K.G. Cassman, 2004. Rice yields decline with higher night temperature from global warming. *Proceedings of the National Academy of Sciences*, 101(27): 9971-9975.

Penuelas, J., S.B. Idso, A. Ribas, and B.A. Kimball, 1997. Effects of long-term atmospheric CO$_2$ enrichment on the mineral concentration of citrus aurantium leaves. *New Phytologist*, 135(3): 439-444.

Peralta, A., and M. Wander, 2008. Soil organic matter dynamics under soybean exposed to elevated CO$_2$. *Plant and Soil*, 303(1): 69-81.

Peri, P., and M. Bloomberg, 2002. Windbreaks in southern Patagonia, Argentina: A review of research on growth models, windspeed reduction, and effects on crops. *Agroforestry Systems*, 56(2): 129-144.

Perry, S.W., and D.R. Krieg, 1981. *Gross net photosynthesis ratios of cotton as affected by environment and genotype*, paper presented at Beltwide Cotton Producers Research Conference. Memphis, TN. National Cotton Council.

Peterson, G., 2009. Ecological limits of adaptation to climate change, *in* W.N. Adger, I. Lorenzoni and K.L. O'Brien (eds.), *Adapting to climate change: Thresholds, values, governance*. Cambridge University Press, Cambridge; New York.

Peterson, G.L., 2009b. Reaction of selected winter wheat cultivars from Europe and United States to karnal bunt. *European Journal of Plant Pathology*, 125(3): 497-507.

Pettigrew, W.T., 1995. Source-to-sink manipulation effects on cotton fiber quality. *Agronomy Journal*, 87(5): 947-952.

Pettigrew, W.T., 2002. Improved yield potential with an early planting cotton production system. *Agronomy Journal*, 94(5): 997-1003.

Pettigrew, W.T., 2004a. Moisture deficit effects on cotton lint yield, yield components, and boll distribution. *Agronomy Journal*, 96(2): 377-383.

Pettigrew, W.T., 2004b. Physiological consequences of moisture deficit stress in cotton. *Crop Science*, 44(4): 1265-1272.

Pettigrew, W.T., 2008. The effect of higher temperatures on cotton lint yield production and fiber quality. *Crop Science*, 48(1): 278-285.

Pfeifer, R.A., and M. Habeck, 2002. Farm level economic impacts of climate change, *in* O.C. Doering (ed.) *Effects of climate change and variability on agricultural production systems*, pp. 159–178. Kluwer Academic Publishers, Boston.

Pfeifer, R.A., J. Southworth, O.C. Doering, and L. Moore, 2002. Climate variability impacts on farm-level risk, *in* O.C. Doering (ed.) *Effects of climate change and variability on agricultural production systems*, pp. 179–193. Kluwer Academic Publishers, Boston.

Pigali, P., and S. Pandey, 2001. World maize needs meeting: Technological opportunities and priorities for the public sector, *in* P.L. Pingali (ed.) *CIMMYT 1999-2000 world maize facts and trends*, pp. 1-24. CIMMYT, Mexico.

Pike, A.C., T.G. Mueller, A. Schörgendorfer, S.A. Shearer, and A.D. Karathanasis, 2009. Erosion index derived from terrain attributes using logistic regression and neural networks. *Agronomy Journal*, 101(5): 1068-1079.

Pino, M.d.l.A., E. Terry, and F. Soto, 2002. Natural shade systems as a phytoclimate modifier in tomato crop (*lycopersicon esculentum mill*). *Cultivos Tropicale*, 23: 5-10.

Polley, H., P. Fay, V. Jin, and G. Combs, 2011. CO$_2$ enrichment increases element concentrations in grass mixtures by changing species abundances. *Plant Ecology*, 212(6): 945-957.

Polley, H.W., 1997. Implications of rising atmospheric carbon dioxide concentration for rangelands. *Journal of Range Management*, 50(6): 562-577.

Polley, H.W., J.A. Morgan, and P.A. Fay, 2011. Application of a conceptual framework to interpret variability in rangeland responses to atmospheric CO$_2$ enrichment. *The Journal of Agricultural Science*, 149(01): 1-14.

Poorter, H., and M.-L. Navas, 2003. Plant growth and competition at elevated CO$_2$: On winners, losers and functional groups. *New Phytologist*, 157(2): 175-198.

Porter, J.R., and M. Gawith, 1999. Temperatures and the growth and development of wheat: A review. *European Journal of Agronomy*, 10(1): 23-36.

Postweiler, K., R. Stösser, and S.F. Anvari, 1985. The effect of different temperatures on the viability of ovules in cherries. *Scientia Horticulturae*, 25(3): 235-239.

Poudel, P.R., R. Mochioka, K. Beppu, and I. Kataoka, 2009. Influence of temperature on berry composition of interspecific hybrid wine grape 'kadainou r-1' (*Vitis ficifolia* var. Ganebu x v. Vinifera 'muscat of alexandria'). *Journal of the Japanese Society for Horticultural Science*, 78(2): 169-174.

Powell, A.A., and G. David, 2011. *Fruit crops. Principles of freeze protection for fruit crops*. Alabama Cooperative Extension System http://www.aces.edu/dept/peaches/frzcritical.html.

Power, A.G., 2010. Ecosystem services and agriculture: Tradeoffs and synergies. *Philosophical Transactions of the Royal Society B: Biological Sciences*, 365(1554): 2959-2971.

Prasad, P.V.V., K.J. Boote, and L.H. Allen Jr, 2006. Adverse high temperature effects on pollen viability, seed-set, seed yield and harvest index of grain-sorghum (*Sorghum bicolor l.* moench) are more severe at elevated carbon dioxide due to higher tissue temperatures. *Agricultural and Forest Meteorology*, 139(3–4): 237-251.

Prasad, P.V.V., K.J. Boote, L.H. Allen, and J.M.G. Thomas, 2002. Effects of elevated temperature and carbon dioxide on seed-set and yield of kidney bean (*Phaseolus vulgaris l.*). *Global Change Biology*, 8(8): 710-721.

Prasad, P.V.V., K.J. Boote, L. Hartwell Allen, and J.M.G. Thomas, 2003. Super-optimal temperatures are detrimental to peanut (*Arachis hypogaea L.*) reproductive processes and yield at both ambient and elevated carbon dioxide. *Global Change Biology*, 9(12): 1775-1787.

Pressman, E., M.M. Peet, and D.M. Pharr, 2002. The effect of heat stress on tomato pollen characteristics is associated with changes in carbohydrate concentration in the developing anthers. *Annals of Botany*, 90(5): 631-636.

Prior, S.A., H.A. Torbert, G.B. Runion, G.L. Mullins, H.H. Rogers, and J.R. Mauney, 1998. Effects of carbon dioxide enrichment on cotton nutrient dynamics. *Journal of Plant Nutrition*, 21(7): 1407-1426.

Pruski, F.F., and M.A. Nearing, 2002a. Runoff and soil-loss responses to changes in precipitation: A computer simulation study. *Journal of Soil and Water Conservation*, 57(1): 7-16.

Pruski, F.F., and M.A. Nearing, 2002b. Climate-induced changes in erosion during the 21st century for eight U.S. locations. *Water Resour. Res.*, 38(12): 1298.

Pryor, S.C., R.J. Barthelmie, D.T. Young, E.S. Takle, R.W. Arritt, D. Flory, W.J. Gutowski, Jr., A. Nunes, and J. Roads, 2009. Wind speed trends over the contiguous United States. *J. Geophys. Res.*, 114(D14): D14105.

Quine, T.A., and Y. Zhang, 2002. An investigation of spatial variation in soil erosion, soil properties, and crop production within an agricultural field in Devon, United Kingdom. *Journal of Soil and Water Conservation*, 57(1): 55-65.

Raj, P.P., M. Ryosuke, B. Kenji, and K. Ikuo, 2009. Influence of temperature on berry composition of interspecific hybrid wine grape 'kadainou r-1' (*Vitis ficifolia* var. Ganebu×v. Vinifera 'muscat of alexandria'). *Journal of the Japanese Society for Horticultural Science*, 78(2): 169-174.

Ramey, H.H., 1986. Stress influence on fiber development, *in* J.R. Mauney and J.M. Stewart (eds.), *Cotton physiology*, pp. 351-360. Cotton Foundation, Memphis, TN.

Rashidi, M., and M. Gholami, 2008. Review of crop water productivity values for tomato, potato, melon, watermelon and cantaloupe in Iran. *International Journal of Agriculture and Biology*, 10: 432-436.

Ravi, S., P. D'Odorico, D.D. Breshears, J.P. Field, A.S. Goudie, T.E. Huxman, J. Li, G.S. Okin, R.J. Swap, A.D. Thomas, S. Van Pelt, J.J. Whicker, and T.M. Zobeck, 2011. Aeolian processes and the biosphere. *Rev. Geophys.*, 49(3): RG3001.

Reckendorf, F., 1995. *RCA III, sedimentation in irrigation water bodies, reservoirs, canals, and ditches.* USDA - Natural Resources Conservation Service http://www.nrcs.usda.gov/wps/portal/nrcs/detail/national/technical/nra/rca/?&cid=nrcs143_014200.

Reddy, K.R., G.H. Davidonis, A.S. Johnson, and B.T. Vinyard, 1999. Temperature regime and carbon dioxide enrichment alter cotton boll development and fiber properties. *Agronomy Journal*, 91(5): 851-858.

Reddy, K.R., H.F. Hodges, and J.M. McKinion, 1995. Carbon dioxide and temperature effects on pima cotton development. *Agronomy Journal*, 87(5): 820-826.

Reddy, K.R., H.F. Hodges, and J.M. McKinion, 1997. A comparison of scenarios for the effect of global climate change on cotton growth and yield. *Australian Journal of Plant Physiology*, 24: 707-713.

Reddy, K.R., H.F. Hodges, J.M. McKinion, and G.W. Wall, 1992a. Temperature effects on pima cotton growth and development. *Agron. J.*, 84(2): 237-243.

Reddy, K.R., H.F. Hodges, and V.R. Reddy, 1992. Temperature effects on cotton fruit retention. *Agronomy Journal*, 84(1): 26-30.

Reddy, K.R., H.F. Hodges, and V.R. Reddy, 1992b. Temperature effects on cotton fruit retention. *Agron. J.*, 84(1): 26-30.

Reddy, K.R., V.R. Reddy, and H.F. Hodges, 1992. Temperature effects on early season cotton growth and development. *Agronomy Journal*, 84(2): 229-237.

Reddy, K.R., P.V. Vara Prasad, and V.G. Kakani, 2005. Crop responses to elevated carbon dioxide and interactions with temperature. *Journal of Crop Improvement*, 13(1-2): 157-191.

Reddy, V.R., D.N. Baker, and H.F. Hodges, 1991. Temperature effects on cotton canopy growth, photosynthesis, and respiration. *Agronomy Journal*, 83(4): 699-704.

Reddy, V.R., H.F. Hodges, W.H. McCarty, and J.M. McKinnon, 1996. *Weather and cotton growth: Present and future.* Mississippi Agr. & Forestry Exp. Sta., Mississippi State University, Starkeville, MS.

Reddy, V.R., K.R. Reddy, and H.F. Hodges, 1995. Carbon dioxide enrichment and temperature effects on cotton canopy photosynthesis, transpiration, and water-use efficiency. *Field Crops Research*, 41(1): 13-23.

Reddy, V.R., K.R. Reddy, and H.F. Hodges, 1995. Carbon dioxide enrichment and temperature effects on cotton canopy photosynthesis, transpiration, and water-use efficiency. *Field Crops Research*, 41(1): 13-23.

Redman, R.S., Y.O. Kim, C.J.D.A. Woodward, C. Greer, L. Espino, S.L. Doty, and R.J. Rodriguez, 2011. Increased fitness of rice plants to abiotic stress via habitat adapted symbiosis: A strategy for mitigating impacts of climate change. *PLoS ONE*, 6(7): e14823.

Reginato, G.H., R.H. Callejas, R.A. Sapiaín, and V. García-de-Cortázar, 2010. Rest completion and growth of 'Thompson seedless' grapes as a function of temperatures. *Acta Hort. (ISHS)*, 872: 427-430.

Reich, P.B., B.A. Hungate, and Y.Q. Luo, 2006. Carbon-nitrogen interactions in terrestrial ecosystems in response to rising atmospheric carbon dioxide, *in Annual review of ecology evolution and systematics*, pp. 611-636. Annual Reviews, Palo Alto.

Reich, P.B., D. Tilman, J. Craine, D. Ellsworth, M.G. Tjoelker, J. Knops, D. Wedin, S. Naeem, D. Bahauddin, J. Goth, W. Bengtson, and T.D. Lee, 2001. Do species and functional groups differ in acquisition and use of C, N and water under varying atmospheric CO_2 and N availability regimes? A field test with 16 grassland species. *New Phytologist*, 150(2): 435-448.

Reicosky, D.C., 1997. Tillage-induced CO_2 emission from soil. *Nutrient Cycling in Agroecosystems*, 49(1): 273-285.

Rhoads, M.L., R.P. Rhoads, M.J. VanBaale, R.J. Collier, S.R. Sanders, W.J. Weber, B.A. Crooker, and L.H. Baumgard, 2009. Effects of heat stress and plane of nutrition on lactating Holstein cows: I. Production, metabolism, and aspects of circulating somatotropin1. *Journal of dairy science*, 92(5): 1986-1997.

Richardson, D.M., J.J. Hellmann, J.S. McLachlan, D.F. Sax, M.W. Schwartz, P. Gonzalez, E.J. Brennan, A. Camacho, T.L. Root, O.E. Sala, S.H. Schneider, D.M. Ashe, J.R. Clark, R. Early, J.R. Etterson, E.D. Fielder, J.L. Gill, B.A. Minteer, S. Polasky, H.D. Safford, A.R. Thompson, and M. Vellend, 2009. Multidimensional evaluation of managed relocation. *Proceedings of the National Academy of Sciences*, 106(24): 9721-9724.

Ro, H.M., P.G. Kim, I.B. Lee, M.S. Yiem, and S.Y. Woo, 2001. Photosynthetic characteristics and growth responses of dwarf apple (*Malus domestica borkh. Cv. Fuji*) saplings after 3 years of exposure to elevated atmospheric carbon dioxide concentration and temperature. *Trees-Structure and Function*, 15(4): 195-203.

Roberts, P., 2008. *The end of food.* Houghton Mifflin Company, Boston.

Robertson, T., J. Zak, and D. Tissue, 2010. Precipitation magnitude and timing differentially affect species richness and plant density in the Sotol grassland of the Chihuahuan desert. *Oecologia*, 162(1): 185-197.

Root, T.L., J.T. Price, K.R. Hall, S.H. Schneider, C. Rosenzweig, and J.A. Pounds, 2003. Fingerprints of global warming on wild animals and plants. *Nature*, 421(6918): 57-60.

Rosegrant, M., M. Sombilla, and N. Perez, 1995. *Food, agriculture and the environment: Discussion paper no. 5*, 1-34 pp. International Food Policy Research Institute, Washington, D.C.

Rosenzweig, C., and D. Hillel, 1998. *Climate change and the global harvest: Potential impacts of the greenhouse effect on agriculture.* Oxford University Press, New York.

Rosenzweig, C., J. Phillips, R. Goldberg, J. Carroll, and T. Hodges, 1996. Potential impacts of climate change on citrus and potato production in the U.S. *Agricultural Systems*, 52(4): 455-479.

167

Rötter, R., and S.C. van de Geijn, 1999. Climate change effects on plant growth, crop yield and livestock. *Climatic Change*, 43(4): 651-681.

Rouse, R.E., and W.B. Sherman, 2002. High night temperatures during bloom affect fruit set in peach. *Proc. FL State Hort. Soc*, 115: 96-97.

Sacks, W.J., and C.J. Kucharik, 2011. Crop management and phenology trends in the U.S. corn belt: Impacts on yields, evapotranspiration and energy balance. *Agricultural and Forest Meteorology*, 151(7): 882-894.

Saini, H.S., and D. Aspinall, 1982. Abnormal sporogenesis in wheat (*Triticum-aestivum l*) induced by short periods of high-temperature. *Annals of Botany*, 49(6): 835-846.

Sala, O.E., W.J. Parton, L.A. Joyce, and W.K. Lauenroth, 1988. Primary production of the central grassland region of the United States. *Ecology*, 69(1): 40-45.

Sanderson, M.A., D. Wedin, and B. Tracy, 2009. Grassland: Definition, origins, extent and future, *in* W.F. Wedin and S.L. Fales (eds.), *Grassland : Quietness and strength for a new American agriculture*, pp. 57-74. American Society of Agronomy, Crop Science Society of America, Soil Society of America, Madison, WI.

Sankey, J.B., M.J. Germino, S.G. Benner, N.F. Glenn, and A.N. Hoover, 2012. Transport of biologically important nutrients by wind in an eroding cold desert. *Aeolian Research*(0).

Santos, C.L.d., J. Seabra, S., J.G. de Lalla, V.C.d.A. Theodoro, and A. Nespoli, 2009. Performance of crispi lettuce cultivars under high temperatures in Cáceres-MT. *Revista Agrarian*, 2: 87-98.

Sanzol, J., and M. Herrero, 2001. The "effective pollination period" in fruit trees. *Scientia Horticulturae*, 90(1–2): 1-17.

Saranga, Y., C.X. Jiang, R.J. Wright, D. Yakir, and A.H. Paterson, 2004. Genetic dissection of cotton physiological responses to arid conditions and their inter-relationships with productivity. *Plant, Cell & Environment*, 27(3): 263-277.

Satake, T., 1995. High temperature injury, *in* T. Matsuo (ed.) *Science of the rice plant*. Food and Agriculture Policy Research Center, Tokyo.

Satake, T., and H. Hayase, 1970. Male sterility caused by cooling treatment at the young microspore stage in rice plants. 5. Estimations of pollen development stage and the most sensitive stage to coolness. *Proceedings of the Crop Science Society of Japan*, 39: 468-473.

Satake, T., and S. Yoshida, 1976. High temperature-induced sterility in indica ice at flowering. *Japanese Journal of Crop Science*, 47: 4-17.

Sato, S., 2006. The effects of moderately elevated temperature stress due to global warming on the yield and the male reproductive development of tomato (*lycopersicon esculentum mill.*). *HortResearch*, 60: 85-89.

Sato, S., M.M. Peet, and R.G. Gardner, 2004. Altered flower retention and developmental patterns in nine tomato cultivars under elevated temperature. *Scientia Horticulturae*, 101(1–2): 95-101.

Sato, S., M.M. Peet, and J.F. Thomas, 2000. Physiological factors limit fruit set of tomato (*Lycopersicon esculentum Mill.*) under chronic, mild heat stress. *Plant, Cell & Environment*, 23(7): 719-726.

Schillinger, W.F., 2001. *Reducing water runoff and erosion from frozen agricultural soils*, paper presented at Soil Erosion Research for the 21st Century: Proceedings of the International Symposium, 3-5 January, 2001. American Society of Agricultural Engineers, Honolulu, Hawaii, USA, 2001.

Schillinger, W.F., S.E. Schofstoll, and J.R. Alldredge, 2008. Available water and wheat grain yield relations in a Mediterranean climate. *Field Crops Research*, 109(1-3): 45-49.

Schumacher, J.A., T.C. Kaspar, J.C. Ritchie, T.E. Schumacher, D.L. Karlen, E.R. Venteris, G.W. McCarty, T.S. Colvin, D.B. Jaynes, M.J. Lindstrom, and T.E. Fenton, 2005. Identifying spatial patterns of erosion for use in precision conservation. *Journal of Soil and Water Conservation*, 60(6): 355-362.

Scott, H.D., J.A. Ferguson, L. Hanson, T. Fugitt, and E. Smith, 1998. Agricultural water management in the Mississippi Delta region of Arkansas. *Arkansas Agricultural Experiment Station Bulletin 959*, http://arkansasagnews.uark.edu/1353.htm.

Seager, R., and G.A. Vecchi, 2010. Greenhouse warming and the 21st century hydroclimate of southwestern North America. *Proceedings of the National Academy of Sciences*, 107(50): 21277-21282.

Segal, M., Z. Pan, R.W. Arritt, and E.S. Takle, 2001. On the potential change in wind power over the U.S. due to increases of atmospheric greenhouse gases. *Renewable Energy*, 24(2): 235-243.

Seif, S., and W. Gruppe, 1985. Chilling requirements of sweet cherries (*Prunus avium*) and interspecific cherry hybrids (*Prunus x ssp.*). *Acta Hort. (ISHS)*, 169(289-294).

Seo, S.N., and B. McCarl, 2011. Managing livestock species under climate change in Australia. *Animals*, 1(4): 343-365.

Serrat-Capdevila, A., J.B. Valdés, J.G. Pérez, K. Baird, L.J. Mata, and T. Maddock Iii, 2007. Modeling climate change impacts – and uncertainty – on the hydrology of a riparian system: The San Pedro basin (Arizona/Sonora). *Journal of Hydrology*, 347(1–2): 48-66.

Shah, F., J. Huang, K. Cui, L. Nie, T. Shah, C. Chen, and K. Wang, 2011. Impact of high-temperature stress on rice plant and its traits related to tolerance. *The Journal of Agricultural Science*, 149(05): 545-556.

Shaw, M.R., E.S. Zavaleta, N.R. Chiariello, E.E. Cleland, H.A. Mooney, and C.B. Field, 2002. Grassland responses to global environmental changes suppressed by elevated CO_2. *Science*, 298(5600): 1987-1990.

Sherry, R.A., X. Zhou, S. Gu, J.A. Arnone, D.S. Schimel, P.S. Verburg, L.L. Wallace, and Y. Luo, 2007. Divergence of reproductive phenology under climate warming. *Proceedings of the National Academy of Sciences*, 104(1): 198-202.

Shipitalo, M.J., and L.B. Owens, 2006. Tillage system, application rate, and extreme event effects on herbicide losses in surface runoff. *J. Environ. Qual.*, 35(6): 2186-2194.

Singh, R.P., P.V.V. Prasad, K. Sunita, S.N. Giri, and K.R. Reddy, 2007. Influence of high temperature and breeding for heat tolerance in cotton: A review, *in* L.S. Donald (ed.) *Advances in agronomy*, pp. 313-385. Academic Press.

Skidmore, E.L., 1965. Assessing wind erosion forces: Directions and relative magnitudes. *Soil Science Society of America Proceedings*, 29(5): 587-590.

Skidmore, E.L., P.S. Fisher, and N.P. Woodruff, 1970. Wind erosion equation: Computer solution and application. *Soil Science Society of America Journal*, 34(6): 931-935.

Snider, J.L., D.M. Oosterhuis, and E.M. Kawakami, 2010. Genotypic differences in thermotolerance are dependent upon prestress capacity for antioxidant protection of the photosynthetic apparatus in *Gossypium hirsutum*. *Physiologia Plantarum*, 138(3): 268-277.

Snider, J.L., D.M. Oosterhuis, and E.M. Kawakami, 2011. Diurnal pollen tube growth rate is slowed by high temperature in field-grown *Gossypium hirsutum* pistils. *Journal of Plant Physiology*, 168(5): 441-448.

Snider, J.L., D.M. Oosterhuis, and E.M. Kawakami, 2011. Mechanisms of reproductive thermotolerance in *Gossypium hirsutum*: The effect of genotype and exogenous calcium application. *Journal of Agronomy & Crop Science*, 197(3): 228-236.

Snider, J.L., D.M. Oosterhuis, B.W. Skulman, and E.M. Kawakami, 2009. Heat stress-induced limitations to reproductive success in *Gossypium hirsutum*. *Physiologia Plantarum*, 137(2): 125-138.

Snyder, R.L.a.J.P.d.M.-A., 2005. *Frost protection: Fundamentals, practice, and economics.* Environment and Natural Resource Series 10. Food and Agriculture Organization of the United Nations, Rome.

Sojka, R.E., D.L. Bjorneberg, T.J. Trout, T.S. Strelkoff, and M.A. Nearing, 2007. The importance and challenge of modeling irrigation-induced erosion. *Journal of Soil and Water Conservation,* 62(3): 153-162.

Sønsteby, A., and O.M. Heide, 2008. Temperature responses, flowering and fruit yield of the June-bearing strawberry cultivars florence, frida and korona. *Scientia Horticulturae,* 119(1): 49-54.

Sorensen, R.B., C.L. Butts, and R.C. Nutti, 2011. Deep subsurface drip irrigation for cotton in the Southeast. *Journal of Cotton Science,* 15: 233-242.

Southworth, J., R.A. Pfeifer, M. Habeck, J.C. Randolph, O.C. Doering, J.J. Johnston, and D.G. Rao, 2002. Changes in soybean yields in the midwestern United States as a result of future changes in climate, climate variability, and CO_2 fertilization. *Climatic Change,* 53(4): 447-475.

Southworth, J., J.C. Randolph, M. Habeck, O.C. Doering, R.A. Pfeifer, D.G. Rao, and J.J. Johnston, 2000. Consequences of future climate change and changing climate variability on maize yields in the midwestern United States. *Agriculture, Ecosystems, and Environment,* 82(1–3): 139-158.

Spayd, S.E., J.M. Tarara, D.L. Mee, and J.C. Ferguson, 2002. Separation of sunlight and temperature effects on the composition of *Vitis vinifera* cv. Merlot berries. *American Journal of Enology and Viticulture,* 53(3): 171-182.

Sprott, L.R., G.E. Selk, and D.C. Adams, 2001. Review: Factors affecting decisions on when to calve beef females. *The Professional Animal Scientist,* 17(4): 238-246.

Srinivasan, C., A.M. Callahan, C. Dardick, and R. Scorza, 2010. Expression of the poplar flowering locus T1 (FT1) gene reduces the generation time in plum (*Prunus domestica l.*), Lisbon, Portugal.

Srinivasan, C., and M.G. Mullins, 1981. Physiology of flowering in the grapevine — a review. *American Journal of Enology and Viticulture,* 32(1): 47-63.

Stanton, M.A., J.C. Scheerens, R.C. Funt, and J.R. Clark, 2007. Floral competence of primocane-fruiting blackberries prime-jan and prime-jim grown at three temperature regimens. *HortScience,* 42(3): 508-513.

Stöckle, C., R. Nelson, S. Higgins, J. Brunner, G. Grove, R. Boydston, M. Whiting, and C. Kruger, 2010. Assessment of climate change impact on eastern Washington agriculture. *Climatic Change,* 102(1): 77-102.

Stockton, J.R., L.D. Doneen, and V.T. Walhood, 1961. Boll shedding and growth of the cotton plant in relation to irrigation frequency. *Agronomy Journal,* 53(4): 272-275.

Stone, L.R., and A.J. Schlegel, 2006. Yield–water supply relationships of grain sorghum and winter wheat. *Agron. J.,* 98(5): 1359-1366.

Stott, G.H., 1981. What is animal stress and how is it measured? *Journal of Animal Science,* 52(1): 150-153.

Stover, E.W., and D.W. Greene, 2005. Environmental effects on the performance of foliar applied plant growth regulators: A review focusing on tree fruits. *HortTechnology,* 15(2): 214-221.

St-Pierre, N.R., B. Cobanov, and G. Schnitkey, 2003. Economic losses from heat stress by U.S. livestock industries. *Journal of Dairy Science,* 86: E52-E77.

Sun, Y.-C., J. Yin, F.-J. Chen, G. Wu, and F. Ge, 2011. How does atmospheric elevated CO_2 affect crop pests and their natural enemies? Case histories from China. *Insect Science,* 18(4): 393-400.

Sutherst, R., R. Baker, S. Coakley, R. Harrington, D. Kriticos, and H. Scherm, 2007. Pests under global change—meeting your future landlords? *in* J.G. Canadell, D.E. Pataki and L. Pitelka (eds.), *Terrestrial ecosystems in a changing world.* Springer, Berlin; New York.

Suttle, K.B., M.A. Thomsen, and M.E. Power, 2007. Species interactions reverse grassland responses to changing climate. *Science,* 315(5812): 640-642.

Swartz, H.J., and L.E. Powell Jr, 1981. The effect of long chilling requirement on time of bud break in apple. *Acta Hort. (ISHS),* 120: 173-178.

Syvertsen, J., and J. Graham, 1999. Phosphorus supply and arbuscular mycorrhizas increase growth and net gas exchange responses of two citrus spp. grown at elevated CO_2. *Plant and Soil,* 208(2): 209-219.

Taha, M.A., M.N.A. Malik, F.I. Chaudhry, and M.I. Makhdum, 1981. Heat-induced sterility in cotton sown during early April in West Punjab. *Experimental Agriculture,* 17(02): 189-194.

Takeda, F., and J. Phillips, 2011. Horizontal cane orientation and rowcover application improve winter survival and yield of trailing 'siskiyou' blackberry. *HortTechnology,* 21(2): 170-175.

Tartachnyk, I.I., and M.M. Blanke, 2007. Photosynthesis and transpiration of tomato and CO_2 fluxes in a greenhouse under changing environmental conditions in winter. *Annals of Applied Biology,* 150(2): 149-156.

Tashiro, T., and I. Wardlaw, 1990. The response to high temperature shock and humidity changes prior to and during the early stages of grain development in wheat. *Functional Plant Biology,* 17(5): 551-561.

Tesfaendrias, M.T., M.R. McDonald, and J. Warland, 2010. Consistency of long-term marketable yield of carrot and onion cultivars in muck (organic) soil in relation to seasonal weather. *Canadian Journal of Plant Science,* 90(5): 755-765.

Tew, M., G. Battel, and C.A. Nelson., 2002. *Implementation of a new wind chill temperature index by the National Weather Service.* American Meteorology Society, Orlando, FL.

Thakur, P., S. Kumar, J.A. Malik, J.D. Berger, and H. Nayyar, 2010. Cold stress effects on reproductive development in grain crops: An overview. *Environmental and Experimental Botany,* 67(3): 429-443.

Thornton, P.K., 2010. Livestock production: Recent trends, future prospects. *Philosophical Transactions of the Royal Society B: Biological Sciences,* 365(1554): 2853-2867.

Tian, X., T. Matsui, S. Li, M. Yoshimoto, K. Kobayasi, and T. Hasegawa, 2010. Heat-induced floret sterility of hybrid rice (*Oryza sativa*) cultivars under humid and low wind conditions in the field of Jianghan Basin, China. *Plant Production Science,* 13(3): 243-251.

Timlin, D., S.M. Lutfor Rahman, J. Baker, V.R. Reddy, D. Fleisher, and B. Quebedeaux, 2006. Whole plant photosynthesis, development, and carbon partitioning in potato as a function of temperature. *Agronomy Journal,* 98(5): 1195-1203.

Tomer, M.D., 2010. How do we identify opportunities to apply new knowledge and improve conservation effectiveness? *Journal of Soil and Water Conservation,* 65(4): 261-265.

Torell, L.A., S. Murugan, and O.A. Ramirez, 2010. Economics of flexible versus conservative stocking strategies to manage climate variability risk. *Rangeland Ecology & Management,* 63(4): 415-425.

Tromp, J., 1976. Flower-bud formation and shoot growth in apple as affected by temperature. *Scientia Horticulturae,* 5(4): 331-338.

Tromp, J., and O. Borsboom, 1994. The effect of autumn and spring temperature on fruit set and the effective pollination period in apple and pear. *Scientia Horticulturae,* 60: 23-30.

Tsvetsinskaya, E.A., L.O. Mearns, T. Mavromatis, W. Gao, L. McDaniel, and M.W. Downton, 2003. The effect of spatial scale of climatic change scenarios on simulated maize, winter wheat, and rice production in the southeastern United States. *Climatic Change*, 60(1-2): 37-71.

Tubiello, F.N., C. Rosenzweig, R.A. Goldberg, S. Jagtap, and J.W. Jones, 2002. Effects of climate change on U.S. crop production: Simulation results using two different GCM scenarios. Part I: Wheat, potato, maize, and citrus. *Climate Research*, 20(3): 259-270.

Turner, N.C., A.B. Hearn, J.E. Begg, and G.A. Constable, 1986. Cotton (*Gossypium hirsutum l.*): Physiological and morphological responses to water deficits and their relationship to yield. *Field Crops Research*, 14(0): 153-170.

Tylianakis, J.M., R.K. Didham, J. Bascompte, and D.A. Wardle, 2008. Global change and species interactions in terrestrial ecosystems. *Ecology Letters*, 11(12): 1351-1363.

U.S. Corps of Army Engineers, 2000. Grand Prairie area demonstration project. Http://www.Mvm.Usace.Army.Mil/grandprairie/.

US EPA, 2010. *Climate change indicators in the United States. EPA 430–r–10–007* www.epa.gov/climatechange/science/indicators/.

USDA, 1956. *Notes of the joint conference on slope-practice.* USDA Soil and Water Conservation Research Branch, Agricultural Research Service and Soil Conservation Service, Purdue University, Feb. 27-March 1, 1956.

USDA, 1990-2010. *Agricultural statistics.* United States Government Printing Office, United States Department of Agriculture, Washington, DC.

USDA, 2008. The synthesis and assessment product 4.3 (SAP 4.3): The effects of climate change on agriculture, land resources, water resources, and biodiversity in the United States.

USDA, 2010. *2007 national resources inventory – soil erosion on croplands.* Natural Resources Conservation Service http://www.nrcs.usda.gov/Internet/FSE_DOCUMENTS/nrcs143_012269.pdf.

Uzun, S., 2007. The effect of temperature and mean cumulative daily light intensity on fruiting behavior of greenhouse-grown tomato. *Journal of the American Society for Horticultural Science*, 132(4): 459-466.

Van Klaveren, R.W., and D.K. McCool, 2010. Freeze–thaw and water tension effects on soil detachment. *Soil Science Society of America Journal*, 74(4): 1327-1338.

Vicens, N., and J. Bosch, 2000. Weather-dependent pollinator activity in an apple orchard, with special reference to *Osmia cornuta* and *Apis mellifera* (hymenoptera: Megachilidae and apidae). *Environmental Entomology*, 29(3): 413-420.

Vidon, P., 2010. Riparian zone management and environmental quality: A multi-contaminant challenge. *Hydrological Processes*, 24(11): 1532-1535.

Villafañe, R., and R. Hernández, 2000. Relationship between furrow orientation and the distribution of moisture and salinity in top soil. *Agronomía Tropical (Maracay)*, 50: 665-673.

Vu, J.C.V., Y.C. Newman, L.H. Allen Jr, M. Gallo-Meagher, and M.-Q. Zhang, 2002. Photosynthetic acclimation of young sweet orange trees to elevated growth CO_2 and temperature. *Journal of Plant Physiology*, 159(2): 147-157.

Wagstaffe, A., and N.H. Battey, 2004. Analysis of shade and temperature effects on assimilate partitioning in the everbearing strawberry 'everest' as the basis for optimised long-season fruit production. *Journal of Horticultural Science & Biotechnology*, 79(6): 917-922.

Wagstaffe, A., and N.H. Battey, 2006. The optimum temperature for long-season cropping in the everbearing strawberry 'everest'. . *Acta Hort. (ISHS)*, 708: 45-50.

Wand, S.J.E., G.F. Midgley, M.H. Jones, and P.S. Curtis, 1999. Responses of wild C4 and C3 grass (poaceae) species to elevated atmospheric CO_2 concentration: A meta-analytic test of current theories and perceptions. *Global Change Biology*, 5(6): 723-741.

Wanjura, D.F., E.B. Hudspeth, and J.D. Bilbro, 1969. Emergence time, seed quality, and planting depth effects on yield and survival of cotton (*Gossypium hirsutum l.*). *Agronomy Journal*, 61(1): 63-65.

Warrington, I.J., T.A. Fulton, E.A. Halligan, and H.N. de Silva, 1999. Apple fruit growth and maturity are affected by early season temperatures. *Journal of the American Society for Horticultural Science*, 124(5): 468-477.

Wassmann, R., S.V.K. Jagadish, K. Sumfleth, H. Pathak, G. Howell, A. Ismail, R. Serraj, E. Redona, R.K. Singh, and S. Heuer, 2009. Regional vulnerability of climate change impacts on Asian rice production and scope for adaptation, *in* D.L. Sparks (ed.) *Advances in agronomy, vol 102*, pp. 91-133.

Way, R.D., 1995. *Pollination and fruit set of fruit crops.* Cornell Cooperative Extension. Infor. Bul. 237.

Weinberger, J.H., 1954. Effects of high temperatures during the breaking of the rest of Sullivan Elberta peach buds. *Journal of the American Society for Horticultural Science*, 63: 157-162.

Welch, J.R., J.R. Vincent, M. Auffhammer, P.F. Moya, A. Dobermann, and D. Dawe, 2010. Rice yields in tropical/subtropical Asia exhibit large but opposing sensitivities to minimum and maximum temperatures. *Proceedings of the National Academy of Sciences*, 107(33): 14562-14567.

Weltz, M.A., G. Dunn, J. Reeder, and G. Frasier, 2003. Ecological sustainability of rangelands. *Arid Land Research and Management*, 17(4): 369-388.

Weltz, M.A., L. Jolley, D. Goodrich, K. Boykin, M. Nearing, J. Stone, P. Guertin, M. Hernandez, K. Spaeth, F. Pierson, C. Morris, and B. Kepner, 2011. Techniques for assessing the environmental outcomes of conservation practices applied to rangeland watersheds. *Journal of Soil and Water Conservation*, 66(5): 154A-162A.

Wert, T.W., J.G. Williamson, J.X. Chaparro, E.P. Miller, and R.E. Rouse, 2009. The influence of climate on fruit development and quality of four low-chill peach cultivars. *HortScience*, 44(3): 666-670.

Wertheim, S.J., 2000. Developments in the chemical thinning of apple and pear. *Plant Growth Regulation*, 31(1): 85-100.

Wheeler, T.R., G.R. Batts, R.H. Ellis, P. Hadley, and J.I.L. Morison, 1996. Growth and yield of winter wheat (*Triticum aestivum*) crops in response to CO_2 and temperature. *The Journal of Agricultural Science*, 127(01): 37-48.

Wheeler, T.R., A.J. Daymond, J.I.L. Morison, R.H. Ellis, and P. Hadley, 2004. Acclimation of photosynthesis to elevated CO2 in onion (*Allium cepa*) grown at a range of temperatures. *Annals of Applied Biology*, 144(1): 103-111.

White, J.W., G. Hoogenboom, B.A. Kimball, and G.W. Wall, 2011. Methodologies for simulating impacts of climate change on crop production. *Field Crops Research*, 124(3): 357-368.

White, M.A., N.S. Diffenbaugh, G.V. Jones, J.S. Pal, and F. Giorgi, 2006. Extreme heat reduces and shifts United States premium wine production in the 21st century. *Proceedings of the National Academy of Sciences*, 103(30): 11217-11222.

Whitford, W.G., and Y. Steinberger, 2011. Effects of simulated storm sizes and nitrogen on three Chihuahuan desert perennial herbs and a grass. *Journal of Arid Environments*, 75(9): 861-864.

Wiatrak, P.J., D.L. Wright, and J.J. Marois, 2006. Development and yields of cotton under two tillage systems and nitrogen application following white lupin grain crop. *Journal of Cotton Science*, 10: 1-8.

Wiatrak, P.J., D.L. Wright, J.J. Marois, W. Koziara, and J.A. Pudelko, 2005. Tillage and nitrogen application impact on cotton following wheat. *Agronomy Journal*, 97(1): 288-293.

Wilkie, J.D., M. Sedgley, and T. Olesen, 2008. Regulation of floral initiation in horticultural trees. *Journal of Experimental Botany*, 59(12): 3215-3228.

Wilks, D.S., and D.W. Wolfe, 1998. Optimal use and economic value of weather forecasts for lettuce irrigation in a humid climate. *Agricultural and Forest Meteorology*, 89(2): 115-129.

Williams, J., M. Nearing, A. Nicks, E. Skidmore, C. Valentin, K. King, and R. Savabi, 1996. Using soil erosion models for global change studies. *Journal of Soil and Water Conservation*, 51(5): 381-385.

Williams, J.H., J.H.H. Wilson, and G.C. Bate., 1975. The growth of groundnuts (*Arachis hypogaea* L. Cv. Makulu red) at three altitudes in Rhodesia. *Rhodesian Journal of Agricultural Research*, 13: 33-43.

Williams, R.S., R.J. Norby, and D.E. Lincoln, 2000. Effects of elevated CO_2 and temperature-grown red and sugar maple on gypsy moth performance. *Global Change Biology*, 6(6): 685-695.

Wilson, B.C., J.L. Sibley, J.E. Altland, E.H. Simonne, and D.J. Eakes, 2002. Chill and heat unit levels affect foliar budbreak of selected red and Freeman maple cultivars. *Arboriculture & Urban Forestry Online*, 28: 148-152.

Wischmeier, W.H., 1959. A rainfall erosion index for a universal soil-loss equation. *Soil Sci. Soc. Am. J.*, 23(3): 246-249.

Wischmeier, W.H., and D.D. Smith, 1965. *Predicting rainfall-erosion losses from cropland east of the Rocky Mountains : Guide for selection of practices for soil and water conservation*. Agricultural Research Service, U.S. Dept of Agriculture in cooperation with Purdue Agricultural Experiment Station, Washington, D.C.

Wischmeier, W.H., and D.D. Smith, 1978. *Predicting rainfall erosion losses : A guide to conservation planning*. Dept. of Agriculture, Science and Education Administration, Washington, D.C.

Wisniewski, M., J. Norelli, C. Bassett, T. Artlip, and D. Macarisin, 2011. Ectopic expression of a novel peach (*Prunus persica*) CBF transcription factor in apple (*Malus × domestica*) results in short-day induced dormancy and increased cold hardiness. *Planta*, 233(5): 971-983.

Wissemeier, A.H., and G. Zuhlke, 2002. Relation between climatic variables, growth and the incidence of tipburn in field-grown lettuce as evaluated by simple, partial and multiple regression analysis. *Scientia Horticulturae*, 93(3-4): 193-204.

Wolf, J., 2002. Comparison of two potato simulation models under climate change. I. Model calibration and sensitivity analyses. *Climate Research*, 21(2): 173-186.

Wolfe, D., L.H. Ziska, C. Petzoldt, A. Seaman, L. Chase, and K. Hayhoe, 2008. Projected change in climate thresholds in the northeastern U.S.: Implications for crops, pests, livestock, and farmers. *Mitigation and Adaptation Strategies for Global Change*, 13(5): 555-575.

Wolfe, D.W., M.D. Schwartz, A.N. Lakso, Y. Otsuki, R.M. Pool, and N.J. Shaulis, 2005. Climate change and shifts in spring phenology of three horticultural woody perennials in northeastern USA. *International Journal of Biometeorology*, 49(5): 303-309.

Woolf, A.B., and I.B. Ferguson, 2000. Postharvest responses to high fruit temperatures in the field. *Postharvest Biology and Technology*, 21(1): 7-20.

World Bank, 2000. *Bangladesh - climate change and sustainable development* http://go.worldbank.org/HMG5ZNLDE0.

World Bank, 2007a. *World development indicators*. World Bank, Washington, DC.

World Bank, 2007b. *World development report 2008, agriculture for development*. World Bank, Washington, DC.

Xia, J., S. Niu, and S. Wan, 2009. Response of ecosystem carbon exchange to warming and nitrogen addition during two hydrologically contrasting growing seasons in a temperate steppe. *Global Change Biology*, 15(6): 1544-1556.

Yang, H., A. Dobermann, K.G. Cassman, and D.T. Walters, 2006. Features, applications, and limitations of the hybrid-maize simulation model. *Agronomy Journal*, 98(3): 737-748.

Young, J.R., and R. Sweeney, 2007. *Arkansas ground water protection and management report for 2006*. Arkansas Natural Resources Commission.

Yuan, R., 2007. Effects of temperature on fruit thinning with ethephon in 'golden delicious' apples. *Scientia Horticulturae*, 113(1): 8-12.

Zavaleta, E.S., B.D. Thomas, N.R. Chiariello, G.P. Asner, M.R. Shaw, and C.B. Field, 2003. Plants reverse warming effect on ecosystem water balance. *Proceedings of the National Academy of Sciences*, 100(17): 9892-9893.

Zhang, X.C., W.Z. Liu, Z. Li, and J. Chen, 2011. Trend and uncertainty analysis of simulated climate change impacts with multiple GCMs and emission scenarios. *Agricultural and Forest Meteorology*, 151(10): 1297-1304.

Zhang, Y.-G., M. Hernandez, E. Anson, M.A. Nearing, H. Wei, J.J. Stone, and P. Heilman, 2012. Modeling climate change effects on runoff and soil erosion in southeastern Arizona rangelands and implications for mitigation with rangeland conservation practices. *Journal of Soil and Water Conservation*: (in press).

Zhao, D., K.R. Reddy, V.G. Kakani, S. Koti, and W. Gao, 2005. Physiological causes of cotton fruit abscission under conditions of high temperature and enhanced ultraviolet-B radiation. *Physiologia Plantarum*, 124(2): 189-199.

Zhu, C., Q. Zeng, L.H. Ziska, J. Zhu, Z. Xie, and G. Liu, 2008. Effect of nitrogen supply on carbon dioxide–induced changes in competition between rice and barnyardgrass (*Echinochloa crus-galli*). *Weed Science*, 56(1): 66-71.

Zinn, K.E., M. Tunc-Ozdemir, and J.F. Harper, 2010. Temperature stress and plant sexual reproduction: Uncovering the weakest links. *Journal of Experimental Botany*.

Ziska, L., and P. Manalo, 1996. Increasing night temperature can reduce seed set and potential yield of tropical rice. *Functional Plant Biology*, 23(6): 791-794.

Ziska, L.H., 2010. Elevated carbon dioxide alters chemical management of Canada thistle in no-till soybean. *Field Crops Research*, 119(2-3): 299-303.

Ziska, L.H., and J.A. Bunce, 2007. Predicting the impact of changing CO_2 on crop yields: Some thoughts on food. *New Phytologist*, 175(4): 607-618.

Ziska, L.H., S. Faulkner, and J. Lydon, 2004. Changes in biomass and root:shoot ratio of field-grown Canada thistle (*Cirsium arvense*), a noxious, invasive weed, with elevated CO_2: Implications for control with glyphosate. *Weed Science*, 52(4): 584-588.

Ziska, L.H., and A. McClung, 2008. Differential response of cultivated and weedy (red) rice to recent and projected increases in atmospheric carbon dioxide. *Agronomy Journal*, 100(5): 1259-1263.

Ziska, L.H., M.B. Tomecek, and D.R. Gealy, 2010. Competitive interactions between cultivated and red rice as a function of recent and projected increases in atmospheric carbon dioxide. *Agronomy Journal*, 102(1): 118-123.

Zuzel, J.F., R.R. Allmaras, and R. Greenwalt, 1982. Runoff and soil erosion on frozen soils in northeastern Oregon. *Journal of Soil and Water Conservation*, 37(6): 351-354.

Chapter 6 References

Acevedo, M.F., 2011. Interdisciplinary progress in food production, food security and environment research. *Environmental Conservation*, 38(02): 151-171.

Adams, R.M., R.A. Fleming, C.-C. Chang, B.A. McCarl, and C. Rosenzweig, 1995. A reassessment of the economic effects of global climate change on U.S. agriculture. *Climatic Change*, 30(2): 147-167.

Adams, R.M., B. Hurd, S. Lenhart, and N. Leary, 1998. The effects of global warming on agriculture: An interpretative review. *Climatic Change*, 11: 19-30.

Adams, R.M., B.A. McCarl, and L.O. Mearns, 2003. The effects of spatial scale of climate scenarios on economic assessments: An example from U.S. agriculture. *Climatic Change*, 60(1): 131-148.

Adams, R.M., B.A. McCarl, K. Segerson, C. Rosenzweig, K.J. Bryant, B.L. Dixon, R. Connor, R.E. Evenson, and D. Ojima, 1999. The economic effects of climate change on U.S. agriculture, *in* R.O. Mendelsohn and J.E. Neumann (eds.), *The impact of climate change on the United States economy*, pp. 19-54. Cambridge University Press, Cambridge; New York.

Adams, R.M., C. Rosenzweig, R.M. Peart, J.T. Ritchie, B.A. McCarl, J.D. Glyer, R.B. Curry, J.W. Jones, K.J. Boote, and L.H. Allen, 1990. Global climate change and U.S. agriculture. *Nature*, 345(6272): 219-224.

Ainsworth, E.A., and D.R. Ort, 2010. How do we improve crop production in a warming world? *Plant Physiology*, 154(2): 526-530.

Alig, R.J., D.M. Adams, and B.A. McCarl, 2002. Projecting impacts of global climate change on the U.S. forest and agriculture sectors and carbon budgets. *Forest Ecology and Management*, 169(1-2): 3-14.

Antle, J.M., and S.M. Capalbo, 2010. Adaptation of agricultural and food systems to climate change: An economic and policy perspective. *Applied Economic Perspectives and Policy*, 32(3): 386-416.

Antle, J.M., S.M. Capalbo, E.T. Elliott, and K.H. Paustian, 2004. Adaptation, spatial heterogeneity, and the vulnerability of agricultural systems to climate change and CO_2 fertilization: An integrated assessment approach. *Climatic Change*, 64(3): 289-315.

Battisti, D.S., and R.L. Naylor, 2009. Historical warnings of future food insecurity with unprecedented seasonal heat. *Science*, 323(5911): 240-244.

Beach, R.H., C. Zhen, A. Thomson, R.M. Rejesus, P. Sinha, A.W. Lentz, D.V. Vedenov, and B.A. McCarl, 2010. *Climate change impacts on crop insurance*. USDA Risk Management Agency, Kansas City, MO.

Burke, M., J. Dykema, D. Lobell, E. Miguel, and S. Satyanath, 2011. *Incorporating climate uncertainty into estimates of climate change impacts, with applications to U.S. and African agriculture*. NBER Working Paper Series, Working Paper 17092.

Burney, J.A., S.J. Davis, and D.B. Lobell, 2010. Greenhouse gas mitigation by agricultural intensification. *Proceedings of the National Academy of Sciences*, 107(26): 12052-12057.

Challinor, A., T. Wheeler, C. Garforth, P. Craufurd, and A. Kassam, 2007. Assessing the vulnerability of food crop systems in Africa to climate change. *Climatic Change*, 83(3): 381-399.

Challinor, A.J., F. Ewert, S. Arnold, E. Simelton, and E. Fraser, 2009. Crops and climate change: Progress, trends, and challenges in simulating impacts and informing adaptation. *Journal of Experimental Botany*, 60(10): 2775-2789.

Chen, C.-C., B.A. McCarl, and D.E. Schimmelpfennig, 2004. Yield variability as influenced by climate: A statistical investigation. *Climatic Change*, 66(1): 239-261.

Chhetri, N.B., W.E. Easterling, A. Terando, and L. Mearns, 2010. Modeling path dependence in agricultural adaptation to climate variability and change. *Annals of the Association of American Geographers*, 100(4): 894-907.

Claessens, L., J.M. Antle, J.J. Stoorvogel, R.O. Valdivia, P.K. Thornton, and M. Herrero, 2012. A method for evaluating climate change adaptation strategies for small-scale farmers using survey, experimental and modeled data. *Agricultural Systems*, 111: 85-95.

Cline, W.R., 2007. *Global warming and agriculture: Impact estimates by country*. Center for Global Development : Peterson Institute for International Economics, Washington, DC.

Cox, C., A. Hug, and N. Buruzelius, 2011. *Losing ground: Executive summary*. Environmental Working Group, Washington, DC.

Darwin, R., 2004. Effects of greenhouse gas emissions on world agriculture, food consumption, and economic welfare. *Climatic Change*, 66(1): 191-238.

Darwin, R., M.E. Tsigas, J. Lewandrowski, and A. Raneses, 1995. *World agriculture and climate change: Economic adaptations*. United States Department of Agriculture, Economic Research Service http://ideas.repec.org/p/ags/uerser/33933.html.

Deschenes, O., and M. Greenstone, 2007. The economic impacts of climate change: Evidence from agricultural output and random fluctuations in weather. *The American Economic Review*, 97(1): 354-385.

Easterling, W.E., P.K. Aggarwal, P. Batima, K.M. Brander, L. Erda, S.M. Howden, A. Kirilenko, J. Morton, J.-F. Soussana, J. Schmidhuber, and F.N. Tubiello, 2007. Food, fibre and forest products, *in* M.L. Parry, O.F. Canziani, J.P. Palutikof, P.J. Van Der Linden and C.E. Hanson (eds.), *Climate change 2007: Impacts, adaptation and vulnerability: Working Group II contribution to the Fourth Assessment Report of the IPCC Intergovernmental Panel on Climate Change*, pp. 273-313. Cambridge University Press, Cambridge, UK.

Economic Research Service, 2012. *Cash receipts, by commodity groups and selected commodities, 1924-2010 [computer file]*. U.S. Department of Agriculture: Washington DC. URL: http://www.ers.usda.gov/data/farmincome/finfidmu.htm#cashrec . Accessed May 31, 2012.

Edgerton, M.D., 2009. Increasing crop productivity to meet global needs for feed, food, and fuel. *Plant Physiology*, 149(1): 7-13.

ERS, 2012. *Wheat data: Yearbook tables* http://www.ers.usda.gov/data/wheat/YBtable21.asp.

Fischer, G., M. Shah, F. N. Tubiello, and H. van Velhuizen, 2005. Socio-economic and climate change impacts on agriculture: An integrated assessment, 1990-2080. *Philosophical Transactions of the Royal Society B: Biological Sciences*, 360(1463): 2067-2083.

Funk, C.C., and M.E. Brown, 2009. Declining global per capita agricultural production and warming oceans threaten food security. *Food Security*, 1(3): 271-289.

Gornall, J., R. Betts, E. Burke, R. Clark, J. Camp, K. Willett, and A. Wiltshire, 2010. Implications of climate change for agricultural productivity in the early twenty-first century. *Philosophical Transactions of the Royal Society B: Biological Sciences*, 365(1554): 2973-2989.

Hare, W., W. Cramer, M. Schaeffer, A. Battaglini, and C. Jaeger, 2011. Climate hotspots: Key vulnerable regions, climate change and limits to warming. *Regional Environmental Change*, 11(0): 1-13.

Hatfield, J.L., K.J. Boote, B.A. Kimball, L.H. Ziska, R.C. Izaurralde, D. Ort, A.M. Thomson, and D. Wolfe, 2011. Climate impacts on agriculture: Implications for crop production. *Agronomy Journal*, 103(2): 351-370.

Hertel, T.W., M.B. Burke, and D.B. Lobell, 2010. The poverty implications of climate-induced crop yield changes by 2030. *Global Environmental Change-Human and Policy Dimensions*, 20(4): 577-585.

Hertel, T.W., and S.D. Rosch, 2010. Climate change, agriculture, and poverty. *Applied Economic Perspectives and Policy*.

Hitz, S., and J. Smith, 2004. Estimating global impacts from climate change. *Global Environmental Change*, 14(3): 201-218.

Howden, S.M., J.F. Soussana, F.N. Tubiello, N. Chhetri, M. Dunlop, and H. Meinke, 2007. Adapting agriculture to climate change. *Proceedings of the National Academy of Sciences*, 104(50): 19691-19696.

Iowa State University, and National Soil Erosion Research Lab, *Iowa daily erosion project*. National Laboratory for Agriculture and the Environment and the University of Iowa, http://wepp.mesonet. agron.iastate.edu/ http://wepp.mesonet.agron.iastate.edu/.

IPCC, 2007b. *Climate change 2007 : Impacts, adaptation and vulnerability: Contribution of Working Group II to the Fourth Assessment Report of the Intergovernmental Panel on Climate Change*. Cambridge University Press, Cambridge, U.K.; New York.

Isik, M., and S. Devadoss, 2006. An analysis of the impact of climate change on crop yields and yield variability. *Applied Economics*, 38(7): 835-844.

Izaurralde, R.C., N.J. Rosenberg, R.A. Brown, and A.M. Thomson, 2003. Integrated assessment of Hadley Center (HadCM2) climate-change impacts on agricultural productivity and irrigation water supply in the conterminous United States: Part II. Regional agricultural production in 2030 and 2095. *Agricultural and Forest Meteorology*, 117(1–2): 97-122.

Izaurralde, R.C., A.M. Thomson, J.A. Morgan, P.A. Fay, H.W. Polley, and J.L. Hatfield, 2011. Climate impacts on agriculture: Implications for forage and rangeland production. *Agronomy Journal*, 103(2): 371-381.

Jarvis, A., C. Lau, S. Cook, E. Wollenberg, J. Hansen, O. Bonilla, and A. Challinor, 2011. An integrated adaptation and mitigation framework for developing agricultural research: Synergies and trade-offs. *Experimental Agriculture*, 47(2): 185-203.

Karl, T., J. Melillo, T. Peterson, and S.J. Hassol (Eds.), 2009. *Global climate change impacts in the United States*. Cambridge University Press.

Kelly, D.L., and C.D. Kolstad, 1999. Integrated assessment models for climate change control, *in* H. Folmer and T. Tietenberg (eds.), *The international yearbook of environmental and resource economics 1999/2000: A survey of current issues*. Edward Elgar, Cheltenham, UK; Northampton, MA.

Knutson, C.L., T. Haigh, M.J. Hayes, M. Widhalm, J. Nothwehr, M. Kleinschmidt, and L. Graf, 2011. Farmer perceptions of sustainable agriculture practices and drought risk reduction in Nebraska, USA. *Renewable Agriculture and Food Systems*, 26(03): 255-266.

Kucharik, C.J., and S.P. Serbin, 2008. Impacts of recent climate change on Wisconsin corn and soybean yield trends. *Environmental Research Letters*, 3(3): 034003.

Lobell, D.B., and G.P. Asner, 2003. Climate and management contributions to recent trends in U.S. Agricultural yields. *Science*, 299(5609): 1032.

Lobell, D.B., and M.B. Burke, 2008. Why are agricultural impacts of climate change so uncertain? The importance of temperature relative to precipitation. *Environmental Research Letters*, 3(3): 034007.

Lobell, D.B., M.B. Burke, C. Tebaldi, M.D. Mastrandrea, W.P. Falcon, and R.L. Naylor, 2008. Prioritizing climate change adaptation needs for food security in 2030. *Science*, 319(5863): 607-610.

Lobell, D.B., K.G. Cassman, and C.B. Field, 2009. Crop yield gaps: Their importance, magnitudes, and causes. *Annual Review of Environment and Resources*, 34(1): 179-204.

Lobell, D.B., and C.B. Field, 2007. Global scale climate crop yield relationships and the impacts of recent warming. *Environmental Research Letters*, 2(1): 014002.

Lobell, D.B., C.B. Field, K.N. Cahill, and C. Bonfils, 2006. Impacts of future climate change on California perennial crop yields: Model projections with climate and crop uncertainties. *Agricultural and Forest Meteorology*, 141(2–4): 208-218.

Lobell, D.B., W. Schlenker, and J. Costa-Roberts, 2011. Climate trends and global crop production since 1980. *Science*, 333: 208-218.

Long, S.P., E.A. Ainsworth, A.D.B. Leakey, and P.B. Morgan, 2005. Global food insecurity. Treatment of major food crops with elevated carbon dioxide or ozone under large-scale fully open-air conditions suggests recent models may have overestimated future yields. *Philosophical Transactions of the Royal Society B: Biological Sciences*, 360(1463): 2011-2020.

Malcolm, S., 2011. Regional economic and environmental impacts of agricultural adaptation to a changing climate in the United States *Forestry and Agriculture Greenhouse Gas Modeling Forum*, Shepherdstown, WV.

Malcolm, S., E. Marshall, M. Aillery, P. Heisey, M. Livingston, and K. Day-Rubenstein, 2012. *Agricultural adaptation to a changing climate: Economic and environmental implications vary by U.S. Region, ERR-136*. U.S. Department of Agriculture, Economic Research Service.

McCarl, B., 2008. *U.S. agriculture in the climate change squeeze: Part 1: Sectoral sensitivity and vulnerability. Report to the National Environmental Trust* http://agecon2.tamu.edu/people/ faculty/mccarl-bruce/papers/1303Agriculture%20in%20the%20 climate%20change%20squeez1.doc.

Mendelsohn, R., and A. Dinar, 1999. Climate change, agriculture, and developing countries: Does adaptation matter? *The World Bank Research Observer*, 14(2): 277-293.

Mendelsohn, R., W.D. Nordhaus, and D. Shaw, 1994. The impact of global warming on agriculture: A Ricardian analysis. *The American Economic Review*, 84(4): 753-771.

Mertz, O., K. Halsnaes, J.E. Olesen, and K. Rasmussen, 2009. Adaptation to climate change in developing countries. *Environmental Management*, 43(5): 743-752.

Moss, R.H., J.A. Edmonds, K.A. Hibbard, M.R. Manning, S.K. Rose, D.P. van Vuuren, T.R. Carter, S. Emori, M. Kainuma, T. Kram, G.A. Meehl, J.F.B. Mitchell, N. Nakicenovic, K. Riahi, S.J. Smith, R.J. Stouffer, A.M. Thomson, J.P. Weyant, and T.J. Wilbanks, 2010. The next generation of scenarios for climate change research and assessment. *Nature*, 463(7282): 747-756.

Munasinghe, L., T. Jun, and D. Rind, 2011. Climate change: A new metric to measure changes in the frequency of extreme temperatures using record data. *Climatic Change*: 1-24.

Nakicenovic et al. 2000. *Special report on emissions scenarios: A special report of Working Group III of the Intergovernmental Panel on Climate Change*. Cambridge University Press, Cambridge; New York.

NOAA NCDC, 2011. *State of the climate: Global analysis for annual 2011*. NOAA National Climatic Data Center http://www.ncdc.noaa. gov/sotc/global/2011/13.

OECD-INEA-FAO, 2010. *Climate change and crop insurance in the United States*. OECD-INEA-FAO workshop on agriculture and adaptation to climate change.

Parry, M., C. Rosenzweig, and M. Livermore, 2005. Climate change, global food supply and risk of hunger. *Philosophical Transactions of the Royal Society B: Biological Sciences*, 360(1463): 2125-2138.

Parry, M.L., C. Rosenzweig, A. Iglesias, M. Livermore, and G. Fischer, 2004. Effects of climate change on global food production under sres emissions and socio-economic scenarios. *Global Environmental Change*, 14(1): 53-67.

Patt, A., D. van Vuuren, F. Berkhout, A. Aaheim, A. Hof, M. Isaac, and R. Mechler, 2010. Adaptation in integrated assessment modeling: Where do we stand? *Climatic Change*, 99(3): 383-402.

173

Pfister, S., P. Bayer, A. Koehler, and S. Hellweg, 2011. Projected water consumption in future global agriculture: Scenarios and related impacts. *Science of the Total Environment*, 409(20): 4206-4216.

Polsky, C., and W.E. Easterling, 2001. Adaptation to climate variability and change in the US Great Plains: A multi-scale analysis of Ricardian climate sensitivities. *Agriculture, Ecosystems & Environment*, 85(1-3): 133-144.

Quiggin, J., and J. Horowitz, 2003. Costs of adjustment to climate change. *Australian Journal of Agricultural and Resource Economics*, 47(4): 429-446.

Ramankutty, N., J.A. Foley, J. Norman, and K. McSweeney, 2002. The global distribution of cultivable lands: Current patterns and sensitivity to possible climate change. *Global Ecology and Biogeography*, 11(5): 377-392.

Reilly, J., S. Paltsev, B. Felzer, X. Wang, D. Kicklighter, J. Melillo, R. Prinn, M. Sarofim, A. Sokolov, and C. Wang, 2007. Global economic effects of changes in crops, pasture, and forests due to changing climate, carbon dioxide, and ozone. *Energy Policy*, 35(11): 5370-5383.

Reilly, J., F. Tubiello, B. McCarl, D. Abler, R. Darwin, K. Fuglie, S. Hollinger, C. Izaurralde, S. Jagtap, J. Jones, L. Mearns, D. Ojima, E. Paul, K. Paustian, S. Riha, N. Rosenberg, and C. Rosenzweig, 2003. U.S. agriculture and climate change: New results. *Climatic Change*, 57(1): 43-67.

Rosenzweig, C., J.W. Jones, J.L. Hatfield, A.C. Ruane, K.J. Boote, P. Thorburn, J.M. Antle, G.C. Nelson, C. Porter, S. Janssen, S. Asseng, B. Basso, F. Ewert, D. Wallach, G. Baigorria, and J.M. Winter, 2012. The agricultural model intercomparison and improvement project (AgMIP): Protocols and pilot studies. *Agricultural and Forest Meteorology*, (in press).

Rosenzweig, C., and M.L. Parry, 1994. Potential impact of climate change on world food supply. *Nature*, 367(6459): 133-138.

Rosenzweig, C., F.N. Tubiello, R. Goldberg, E. Mills, and J. Bloomfield, 2002. Increased crop damage in the U.S. from excess precipitation under climate change. *Global Environmental Change-Human and Policy Dimensions*, 12(3): 197-202.

Sands, R.D., and J.A. Edmonds, 2005. Climate change impacts for the conterminous USA: An integrated assessment. *Climatic Change*, 69(1): 127-150.

Schlenker, W., W. Hanemann, and A. Fisher, 2007. Water availability, degree days, and the potential impact of climate change on irrigated agriculture in California. *Climatic Change*, 81(1): 19-38.

Schlenker, W., W.M. Hanemann, and A.C. Fisher, 2005. Will U.S. agriculture really benefit from global warming? Accounting for irrigation in the hedonic approach. *American Economic Review*, 95(1): 395-406.

Schlenker, W., and D.B. Lobell, 2010. Robust negative impacts of climate change on African agriculture. *Environmental Research Letters*, 5(1): 014010.

Schlenker, W., and M.J. Roberts, 2009. Nonlinear temperature effects indicate severe damages to U.S. crop yields under climate change. *Proceedings of the National Academy of Sciences*, 106(37): 15594-15598.

Schmidhuber, J., and F.N. Tubiello, 2007. Global food security under climate change. *Proceedings of the National Academy of Sciences*, 104(50): 19703-19708.

Smit, B., and M.W. Skinner, 2002. Adaptation options in agriculture to climate change: A typology. *Mitigation and Adaptation Strategies for Global Change*, 7(1): 85-114.

Soussana, J.-F., A.-I. Graux, and F.N. Tubiello, 2010. Improving the use of modelling for projections of climate change impacts on crops and pastures. *Journal of Experimental Botany*, 61(8): 2217-2228.

Tilman, D., C. Balzer, J. Hill, and B.L. Befort, 2011. Global food demand and the sustainable intensification of agriculture. *Proceedings of the National Academy of Sciences*, 108(50): 20260-20264.

Tol, R.S.J., 2009. The economic effects of climate change. *The Journal of Economic Perspectives*, 23(2): 29-51.

Tubiello, F.N., J.S. Amthor, K.J. Boote, M. Donatelli, W. Easterling, G. Fischer, R.M. Gifford, M. Howden, J. Reilly, and C. Rosenzweig, 2007. Crop response to elevated CO2 and world food supply: A comment on "food for thought..." by Long et al. Science 312:1918–1921, 2006. *European Journal of Agronomy*, 26(3): 215-223.

Tubiello, F.N., J.F. Soussana, and S.M. Howden, 2007. Crop and pasture response to climate change. *Proceedings of the National Academy of Sciences*, 104(50): 19686-19690.

UNDP, 2007. *Human development report 2007/2008. Fighting climate change: Human solidarity in a divided world.* United Nations Development Program, New York.

USDA, 2009. *National resources inventory.* Natural Resources Conservation Service, Washington, D.C., and Center for Survey Statistics and Methodology, Iowa State University, Ames, Iowa http://www.nrcs.usda.gov/Internet/FSE_DOCUMENTS//stelprdb1041379.pdf.

van der Velde, M., F. Tubiello, A. Vrieling, and F. Bouraoui, 2011. Impacts of extreme weather on wheat and maize in France: Evaluating regional crop simulations against observed data. *Climatic Change*: 1-15.

Vanloqueren, G., and P.V. Baret, 2009. How agricultural research systems shape a technological regime that develops genetic engineering but locks out agroecological innovations. *Research Policy*, 38(6): 971-983.

Winkler, J., S. Thornsbury, M. Artavia, F.-M. Chmielewski, D. Kirschke, S. Lee, M. Liszewska, S. Loveridge, P.-N. Tan, S. Zhong, J. Andresen, J. Black, R. Kurlus, D. Nizalov, N. Olynk, Z. Ustrnul, C. Zavalloni, J. Bisanz, G. Bujdosó, L. Fusina, Y. Henniges, P. Hilsendegen, K. Lar, L. Malarzewski, T. Moeller, R. Murmylo, T. Niedzwiedz, O. Nizalova, H. Prawiranata, N. Rothwell, J. van Ravensway, H. von Witzke, and M. Woods, 2010. A conceptual framework for multi-regional climate change assessments for international market systems with long-term investments. *Climatic Change*, 103(3): 445-470.

Winters, P., R. Murgai, E. Sadoulet, A. de Janvry, and G. Frisvold, 1998. Economic and welfare impacts of climate change on developing countries. *Environmental and Resource Economics*, 12(1): 1-24.

Wolfe, D., L.H. Ziska, C. Petzoldt, A. Seaman, L. Chase, and K. Hayhoe, 2008. Projected change in climate thresholds in the northeastern U.S.: Implications for crops, pests, livestock, and farmers. *Mitigation and Adaptation Strategies for Global Change*, 13(5): 555-575.

Yu, T.-H., S. Tokgoz, E. Wailes, and E. Chavez, 2011. A quantitative analysis of trade policy responses to higher world agricultural commodity prices. *Food Policy*, 36(5): 545-561.

Zhang, X.A., and X.M. Cai, 2011. Climate change impacts on global agricultural land availability. *Environmental Research Letters*, 6(1).

Chapter 7 References

Adger, W., S. Agrawala, M.M.Q. Mirza, C. Conde, K. O'Brien, J. Pulhin, R. Pulwarty, B. Smit, and K. Takahashi, 2007. Assessment of adaptation practices, options, constraints and capacity, *in* M.L. Parry, O.F. Canziani, J.P. Palutikof, P.J. Van Der Linden and C.E. Hanson (eds.), *Climate change 2007: Impacts, adaptation and vulnerability: Working Group II contribution to the Fourth Assessment Report of the IPCC Intergovernmental Panel on Climate Change*, pp. 717-743. Cambridge University Press, Cambridge, UK.

Adger, W., S. Dessai, M. Goulden, M. Hulme, I. Lorenzoni, D. Nelson, L. Naess, J. Wolf, and A. Wreford, 2009. Are there social limits to adaptation to climate change? *Climatic Change*, 93(3): 335-354.

Adger, W.N., and K. Vincent, 2005. Uncertainty in adaptive capacity. *Comptes Rendus Geoscience*, 337(4): 399-410.

Agroclimate, 2011. Agroclimate is an interactive website with climate, agriculture, and forestry information that allows users to assess resource management options with respect to their probable outcomes under forecast climate conditions. http://agroclimate.org/. Southeast Climate Consortium.

Alberini, A., A. Chiabai, and L. Muehlenbachs, 2006. Using expert judgment to assess adaptive capacity to climate change: Evidence from a conjoint choice survey. *Global Environmental Change*, 16(2): 123-144.

Allison, H.E., and R.J. Hobbs., 2004. Resilience, adaptive capacity, and the "lock-in trap" of the Western Australian agricultural region. *Ecology and Society*, 9(1): 3.

Antle, J.M., 2009. *Adaptation of agriculture and the food system to climate change: Policy issues. Issues brief 10-03. Testimony to house agriculture subcommittee, 2009 - 206.205.47.99.* Resources for the Future www.rff.org.

Antle, J.M., and S.M. Capalbo, 2010. Adaptation of agricultural and food systems to climate change: An economic and policy perspective. *Applied Economic Perspectives and Policy*, 32(3): 386-416.

Arbuckle, J.G.J., 2011. *Farmer perspectives on climate change and agriculture*, paper presented at Climate Change Conference, November 1-3, 2011. Heartland Regional Water Coordination Initiative, Lied Conference Center, Nebraska City, NE.

Batie, S.S., 2009. Green payments and the U.S. farm bill: Information and policy challenges. *Frontiers in Ecology and the Environment*, 7(7): 380-388.

BC Agriculture, 2012. *Climate change adaptation risk and opportunity assessment. Provincial report.* . British Columbia Agriculture & Food Climate Action Initiative.

Belliveau, S., B. Smit, and B. Bradshaw, 2006. Multiple exposures and dynamic vulnerability: Evidence from the grape industry in the Okanagan Valley, Canada. *Global Environmental Change*, 16(4): 364-378.

Below, T.B., K.D. Mutabazi, D. Kirschke, C. Franke, S. Sieber, R. Siebert, and K. Tscherning, 2012. Can farmers' adaptation to climate change be explained by socio-economic household-level variables? *Global Environmental Change*, 22(1): 223-235.

Biesbroek, G.R., R.J. Swart, T.R. Carter, C. Cowan, T. Henrichs, H. Mela, M.D. Morecroft, and D. Rey, 2010. Europe adapts to climate change: Comparing national adaptation strategies. *Global Environmental Change*, 20(3): 440-450.

Blackstock, K.L., J. Ingram, R. Burton, K.M. Brown, and B. Slee, 2010. Understanding and influencing behaviour change by farmers to improve water quality. *Science of the Total Environment*, 408(23): 5631-5638.

Borick, C., E. Lachapelle, and B. Rabe, 2011. *Climate compared: Public opinion on climate change in the United States and Canada.* Issues in Governance Studies 39, The Brookings Institution, Washington, DC http://www.brookings.edu/papers/2011/04_climate_change_opinion.aspx.

Brooks, N., W. Neil Adger, and P. Mick Kelly, 2005. The determinants of vulnerability and adaptive capacity at the national level and the implications for adaptation. *Global Environmental Change*, 15(2): 151-163.

Brown, A., M. Gawith, K. Lonsdale, and P. Pringle, 2011. *Managing adaptation: Linking theory and practice.* UK Climate Impacts Programme, Oxford, UK. Retrieved from http://www.ukcip.org.uk/wordpress/wp-content/PDFs/UKCIP_Managing_adaptation.pdf.

Brown, P.R., R. Nelson, B. Jacobs, P. Kokic, J. Tracey, M. Ahmed, and P. DeVoil, 2010. Enabling natural resource managers to self-assess their adaptive capacity. *Agricultural Systems*, 103(8): 562-568.

Bryant, C., B. Singh, P. Thomassin, L. Baker, S. Desroches, M. Sovoie, K. Delusca, M. Doyon, and E. Seyoum, 2008. *Evaluation of agricultural adaptation processes and adaptive capacity to climate change and variability: The co-construction of new adaptation planning tools with stakeholders and farming communities in the Saguenay-Lac-Saint-Jean and Montérégie regions of Québec. Final synthesis report project a1332.* Submitted to Natural Resources Canada Climate Change Impacts and Adaptation Program adaptation.nrcan.gc.ca.

Bryant, C.R., B. Smit, M. Brklacich, T.R. Johnston, J. Smithers, Q. Chjotti, and B. Singh, 2000. Adaptation in Canadian agriculture to climatic variability and change. *Climatic Change*, 45(1): 181-201.

Cabrera, D., J.T. Mandel, J.P. Andras, and M.L. Nydam, 2008. What is the crisis? Defining and prioritizing the world's most pressing problems. *Frontiers in Ecology and the Environment*, 6(9): 469-475.

Carpenter, S.R., H.A. Mooney, J. Agard, D. Capistrano, R.S. DeFries, S. Díaz, T. Dietz, A.K. Duraiappah, A. Oteng-Yeboah, H.M. Pereira, C. Perrings, W.V. Reid, J. Sarukhan, R.J. Scholes, and A. Whyte, 2009. Science for managing ecosystem services: Beyond the millennium ecosystem assessment. *Proceedings of the National Academy of Sciences*, 106(5): 1305-1312.

Chhetri, N.B., W.E. Easterling, A. Terando, and L. Mearns, 2010. Modeling path dependence in agricultural adaptation to climate variability and change. *Annals of the Association of American Geographers*, 100(4): 894-907.

Council of Australian Governments, 2007. *National climate change adaptation framework.* Retrieved from http://www.coag.gov.au/coag_meeting_outcomes/2007-04-13/docs/national_climate_change_adaption_framework.pdf.

Eakin, H., and L.A. Bojórquez-Tapia, 2008. Insights into the composition of household vulnerability from multicriteria decision analysis. *Global Environmental Change*, 18(1): 112-127.

Easterling, W.E., 2009. Guidelines for adapting agriculture to climate change, *in* D. Hillel and C. Rosenzweig (eds.), *Handbook of climate change and agroecosystems: Impacts, adaptation, and mitigation.* Imperial College Press; Distributed by World Scientific Publishing Co., London; Singapore; Hackensack, NJ.

Easterling, W.E., P.K. Aggarwal, P. Batima, K.M. Brander, L. Erda, S.M. Howden, A. Kirilenko, J. Morton, J.-F. Soussana, J. Schmidhuber, and F.N. Tubiello, 2007. Food, fibre and forest products, *in* M.L. Parry, O.F. Canziani, J.P. Palutikof, P.J. Van Der Linden and C.E. Hanson (eds.), *Climate change 2007: Impacts, adaptation and vulnerability: Working Group II contribution to the Fourth Assessment Report of the IPCC Intergovernmental Panel on Climate Change*, pp. 273-313. Cambridge University Press, Cambridge, UK.

FAO, 2010. *"Climate-smart" agriculture policies, practices and financing for food security, adaptation and mitigation. Technical input prepared for the Hague conference on agriculture food security and climate change.* FAO Communications Division, Rome.

Field, C.B., L.D. Mortsch, M. Brklacich, D.L. Forbes, P. Kovacs, J.A. Patz, S.W. Running, and M.J. Scott, 2007. North America, *in* M.L. Parry, O.F. Canziani, J.P. Palutikof, P.J.V.D. Linden and C.E. Hanson (eds.), *Climate change 2007: Impacts, adaptation and vulnerability: Contribution of Working Group II to the Fourth Assessment Report of the Intergovernmental Panel on Climate Change.* Cambridge University Press, Cambridge, UK.

Folke, C., S. Carpenter, B. Walker, M. Scheffer, T. Elmqvist, L. Gunderson, and C.S. Holling, 2004. Regime shifts, resilience, and biodiversity in ecosystem management. *Annual Review of Ecology, Evolution, and Systematics*, 35: 557-581.

175

Folke, C., T. Hahn, P. Olsson, and J. Norberg, 2005. Adaptive governance of social-ecological systems, *in Annual review of environment and resources*, pp. 441-473. Annual Reviews, Palo Alto.

GAO, 2009. *Climate change adaptation: Information on selected federal efforts to adapt to a changing climate (GAO-10-114SP)* Government Accounting Office http://www.gao.gov/assets/210/203899. pdf.

Godfray, H.C.J., J.R. Beddington, I.R. Crute, L. Haddad, D. Lawrence, J.F. Muir, J. Pretty, S. Robinson, S.M. Thomas, and C. Toulmin, 2010. Food security: The challenge of feeding 9 billion people. *Science*, 327(5967): 812-818.

Goklany, I., 2007. Integrated strategies to reduce vulnerability and advance adaptation, mitigation, and sustainable development. *Mitigation and Adaptation Strategies for Global Change*, 12(5): 755-786.

Gunderson, L.H., and C.S. Holling, 2002. *Panarchy: Understanding transformations in human and natural systems.* Island Press, Washington, DC.

Gunderson, L.H., and C.S. Holling, 2002. Resilience and adaptive cycles, *in* L.H. Gunderson and C.S. Holling (eds.), *Panarchy.* Island Press, Washington.

Hansen, J.E., R. Ruedy, M. Sato, and K. Lo, 2012. NASA GISS surface temperature (GISTEMP) analysis, *in Trends: A compendium of data on global change.* Carbon Dioxide Information Analysis Center, Oak Ridge National Laboratory, U.S. Department of Energy, Oak Ridge, Tenn., U.S.A.

Harrington, L.M.B., and M. Lu, 2002. Beef feedlots in southwestern Kansas: Local change, perceptions, and the global change context. *Global Environmental Change*, 12(4): 273-282.

Harwood, J., R. Heifner, K. Coble, J. Perry, and A. Somwaru, 1999. *Managing risk in farming: Concepts, research, and analysis.* Market and Trade Economics Division and Resource Economics Division, Economic Research Service, U.S. Department of Agriculture. Agricultural Economic Report No. 774 http://www. ers.usda.gov/publications/aer774/.

Hatfield, J.L., K.J. Boote, B.A. Kimball, L.H. Ziska, R.C. Izaurralde, D. Ort, A.M. Thomson, and D. Wolfe, 2011. Climate impacts on agriculture: Implications for crop production. *Agronomy Journal*, 103(2): 351-370.

Hendrickson, J., G.F. Sassenrath, D. Archer, J. Hanson, and J. Halloran, 2008. Interactions in integrated U.S. agricultural systems: The past, present and future. *Renewable Agriculture and Food Systems*, 23(Special Issue 04): 314-324.

Hills, D., and A. Bennett, 2010. *Framework for developing climate change adaptation strategies and action plans for agriculture in western Australia.* Department of Agriculture and Food, Western Australia Agriculture Authority.

Howden, M., A. Ash, S. Barlow, T. Booth, S. Charles, B. Cechet, S. Crimp, R. Gifford, K. Hennessy, R. Jones, M. Kirschbaum, G. McKeon, H. Meinke, S. Park, B. Sutherst, L. Webb, and P. Whetton, 2003. *An overview of the adaptive capacity of the Australian agricultural sector to climate change – options, costs and benefits.* Scientific and Industrial Research Organisation, Collingwood, Victoria http://www.cse.csiro.au/publications/2003/AGOAgClimateAdaptationReport.pdf.

Howden, S.M., J.F. Soussana, F.N. Tubiello, N. Chhetri, M. Dunlop, and H. Meinke, 2007. Adapting agriculture to climate change. *Proceedings of the National Academy of Sciences*, 104(50): 19691-19696.

Hua, W., C. Zulauf, and B. Sohngen, 2004. *To adopt or not to adopt: Conservation decisions and participation in watershed groups*, paper presented at American Agricultural Economics Association Annual Meeting. Denver CO, July.

IAASTD, 2009. *Agriculture at a crossroads: Synthesis report.* International Assessment of Agricultural Knowledge, Science and Technology for Development (IAASTD), Island Press, Washington, D.C.

Iglesias, A., K. Avis, M.Benzie, P. Fisher, M. Harley, N. Hodgson, L. Horrocks, M. Moneo, and J. Webb, 2007. *Adaptation to climate change in the agricultural sector. Report to European Commission Directorate - General for Agriculture and Rural Development. ED05334* http://ec.europa.eu/agriculture/analysis/external/climate/.

Iglesias, A., S. Quiroga, M. Moneo, and L. Garrote, 2012. From climate change impacts to the development of adaptation strategies: Challenges for agriculture in Europe. *Climatic Change*, 112(1): 143-168.

Izaurralde, R.C., A.M. Thomson, J.A. Morgan, P.A. Fay, H.W. Polley, and J.L. Hatfield, 2011. Climate impacts on agriculture: Implications for forage and rangeland production. *Agronomy Journal*, 103(2): 371-381.

Jackson, L., M. van Noordwijk, J. Bengtsson, W. Foster, L. Lipper, M. Pulleman, M. Said, J. Snaddon, and R. Vodouhe, 2010. Biodiversity and agricultural sustainability: From assessment to adaptive management. *Current Opinion in Environmental Sustainability*, 2(1-2): 80-87.

Jackson, L., S. Wheeler, A. Hollander, A. O'Geen, B. Orlove, J. Six, D. Sumner, F. Santos-Martin, J. Kramer, W. Horwath, R. Howitt, and T. Tomich, 2011. Case study on potential agricultural responses to climate change in a California landscape. *Climatic Change*, 109(Suppl 1): 407-427.

Jackson, L.E., F. Santos-Martin, Hollander, W.R. Horwath, R.E. Howitt, J.B. Kramer, A.T. O'Geen, B. S.Orlove, J.W. Six, S.K. Sokolow, D.A. Sumner, T.P. Tomich, and S.M. Wheeler, 2009. *Potential for adaptation to climate change in an agricultural landscape in the Central Valley of California final paper.* California Energy Commission, Sacramento, California http://bibpurl.oclc. org/web/39491http://www.energy.ca.gov/2009publications/CEC-500-2009-044/CEC-500-2009-044-F.PDF.

Jones, R., P. Dettmann, G. Park, M. Rogers, and T. White, 2007. The relationship between adaptation and mitigation in managing climate change risks: A regional response from North Central Victoria, Australia. *Mitigation and Adaptation Strategies for Global Change*, 12(5): 685-712.

Jones, R.N., and B.L. Preston, 2011. Adaptation and risk management. *Wiley Interdisciplinary Reviews: Climate Change*, 2(2): 296-308.

Kahan, D., 2012. Cultural cognition as a conception of the cultural theory of risk, *in* R. Hillerbrand, P. Sandin, M. Peterson and S. Roeser (eds.), *Handbook of risk theory epistemology, decision theory, ethics and social implications of risk.* Springer Verlag, New York.

Kenny, G., 2011. Adaptation in agriculture: Lessons for resilience from eastern regions of New Zealand. *Climatic Change*, 106(3): 441-462.

Klein, R.J.T., S. Huq, F. Denton, T.E. Downing, R.G. Richels, J.B. Robinson, and F.L. Toth, 2007. Inter-relationships between adaptation and mitigation, *in* M.L. Parry, O.F. Canziani, J.P. Palutikof, P.J. Van Der Linden and C.E. Hanson (eds.), *Climate change 2007: Impacts, adaptation and vulnerability: Working Group II contribution to the Fourth Assessment Report of the IPCC Intergovernmental Panel on Climate Change*, pp. 745-777. Cambridge University Press, Cambridge, UK.

Klein, R.J.T., E.L.F. Schipper, and S. Dessai, 2005. Integrating mitigation and adaptation into climate and development policy: Three research questions. *Environmental Science & Policy*, 8(6): 579-588.

Knutson, C.L., T. Haigh, M.J. Hayes, M. Widhalm, J. Nothwehr, M. Kleinschmidt, and L. Graf, 2011. Farmer perceptions of sustainable agriculture practices and drought risk reduction in Nebraska, USA. *Renewable Agriculture and Food Systems*, 26(03): 255-266.

Krishnamurthy, P.K., J.B. Fisher, and C. Johnson, 2011. Mainstreaming local perceptions of hurricane risk into policymaking: A case study of community GIS in Mexico. *Global Environmental Change*, 21(1): 143-153.

Lal, R., 2011. Sequestering carbon in soils of agro-ecosystems. *Food Policy*, 36, Supplement 1: S33-S39.

Lin, B.B., 2011. Resilience in agriculture through crop diversification: Adaptive management for environmental change. *BioScience*, 61(3): 183-193.

Luers, A.L., and S.C. Moser, 2006. *Preparing for the impacts of climate change in California: Advancing the debate on adaptation.* . Report prepared for the California Energy Commission, Public Interest Energy Research Program and the California Environmental Protection Agency, Sacramento, CA, CEC-500-2005-198-SF.

Malka, A., J.A. Krosnick, and G. Langer, 2009. The association of knowledge with concern about global warming: Trusted information sources shape public thinking. *Risk Analysis*, 29(5): 633-647.

Marshall, N., C. Stokes, S. Howden, and R. Nelson, 2009. Chapter 15 enhancing adaptive capacity, *in* C. Stokes and M. Howden (eds.), *Adapting agriculture to climate change: Preparing Australian agriculture, forestry and fisheries for the future.* CSIRO Publishing, Collingwood, Vic.

Marshall, N.A., 2010. Understanding social resilience to climate variability in primary enterprises and industries. *Global Environmental Change*, 20(1): 36-43.

McCarthy, J., O. Canziani, N. Leary, D. Kokken, and K. White, 2001. Climate change 2001: Impacts, adaptation and vulnerability. Annex B: Glossary of terms, *in Contribution of Working Group II to the Third Assessment Report of the Intergovernmental Panel on Climate Change*. Cambridge University Press Cambridge and New York.

McCown, R.L., 2005. New thinking about farmer decision makers, *in* J.L. Hatfield (ed.) *The farmer's decision: Balancing economic agriculture production with environmental quality*, pp. 11-44. Soil and Water Conservation Society, Ankeny, Iowa.

McCright, A.M., and R.E. Dunlap, 2011. The politicization of climate change and polarization in the American public's views of global warming, 2001–2010. *Sociological Quarterly*, 52(2): 155-194.

McDaniels, T.L., L.J. Axelrod, N.S. Cavanagh, and P. Slovic, 1997. Perception of ecological risk to water environments. *Risk Analysis*, 17(3): 341-352.

Merrill, J., R. Brillinger, and A. Heartwell, 2011. *Ready or not? An assessment of California agriculture's readiness for climate change*. California Climate and Agriculture Network, Sebastopol, CA.

Miles, E., M. Elsner, J. Littell, L. Binder, and D. Lettenmaier, 2010. Assessing regional impacts and adaptation strategies for climate change: The Washington climate change impacts assessment. *Climatic Change*, 102(1): 9-27.

Millar, C.I., N.L. Stephenson, and S.L. Stephens, 2007. Climate change and forests of the future: Managing in the FACE of uncertainty. *Ecological Applications*, 17(8): 2145-2151.

Moebius-Clune, B., H.van Es, and J. Melkonian, 2011. *Adapt-N: An adaptive nitrogen management tool for corn production. Adaptive N management fact sheet 1.* Cornell University http://adapt-n.cals.cornell.edu/pubs/pdfs/Factsheet1_IntroToAdaptN.pdf.

Morton, L.W., 2008. The role of civic structure in achieving performance-based watershed management. *Society & Natural Resources*, 21(9): 751-766.

Moser, S., R. Kasperson, G. Yohe, and J. Agyeman, 2008. Adaptation to climate change in the northeast United States: Opportunities, processes, constraints. *Mitigation and Adaptation Strategies for Global Change*, 13(5): 643-659.

Moser, S.C., and J.A. Ekstrom, 2010. A framework to diagnose barriers to climate change adaptation. *Proceedings of the National Academy of Sciences*, 107(51): 22026-22031.

Nelson, D.R., W.N. Adger, and K. Brown, 2007. Adaptation to environmental change: Contributions of a resilience framework. *Annual Review of Environment and Resources*, 32: 395-419.

NAS, 2010. *Toward sustainable agricultural systems in the 21st century. Committee on twenty-first century systems, agriculture.* National Research Council, National Academies Press, Washington, D.C.

Nelson, R., P. Kokic, S. Crimp, P. Martin, H. Meinke, S.M. Howden, P. de Voil, and U. Nidumolu, 2010. The vulnerability of Australian rural communities to climate variability and change: Part II—integrating impacts with adaptive capacity. *Environmental Science & Policy*, 13(1): 18-27.

Newton, A., S. Johnson, and P. Gregory, 2011. Implications of climate change for diseases, crop yields and food security. *Euphytica*, 179(1): 3-18.

NRC, 2009. *Restructuring federal climate research to meet the challenges of climate change.* National Academies Press, Washington, D.C.

NRC, 2010. *Toward sustainable agricultural systems in the 21st century.* National Research Council Committee on Twenty-First Century Systems, Agriculture, National Academies Press, Washington, D.C.

Olesen, J.E., and M. Bindi, 2002. Consequences of climate change for European agricultural productivity, land use and policy. *European Journal of Agronomy*, 16(4): 239-262.

Olesen, J.E., M. Trnka, K.C. Kersebaum, A.O. Skjelvåg, B. Seguin, P. Peltonen-Sainio, F. Rossi, J. Kozyra, and F. Micale, 2011. Impacts and adaptation of European crop production systems to climate change. *European Journal of Agronomy*, 34(2): 96-112.

Parry, M., N. Arnell, P. Berry, D. Dodman, S. Fankhauser, C. Hope, S. Kovats, R. Nicholls, D. Satterthwaite, R. Tiffin, and T. Wheeler, 2009. *Assessing the costs of adaptation to climate change: A review of the UNFCCC and other recent estimates.* International Institute for Environment and Development and Grantham Institute for Climate Change, London.

Parry, M., O.F. Canziani, J.P. Palutikof, P.J.v.d. Linden, and C.E. Hanson, 2007. Climate change 2007: Impacts, adaptation and vulnerability appendix 1: Glossary, *in Contribution of Working Group II to the Fourth Assessment Report of the Intergovernmental Panel on Climate Change*. Cambridge Univeristy Press, Cambridge, UK, and New York.

Pelling, M., 2011. Conclusion: Adapting with climate change, *in* M. Pelling (ed.) *Adaptation to climate change: From resilience to transformation.* Routledge, London; New York.

Peterson, G., 2009. Ecological limits of adaptation to climate change, *in* W.N. Adger, I. Lorenzoni and K.L. O'brien (eds.), *Adapting to climate change: Thresholds, values, governance.* Cambridge University Press, Cambridge; New York.

Petheram, L., K.K. Zander, B.M. Campbell, C. High, and N. Stacey, 2010. 'Strange changes': Indigenous perspectives of climate change and adaptation in NE Arnhem Land (Australia). *Global Environmental Change*, 20(4): 681-692.

Preston, B., R. Westaway, and E. Yuen, 2011. Climate adaptation planning in practice: An evaluation of adaptation plans from three developed nations. *Mitigation and Adaptation Strategies for Global Change*, 16(4): 407-438.

Prokopy, L.S., K. Floress, D. Klotthor-Weinkauf, and A. Baumgart-Getz, 2008. Determinants of agricultural best management practice adoption: Evidence from the literature. *Journal of Soil and Water Conservation*, 63(5): 300 311.

177

Reid, G., 2009. Building resilience to climate change in rain-fed agricultural enterprises: An integrated property planning tool. *Agriculture and Human Values*, 26(4): 391-397.

Reid, S., B. Smit, W. Caldwell, and S. Belliveau, 2007. Vulnerability and adaptation to climate risks in Ontario agriculture. *Mitigation and Adaptation Strategies for Global Change*, 12(4): 609-637.

Reidsma, P., F. Ewert, A.O. Lansink, and R. Leemans, 2010. Adaptation to climate change and climate variability in European agriculture: The importance of farm level responses. *European Journal of Agronomy*, 32(1): 91-102.

Reilly, J.M., and E. Blanc, 2009. Economics of agricultural impacts, adaptation and mitigation, *in* D. Hillel and C. Rosenzweig (eds.), *Handbook of climate change and agroecosystems*. Imperial College Press, London.

Rogovska, N., and R. Cruse, 2011. *Consequences for agriculture in Iowa*. In, Climate Change Impacts on Iowa 2010, Report to the Governor and the Iowa General Assembly, Iowa Climate Change Impacts Committee.

Rosenzweig, C., and F. Tubiello, 2007. Adaptation and mitigation strategies in agriculture: An analysis of potential synergies. *Mitigation and Adaptation Strategies for Global Change*, 12(5): 855-873.

Roser-Renouf, C., and M. Nisbet, 2008. The measurement of key behavioural science constructs in climate change research. *International Journal of Sustainability Communications*, 3: 37-95.

Simões, A.F., D.C. Kligerman, E.L.L. Rovere, M.R. Maroun, M. Barata, and M. Obermaier, 2010. Enhancing adaptive capacity to climate change: The case of smallholder farmers in the Brazilian semi-arid region. *Environmental Science & Policy*, 13(8): 801-808.

Slovic, P., and International Institute for Environment and Development, 2010. *The feeling of risk: New perspectives on risk perception*. Earthscan: In association with the International Institute for Environment and Development, London; Washington, DC.

Smit, B., I. Burton, R.J.T. Klein, and R. Street, 1999. The science of adaptation: A framework for assessment. *Mitigation and Adaptation Strategies for Global Change*, 4(3): 199-213.

Smit, B., I. Burton, R.J.T. Klein, and J. Wandel, 2000. An anatomy of adaptation to climate change and variability. *Climatic Change*, 45(1): 223-251.

Smit, B., and M.W. Skinner, 2002. Adaptation options in agriculture to climate change: A typology. *Mitigation and Adaptation Strategies for Global Change*, 7(1): 85-114.

Smit, B., and J. Wandel, 2006. Adaptation, adaptive capacity and vulnerability. *Global Environmental Change*, 16(3): 282-292.

Smith, C.M., J.M. Peterson, and J.C. Leatherman, 2007. Attitudes of Great Plains producers about best management practices, conservation programs, and water quality. *Journal of Soil and Water Conservation*, 62(5): 97A-103A.

Smith, J., J. Vogel, and J.E. Cromwell III, 2009. An architecture for government action on adaptation to climate change. An editorial comment. *Climatic Change*, 95(1): 53-61.

Spence, A., W. Poortinga, C. Butler, and N.F. Pidgeon, 2011. Perceptions of climate change and willingness to save energy related to flood experience. *Nature Climate Change*, 1(1): 46-49.

Spies, T., T. Giesen, F. Swanson, J. Franklin, D. Lach, and K. Johnson, 2010. Climate change adaptation strategies for federal forests of the Pacific Northwest, USA: Ecological, policy, and socio-economic perspectives. *Landscape Ecology*, 25(8): 1185-1199.

Stöckle, C., R. Nelson, S. Higgins, J. Brunner, G. Grove, R. Boydston, M. Whiting, and C. Kruger, 2010. Assessment of climate change impact on eastern Washington agriculture. *Climatic Change*, 102(1): 77-102.

Strategy, N.C.A., 2012. *National fish, wildlife and plants climate adaptation strategy: Public review draft* http://www.wildlifeadaptationstrategy.gov/pdf/public_review_draft.pdf.

Swanson, D.A., J.C. Hiley, H.D. Venema, and R. Grosshans, 2009. *Indicators of adaptive capacity to climate change for agriculture in the prairie region of Canada: Comparison with field observations*. Working Paper for the Prairie Climate Resilience Project, Winnipeg: International Institute for Sustainable Development.

Tarnoczi, T., 2011. Transformative learning and adaptation to climate change in the Canadian prairie agro-ecosystem. *Mitigation and Adaptation Strategies for Global Change*, 16(4): 387-406.

Tarnoczi, T., and F. Berkes, 2010. Sources of information for farmers' adaptation practices in Canada's prairie agro-ecosystem. *Climatic Change*, 98(1): 299-305.

Tomich, T.P., S. Brodt, H. Ferris, R. Galt, W.R. Horwath, E. Kebreab, J.H.J. Leveau, D. Liptzin, M. Lubell, P. Merel, R. Michelmore, T. Rosenstock, K. Scow, J. Six, N. Williams, and L. Yang, 2011. Agroecology: A review from a global-change perspective. *Annual Review of Environment and Resources*, 36(1): 193-222.

USDA Forest Service, 2010. *National roadmap for responding to climate change* www.fs.fed.us/climatechange/pdf/roadmap.pdf.

Valentin, L., D.J. Bernardo, and T.L. Kastens, 2004. Testing the empirical relationship between best management practice adoption and farm profitability. *Applied Economic Perspectives and Policy*, 26(4): 489-504.

van Apeldoorn, D.F., K. Kok, M.P.W. Sonneveld, and T. Veldkamp, 2011. Panarchy rules: Rethinking resilience of agroecosystems, evidence from Dutch dairy-farming. *Ecology and Society*, 16(1).

Vincent, K., 2007. Uncertainty in adaptive capacity and the importance of scale. *Global Environmental Change*, 17(1): 12-24.

Walker, B., C.S. Hollin, S.R. Carpenter, and A. Kinzig, 2004. Resilience, adaptability and transformability in social-ecological systems. *Ecology and Society*, 9(2).

Wall, E., and B. Smit, 2005. Climate change adaptation in light of sustainable agriculture. *Journal of Sustainable Agriculture*, 27(1): 113-123.

Wardekker, J.A., A.C. Petersen, and J.P. van der Sluijs, 2009. Ethics and public perception of climate change: Exploring the Christian voices in the U.S. public debate. *Global Environmental Change-Human and Policy Dimensions*, 19(4): 512-521.

Weber, E.U., and P.C. Stern, 2011. Public understanding of climate change in the United States. *American Psychologist*, 66(4): 315-328.

Wery, J., and J.W.A. Langeveld, 2010. Introduction to the EJA special issue on "Cropping systems design: New methods for new challenges". *European Journal of Agronomy*, 32(1): 1-2.

Wilbanks, T., P. Leiby, R. Perlack, J. Ensminger, and S. Wright, 2007. Toward an integrated analysis of mitigation and adaptation: Some preliminary findings. *Mitigation and Adaptation Strategies for Global Change*, 12(5): 713-725.

Winsten, J.R., 2009. Improving the cost-effectiveness of agricultural pollution control: The use of performance-based incentives. *Journal of Soil and Water Conservation*, 64(3): 88A-93A.

Wolfe, D., L.H. Ziska, C. Petzoldt, A. Seaman, L. Chase, and K. Hayhoe, 2008. Projected change in climate thresholds in the northeastern U.S.: Implications for crops, pests, livestock, and farmers. *Mitigation and Adaptation Strategies for Global Change*, 13(5): 555-575.

World Bank, 2010. *The economics of adaptation to climate change: A synthesis report*. World Bank Group, Washington http://climatechange.worldbank.org/sites/default/files/documents/EACCSynthesisReport.pdf.

World Bank, 2011. *Climate smart agriculture: Increased productivity and food security, enhanced resilience and reduced carbon emissions for sustainable development*. The International Bank for Reconstruction and Development/The World Bank, Washington, DC.

Yohe, G., and R.S.J. Tol, 2002. Indicators for social and economic coping capacity - moving toward a working definition of adaptive capacity. *Global Environmental Change-Human and Policy Dimensions*, 12(1): 25-40.

Yohe, G.W., R.D. Lasco, Q.K. Ahmad, N.W. Arnell, S.J. Cohen, C. Hope, A.C. Janetos, and R.T. Perez, 2007. Perspectives on climate change and sustainability, *in* M.L. Parry, O.F. Canziani, J.P. Palutikof, P.J. Van Der Linden and C.E. Hanson (eds.), *Climate change 2007: Impacts, adaptation and vulnerability: Working Group II contribution to the Fourth Assessment Report of the IPCC Intergovernmental Panel on Climate Change*, pp. 811-841. Cambridge University Press, Cambridge, UK.

Chapter 8 References

Adger, W., S. Agrawala, M.M.Q. Mirza, C. Conde, K. O'Brien, J. Pulhin, R. Pulwarty, B. Smit, and K. Takahashi, 2007. Assessment of adaptation practices, options, constraints and capacity, *in* M.L. Parry, O.F. Canziani, J.P. Palutikof, P.J. Van Der Linden and C.E. Hanson (eds.), *Climate change 2007 : Impacts, adaptation and vulnerability : Working Group II contribution to the Fourth Assessment Report of the IPCC Intergovernmental Panel on Climate Change*, pp. 717-743. Cambridge University Press, Cambridge, UK.

Amundson, J.L., T.L. Mader, R.J. Rasby, and Q.S. Hu, 2006. Environmental effects on pregnancy rate in beef cattle. *Journal of Animal Science*, 84(12): 3415-3420.

Baylis, M., and A.K. Githeko, 2006. *T7.3: The effects of climate change on infectious diseases of animals.* Foresight http://www.bis.gov.uk/foresight/our-work/projects/published-projects/infectious-diseases/reports-and-publications.

Beach, R.H., C. Zhen, A. Thomson, R.M. Rejeseus, P. Sinha, A.W. Lentz, D.V. Vedenov, and B.A. McCarl, 2010. *Climate change impacts on crop insurance.* USDA Risk Managment Agency, Kansas City, MO.

Burton, I., and B. Lim, 2005. Achieving adequate adaptation in agriculture. *Climatic Change*, 70(1): 191-200.

Canto, T., M.A. Aranda, and A. Fereres, 2009. Climate change effects on physiology and population processes of hosts and vectors that influence the spread of hemipteran-borne plant viruses. *Global Change Biology*, 15(8): 1884-1894.

Diffenbaugh, N.S., C.H. Krupke, M.A. White, and C.E. Alexander, 2008. Global warming presents new challenges for maize pest management. *Environment Research Letters*, 3.

Easterling, W.E., 2009. Guidelines for adapting agriculture to climate change, *in* D. Hillel and C. Rosenzweig (eds.), *Handbook of climate change and agroecosystems: Impacts, adaptation, and mitigation.* Imperial College Press; Distributed by World Scientific Publishing Co., London; Singapore; Hackensack, NJ.

Easterling, W.E., P.K. Aggarwal, P. Batima, K.M. Brander, L. Erda, S.M. Howden, A. Kirilenko, J. Morton, J.-F. Soussana, J. Schmidhuber, and F.N. Tubiello, 2007. Food, fibre and forest products, *in* M.L. Parry, O.F. Canziani, J.P. Palutikof, P.J. Van Der Linden and C.E. Hanson (eds.), *Climate change 2007: Impacts, adaptation and vulnerability: Working Group II contribution to the Fourth Assessment Report of the IPCC Intergovernmental Panel on Climate Change*, pp. 273-313. Cambridge University Press, Cambridge, UK.

Field, C.B., L.D. Mortsch, M. Brklacich, D.L. Forbes, P. Kovacs, J.A. Patz, S.W. Running, and M.J. Scott, 2007. North America, *in* M.L. Parry, O.F. Canziani, J.P. Palutikof, P.J.V.D. Linden and C.E. Hanson (eds.), *Climate change 2007 : Impacts, adaptation and vulnerability : Contribution of Working Group II to the Fourth Assessment Report of the Intergovernmental Panel on Climate Change.* Cambridge University Press, Cambridge, UK.

Frank, K.L., T.L. Mader, J.A. Harrington, G.L. Hahn, and M.S. Davis, 2001. *Climate change effects on livestock production in the Great Plains*, paper presented at 6th International Livestock Environment Symposium, American Society of Agricultural Engineers, St. Joseph, MI.

Fuglie, K.O., and P.W. Heisey, 2007. *Economic Returns to Public Agricultural Research.* USDA EB 10, U.S. Dept. of Agriculture, Economic Research Service.

Gaughan, J.B., N. Lacetera, S.E. Valtorta, H.H. Khalifa, L. Hahn, and T. Mader, 2009. Response of domestic animals to animal challenges, *in* K.L. Ebi, I. Burton and G.R. Mcgregor (eds.), *Biometeorology for adaptation to climate variability and change.* Springer, Dordrecht; London.

Gutierrez, A., L. Ponti, and Q. Cossu, 2009. Effects of climate warming on olive and olive fly (*Bactrocera oleae* (Gmelin)) in California and Italy. *Climatic Change*, 95(1): 195-217.

Gutierrez, A.P., T. D'Oultremont, C.K. Ellis, and L. Ponti, 2006. Climatic limits of pink bollworm in Arizona and California: Effects of climate warming. *Acta Oecologica*, 30(3): 353-364.

Hahn, G.L., 1995. Environmental management for improved livestock performance, health and well-being. *Japanese journal of livestock management*, 30(3): 113-127 /.

Hatfield, J.L., K.J. Boote, B.A. Kimball, L.H. Ziska, R.C. Izaurralde, D. Ort, A.M. Thomson, and D. Wolfe, 2011. Climate impacts on agriculture: Implications for crop production. *Agronomy Journal*, 103(2): 351-370.

Howden, M., A. Ash, S. Barlow, T. Booth, S. Charles, B. Cechet, S. Crimp, R. Gifford, K. Hennessy, R. Jones, M. Kirschbaum, G. McKeon, H. Meinke, S. Park, B. Sutherst, L. Webb, and P. Whetton, 2003. *An overview of the adaptive capacity of the Australian agricultural sector to climate change – options, costs and benefits.* Scientific and Industrial Research Organisation, Collingwood, Victoria http://www.cse.csiro.au/publications/2003/AGOAgClimateAdaptationReport.pdf.

Howden, S.M., J.F. Soussana, F.N. Tubiello, N. Chhetri, M. Dunlop, and H. Meinke, 2007. Adapting agriculture to climate change. *Proceedings of the National Academy of Sciences*, 104(50): 19691-19696.

Izaurralde, R.C., A.M. Thomson, J.A. Morgan, P.A. Fay, H.W. Polley, and J.L. Hatfield, 2011. Climate impacts on agriculture: Implications for forage and rangeland production. *Agronomy Journal*, 103(2): 371-381.

Kimball, B.A., 2011. Effects and interactions with water, nitrogen and temperature, *in* D. Hillel and C. Rosenzweig (eds.), *Handbook of climate change and agroecosystems : Impacts, adaptation, and mitigation*, pp. 87-107. Imperial College Press ; Distributed by World Scientific Publishing Co., London; Singapore; Hackensack, NJ.

Mika, A.M., R.M. Weiss, O. Olfert, R.H. Hallett, and J.A. Newman, 2008. Will climate change be beneficial or detrimental to the invasive swede midge in North America? Contrasting predictions using climate projections from different general circulation models. *Global Change Biology*, 14(8): 1721-1733.

Morgan, J.A., 2005. Rising atmospheric CO_2 and global climate change: Responses and management implications for grazing lands, *in* S.G. Reynolds and J. Frame (eds.), *Grasslands : Developments, opportunities, perspectives.* Food and Agricultural Organization of the United Nations; Science Publishers, Inc., Rome; Enfield, NH.

Morgan, J.A., D.E. Pataki, C. Körner, H. Clark, S.J. Grosso, J.M. Grünzweig, A.K. Knapp, A.R. Mosier, P.C.D. Newton, P.A. Niklaus, J.B. Nippert, R.S. Nowak, W.J. Parton, H.W. Polley, and M.R. Shaw, 2004. Water relations in grassland and desert ecosystems exposed to elevated atmospheric CO_2. *Oecologia*, 140(1). 11-25.

Navas-Castillo, J., E. Fiallo-Olivé, and S. Sánchez-Campos, 2011. Emerging virus diseases transmitted by whiteflies. *Annual Review of Phytopathology*, 49(1): 219-248.

Olesen, J.E., M. Trnka, K.C. Kersebaum, A.O. Skjelvåg, B. Seguin, P. Peltonen-Sainio, F. Rossi, J. Kozyra, and F. Micale, 2011. Impacts and adaptation of European crop production systems to climate change. *European Journal of Agronomy*, 34(2): 96-112.

Patterson, D.T., J.K. Westbrook, R.J.V. Joyce, P.D. Lingren, and J. Rogasik, 1999. Weeds, insects, and diseases. *Climatic Change*, 43(4): 711-727.

Polley, H.W., H.B. Johnson, and J.D. Derner, 2003. Increasing CO_2 from subambient to superambient concentrations alters species composition and increases above-ground biomass in a C3/C4 grassland. *New Phytologist*, 160(2): 319-327.

Savary, S., A. Nelson, A.H. Sparks, L. Willocquet, E. Duveiller, G. Mahuku, G. Forbes, K.A. Garrett, D. Hodson, J. Padgham, S. Pande, M. Sharma, J. Yuen, and A. Djurle, 2011. International agricultural research tackling the effects of global and climate changes on plant diseases in the developing world. *Plant Disease*, 95(10): 1204-1216.

Smit, B., and M.W. Skinner, 2002. Adaptation options in agriculture to climate change: A typology. *Mitigation and Adaptation Strategies for Global Change*, 7(1): 85-114.

Thornton, P.K., 2010. Livestock production: Recent trends, future prospects. *Philosophical Transactions of the Royal Society B: Biological Sciences*, 365(1554): 2853-2867.

Tubiello, F.N., J.S. Amthor, K.J. Boote, M. Donatelli, W. Easterling, G. Fischer, R.M. Gifford, M. Howden, J. Reilly, and C. Rosenzweig, 2007. Crop response to elevated CO_2 and world food supply: A comment on "food for thought..." by Long et al. Science 312:1918–1921, 2006. *European Journal of Agronomy*, 26(3): 215-223.

Ziska, L.H., and J.A. Bunce, 1998. The influence of increasing growth temperature and CO_2 concentration on the ratio of respiration to photosynthesis in soybean seedlings. *Global Change Biology*, 4(6): 637-643.

Ziska, L.H., and J.R. Teasdale, 2000. Sustained growth and increased tolerance to glyphosate observed in a C3, quackgrass (*Elytrigia repens*), grown at elevated carbon dioxide. *Functional Plant Biology*, 27(2): 159-166.

Appendix B

Glossary of Commonly Used Terms

Abscission – The process by which a plant drops one or more of its parts, such as a leaf, fruit, flower, or seed.

Above-ground Net Primary Productivity (ANPP) – The measure of the net rate of photosynthetic carbon sequestration by plants into above-ground components (e.g., leaves, stalks, stems, etc.). ANPP can provide insight on key ecological processes.

Adaptive management – A decision process that promotes flexible decision-making that can be adjusted in the face of uncertainties as outcomes from management actions and other events become better understood. Careful monitoring of these outcomes both advances scientific understanding and helps adjust policies or operations as part of an iterative learning process.

Adaptive capacity – The ability of a system – such as a crop or rangelands, a rural community, a social-ecological system – to weather the effects of and adjust to changing climate.

Aerosols – Extremely fine solid particles that remain suspended in the air.

Agroecology – Loosely defined, agroecology often incorporates ideas about a more environmentally and socially sensitive approach to agriculture, one that focuses not only on production, but also on the ecological sustainability of the productive system. This definition implies a number of features about society and production that go well beyond the limits of the agricultural field. Agroecology is integrative because it uses systems thinking to apply methods and knowledge from multiple scientific disciplines such as agronomy, ecology, sociology, and economics to study interactions between plants, animals, humans, and the environment in agricultural systems.

Anthesis – The period during which a flower is fully open and functional.

Biological (indirect) effects of climate change – The secondary influences on agricultural production, such as weeds, insects, and disease-vector distribution. Changing climate may cause shifts in weed, insect, or disease infestation range, frequency, and intensity, which, in turn, may affect agriculture. Indirect effects of climate change may amplify or counteract the direct effects of climate change.

Biotroph – An organism that derives nutrients from the living tissues of another organism.

Boll – The seed-bearing capsule of certain plants, especially cotton and flax.

C_3 species – Almost all plant life on Earth can be divided into two categories based on the way they assimilate carbon dioxide into their systems. During the first steps in CO_2 assimilation, C_3 plants form a pair of three carbon-atom molecules. C_3 species continue to increase photosynthesis with rising CO_2. C_3 plants include more than 95 percent of the plant species on Earth.

C_4 species – C_4 plants initially form four carbon-atom molecules. C_4 plants include such crop plants as sugar cane and corn. They are the second-most prevalent photosynthetic type, and do not assimilate CO_2 as well as C_3 plants.

Carbon dioxide (CO_2) fertilization – The enhancement of the growth of plants as a result of increased atmospheric CO_2 concentration.

Carbon sink – A carbon reservoir, carbon sinks include the oceans, plants, and other organisms that remove carbon from the atmosphere via photosynthetic processes.

Carbon source – The term describing processes that add CO_2 to the atmosphere.

Carbon sequestration – The term describing processes that remove CO_2 from the atmosphere.

Cardinal temperatures – The critical temperature range for ideal lifecycle development; these vary by species and between vegetative and reproductive growth stages.

Chill requirement – The minimum period of cold weather after which a fruit-bearing tree will blossom.

Climate – Climate, in a narrow sense, is usually defined as the "average weather" or more rigorously as the statistical description in terms of the mean and variability of relevant quantities over a period of time ranging from months to thousands or millions of years. The classical period is 30 years, as defined by the World Meteorological Organization (WMO). The relevant quantities are most often surface variables such as temperature, precipitation, and wind. Climate in a wider sense is the state, including a statistical description, of the climate system.

Climate change – Climate change refers to a statistically significant variation in either the mean state of the *climate* or in its variability, persisting for an extended period (typically decades or longer). Climate change may be due to natural internal processes or external forcings, or to persistent anthropogenic changes in the composition of the atmosphere or in land use. Note that the United Nations Framework Convention on Climate Change (UNFCCC), its Article 1 defines "climate change" as "…a change of climate which is attributed directly or indirectly to human activity that alters the composition of the global atmosphere and which is in addition to natural climate variability observed over comparable time periods." The UNFCCC thus makes a distinction between "climate change" attributable to human activities altering the atmospheric composition, and "climate variability" attributable to natural causes. See also climate variability.

Climate change adaptation – Adjustment in natural or human systems in response to actual or expected climatic stimuli or their effects to moderate harm or exploit beneficial opportunities.

Climate variability – Climate variability refers to variations in the mean state and other statistics (such as standard deviations, the occurrence of extremes, etc.) of the climate on all temporal and spatial scales beyond that of individual weather events. Variability may be due to natural internal processes within the climate system (internal variability), or to variations in natural or anthropogenic external forcing (external variability). See also climate change.

CO_2 enrichment – Addition of CO_2 to the atmosphere.

Coefficient of variation of annual runoff – A measure of the variability of runoff.

Diapause – A suspension of development that can occur at the embryonic, larval, pupal, or adult stage, depending on the species during adverse conditions. **Insect diapause** is triggered by environmental cues, like daylight and temperature.

Direct (physical) climate change effects – Direct effects of climate change effects relate to changes in temperature, precipitation, humidity, and CO_2 concentrations. Each of these variables has a direct influence on the growth and development of crops, rangelands, and livestock.

Dehiscence – Failure of pollen to form normally and be released.

Dryland Production – Farming occurring in semi-arid areas.

Edaphic – Resulting from or influenced by the soil rather than the climate.

Emissions trajectory – The rate and timing of reductions in carbon emissions to meet a given carbon-reduction target.

Endophyte – A plant living within another plant, usually as a parasite.

Evapotranspiration – The sum of evaporation and plant transpiration. Evaporation accounts for the movement of water to the air from sources such as the soil, canopy interception, and water bodies. Transpiration accounts for the movement of water within a plant and the subsequent loss of water as vapor through stomata in its leaves.

Free-Air CO_2 Enrichment (FACE) – FACE is a method and infrastructure used to experimentally enrich the atmosphere enveloping portions of a terrestrial ecosystem with controlled amounts of CO_2 (and in some cases, other gases), without using chambers or walls.

Forb – A broad-leaved herb (not a grass), especially one growing in a field, prairie, or meadow.

Fossil fuels – A natural fuel such as coal or gas, formed in the geological past from the remains of living organisms.

Frost day – A day – or span of days – in which the minimum daily temperature dips below freezing.

Germplasm – The living tissue from which new plants can be grown. It can be a seed or another plant part – e.g., a leaf, a piece of stem, pollen or even just a few cells that can be turned into a whole plant.

General circulation model (GCM) – These numerical models represent physical processes in the atmosphere, ocean, cryosphere and land surface, providing a method for simulating the global climate system's response to increasing greenhouse gas concentrations.

Global dimming – The gradual reduction in the amount of global direct irradiance at the Earth's surface that was observed for several decades after the start of systematic measurements in 1950s.

Greenhouse Gas (GHG) – A gas that traps heat in the atmosphere is called a greenhouse gas. Some GHGs, such as CO_2, may be emitted or drawn from the atmosphere through natural processes or human activities. Other GHGs, such as certain fluorinated gaseous compounds, are created and emitted solely through human activities. The principal GHGs that enter the atmosphere because of human activities are carbon dioxide (CO_2), water vapor (H_2O), methane (CH_4),and nitrogen oxide (NO), as well as fluorinated gases, such as hydrofluorocarbons, perfluorocarbons, and sulfur hexafluoride.

Greenhouse gas mitigation – Reduction in greenhouse gas emissions (also often referred to as climate change mitigation).

Hedonic estimation method – A widely used statistical methodology for impact assessment.

Hectare – A 100 x 100 meter area; often used as a measure for an area of land.

Herbivores – Animals that feed chiefly on plants.

Homeostasis – The scientific study of periodic biological phenomena, such as flowering, breeding, and migration, in relation to climatic conditions.

Integrated assessment model (IAM) – A model that combines scientific and socio-economic aspects of climate change primarily for the purpose of assessing policy options for climate change control.

Irrigation Modes –

Drip irrigation allows water to drip slowly to the roots of plants through a network of valves, pipes, tubing, and emitters.

Flood irrigation pumps water onto the fields. The water then flows freely along the ground among the crops.

Spray irrigation relies on machinery to spray water in all directions.

Leaf area index (LAI) – The ratio of total upper leaf surface of a crop divided by the surface area of the land on which the crop grows.

Lignin – An organic substance that, with cellulose, forms the chief part of woody tissue.

Lysimeter – A device for collecting water from the pore spaces of soils, and for determining the soluble constituents removed in the drainage.

Mutualistic relationship – A positive, reciprocal relationship between two species. Through this relationship, both species enhance their survival, growth or fitness.

Necrosis – The premature death of cells in living tissue.

Necrotroph – An organism that causes the death of host tissues as it grows through them such that it is always colonizing the dead substrate.

Net primary productivity (NPP) – The ratio of all biomass accumulation and biomass losses in units of carbon, weight or energy, per land surface unit, over a set time interval (usually a year).

Panicle – The complete assembly of spikelets found in small grains, e.g., rice, wheat, or barley.

Pathogen – A microorganism that causes disease in its host.

Phenology – The study of periodic biological phenomena (e.g., flowering of plants, breeding patterns, and species migration) in relation to climatic conditions.

Phloem – The living tissue that carries organic nutrients (known as photosynthate), e.g, sucrose, a sugar, to all parts of the plant.

Physical (direct) climate change effects – The physical effects of climate change relate to changes in temperature, precipitation, humidity, and CO_2 concentrations. Each of these variables directly influences the growth and development of crops, rangelands, and livestock.

Potential evapotranspiration – A representation of the environmental demand for evapotranspiration and represents the evapotranspiration rate of a short green crop, completely shading the ground, of uniform height and with adequate water status in the soil profile. It is a reflection of the energy available to evaporate water, and of the wind available to transport the water vapor from the ground up into the lower atmosphere.

Ruminant – Even-toed, cud-chewing, hoofed mammals of the suborder *Ruminantia*, such as domestic cattle.

Sidedress – An application of fertilizer between the rows of growing crops.

Social-Ecological System (SES) – A linked system of humans and nature in which the flow and use of resources (ecological, socioeconomic, and cultural) are regulated by the interaction of ecological and social systems. Agricultural social-ecological systems are ecosystems managed by humans to produce food and fiber for a set of interconnected markets

Spikelet – The individual places on a rice or other similar plant where a grain develops.

Sporulation – Formation of spores.

Stomatal – One of the minute pores in the epidermis of a leaf or stem through which gases and water vapor pass.

Tiller – New shoots that develop at the base of the plant.

Threshhold – A point in a system after which any change that is described as abrupt is one where the change in the response is much larger than the change in the forcing. The changes at the threshold are therefore abrupt relative to the changes that occur before or after the threshold and can lead to a transition to a new state.

Transdisciplinary research – Transdisciplinary research integrates disciplinary knowledge with local knowledge of land managers, technical advisors, and the general public in participatory research and development.

Troposphere – The lowest region of the atmosphere, extending from the Earth's surface to a height of about 6 km to 10 km.

Uncertainty – An expression of the degree to which a value (e.g., the future state of the climate system) is unknown.

Vernalization – Plants requiring a cold period to flower.

Vulnerability – The degree to which a system is susceptible to, or unable to cope with, adverse effects of climate and global change, including climate variability and extremes, as well as climate change in conjunction with other stressors.

Weather – The specific condition of the atmosphere at a particular place and time. It is measured in terms of parameters such as wind, temperature, humidity, atmospheric pressure, cloudiness, and precipitation.

Report Authors and Affiliations

Lead Author Team

Charles L. Walthall
National Program Leader
USDA Agricultural Research Service
Beltsville, MD

Jerry Hatfield
Laboratory Director and Supervisory Plant
Physiologist
USDA Agricultural Research Service
Ames, IA

Peter Backlund
Director, Integrated Science Program and External
Relations
National Center for Atmospheric Research
Boulder, CO

Laura Lengnick
Visiting Scientist
USDA Agricultural Research Service
Beltsville, MD

Elizabeth Marshall
Economist
USDA Economic Research Service
Washington, DC

Margaret Walsh
Ecologist
U.S. Department of Agriculture Climate Change
Program Office
Washington, DC

Section Authors

Scott Adkins
Research Plant Pathologist
USDA Agricultural Research Service
Fort Pierce, FL

Marcel Aillery
Economist
USDA Economic Research Service
Washington, DC

Elizabeth A. Ainsworth
Research Molecular Biologist
USDA Agricultural Research Service
Urbana, IL

Caspar Ammann
Climate Scientist
National Center for Atmospheric Research
Boulder, CO

Christopher J. Anderson
Assistant Director Climate Science Program
Iowa State University
Ames, IA

Ignasi (Nacho) Bartomeus
Post-doctoral Fellow
Rutgers University
New Brunswick, NJ

Lance H. Baumgard
Associate Professor
Iowa State University
Ames, IA

Fitzgerald Booker
Plant Physiologist
USDA Agricultural Research Service
Raleigh, NC

Bethany Bradley
Assistant Professor
University of Massachusetts
Amherst, MA

Dana M. Blumenthal
Ecologist
USDA Agricultural Research Service
Ft. Collins, CO

James Bunce
Research Plant Physiologist
USDA Agricultural Research Service
Beltsville, MD

Kent Burkey
Plant Physiologist
USDA Agricultural Research Service
Raleigh, NC

Seth M. Dabney
Supervisory Research Agronomist
USDA Agricultural Research Service
Oxford, MS

Jorge A. Delgado
Research Soil Scientist
USDA Agricultural Research Service
Fort Collins, CO

Jeffrey Dukes
Associate Professor
Purdue University
West Lafayette, IN

Andy Funk
Agricultural Science Technician
USDA Agricultural Research Service
Beltsville, MD

Karen Garrett
Professor
Kansas State University
Manhattan, KS

Michael Glenn
Plant Physiologist, Director and Research Leader
USDA Agricultural Research Service
Kearneysville, WV

David A. Grantz
Plant Physiologist and Cooperative Extension Air
   Quality Specialist
University of California - Riverside
Parlier, CA

David Goodrich
Civil Engineer
USDA Agricultural Research Service
Tucson, AZ

Shuijin Hu
Associate Professor
Department of Plant Pathology, NC State
   University
Raleigh, NC

R. Cesar Izaurralde
Laboratory Fellow
Pacific Northwest National Laboratory-Joint
   Global Change Research Institute, University of
   Maryland
College Park, MD

Roger A. C. Jones
Principal Plant Virologist
Crop Protection Branch, Department of
   Agriculture and Food Western Australia
Perth, Australia

Soo-Hyung Kim
Assistant Professor
University of Washington
Seattle, WA

Andrew D.B. Leakey
Assistant Professor
University of Illinois, Urbana-Champaign
Urbana, IL

Kim Lewers
Plant Geneticist
USDA Agricultural Research Service
Beltsville, MD

Terry L. Mader
Professor
Animal Science, University of Nebraska
Concord, NE

Anna McClung
Research Plant Geneticist
USDA Agricultural Research Service
Stuttgardt, AR

Jack Morgan
Research Plant Physiologist
USDA Agricultural Research Service
Fort Collins, CO

David J. Muth
Research Engineer
Idaho National Laboratory
Idaho Falls, ID

Mark Nearing
Research Agricultural Engineer
USDA Agricultural Research Service
Tucson, AZ

Derrick M. Oosterhuis
Professor
University of Arkansas
Fayetteville, AR

Donald Ort
Research Plant Physiologist
USDA Agricultural Research Service
Urbana, IL

Camille Parmesan
Chair, Marine Institute
Plymouth University
Plymouth, U.K.

William T. Pettigrew
Plant Physiologist
USDA Agricultural Research Service
Stoneville, MS

Wayne Polley
Research Ecologist
USDA Agricultural Research Service
Temple, TX

Romina Rader
Post-doctoral Fellow
Stockholm University
Stockholm, Sweden

Charles Rice
Professor
Kansas State University
Manhattan, KS

Mike Rivington
Research Scientist
The James Hutton Institute
Aberdeen, Scotland UK

Erin Rosskopf
Microbiologist/Weed Scientist
USDA Agricultural Research Service
Ft. Pierce, FL

William A. Salas
President and Chief Scientist
Applied GeoSolutions, LLC
Durham, NH

Lynn E. Sollenberger
Professor
University of Florida
Gainesville, FL

Robert Srygley
Research Ecologist
USDA Agricultural Research Service
Sidney, MT

Claudio Stöckle
Professor
Washington State University
Pullman, WA

Eugene S. Takle
Director, Climate Science Program
Iowa State University
Ames, IA

Dennis Timlin
Research Soil Scientist
USDA Agricultural Research Service
Beltsville, MD

Jeffrey W. White
Plant Physiologist
USDA Agricultural Research Service
Maricopa, AZ

Rachael Winfree
Assistant Professor
Rutgers University
New Brunswick, NJ

Lois Wright-Morton
Professor
Iowa State University
Ames, IA

Lewis H. Ziska
Research Plant Physiologist
USDA Agricultural Research Service
Beltsville, MD

Staff Support

Brian Bevirt
Proofreader
National Center for Atmospheric Research
Boulder, CO

Chelcy Ford
National Climate Assessment Liaison, U.S. Global
Change Research Program
USDA Forest Service
Washington, DC

Rachel Hauser
External Relations Specialist/Writer
National Center for Atmospheric Research
Boulder, CO

Michelle McCambridge
Program Assistant
National Center for Atmospheric Research
Boulder, CO

Carol Park
Program Assistant
National Center for Atmospheric Research
Boulder, CO

Paula M. Robinson
Program Specialist and Administrator
National Center for Atmospheric Research
Boulder, CO

Michael Shibao
Graphic Design
Smudgeworks
Louisville, CO